Linear Algebra and Matrix Theory

By

Nelson M. Andrews

Published

By

Nutshell Research & Development Enterprises, L.L.C

Published in Caddo Mills, Texas **United States of America**

Library of Congress Control Number: 2008903702

ISBN 978-0-982036-82-2

Andrews, Nelson M.
Linear Algebra And Matrix Theory

Preamble

My original intent for undertaking the writing of this book was to create an eletronic course in linear algebra that utilizes the computer to implement the tedious work associated with implementing computations of linear algebra and matrix theory. In my zeal to computerized these computations, I unwitingly became part of the now-generation. With each chapter of the book, I designed a computer module for implementing the concepts taught in the chapter. Hence for each chapter, the user is empowered to instantly practise the lessons learned by using the Linear Algebra Learning Management System (LMS). The Linear Algebra LMS has two implementations. The first implementation is a personalized application that allows the student to perform linear algebra/matrix operations in a personal computer environment that is independent of the classroom setting. The personalized application provides a convenient aid for assisting with homework, take-home tests and independent research. The second implementation is a web-centric version of the LMS that facilitates teacher-to-student interaction. Both implementations have identical functionality, but the web-centric implementation is administered from a central computer under the control of the teacher or a school appointed administrator. This central control allows the teacher to record assigned lesson responses and the ability to evaluate and compare student responses.

Chapter one is a basic review of the number system, functions, sequences and the limits of functions/sequences. It serves as a review of some basic concepts and allows the reader to refamiliarize themselves with basic numerical concepts. This chapter is optional and may be omitted, at teacher's discretion, without loss of understanding of the ensuing concepts of linear algebra and matrix theory. The remaining chapters build on basic concepts to prove advanced concepts such as the Cayley-Hamilton theorem and the Gram-Smidt algorithm. Unlike other contemporary linear algebra and matrix theory text books, this book elevates the importance of the idempotent matrices and show the development of more advanced concepts in matrix theory. For instance, the proof of the Gram-Smidt algorithm is proved using a sequence of idempotent matrices. Furthermore idempotent matrices are important in the discussion of rings associated with a single matrix. In the discussion of rings associated with singular matrix, we refine the definitions of terms such as generalized inverse and generalized identity elements of the defined ring.

Dedication

This book is dedicated to my one and only fan; my wife Sharon Parrish Andrews. For the previous four years, the duration of time I have been working on the book, the phrase "It's time to rest your eyes and get some sleep" has been her salutation to me, to pry me away from the personal desktop computer. Her caring attitude and echoes of encouragement provided me with additional incentive to successfully complete this project. She also kept me grounded by asking me, from time-to-time, "What are you doing now and do you really need to include it?". As I developed each chapter and the corresponding computer vignette; she critiqued the appearance of the graphical user interface (gui) for the vignette and its relevance to the corresponding chapter in the book. She was also instrumental in designing the book cover. My lack of color coordination resulted in a collage of colors for the original book cover that was very distracting. Sharon came to my rescue and the results was a more appealing color combination and not distracting from the purpose of the book. So Sharon, thanks for keeping it real.

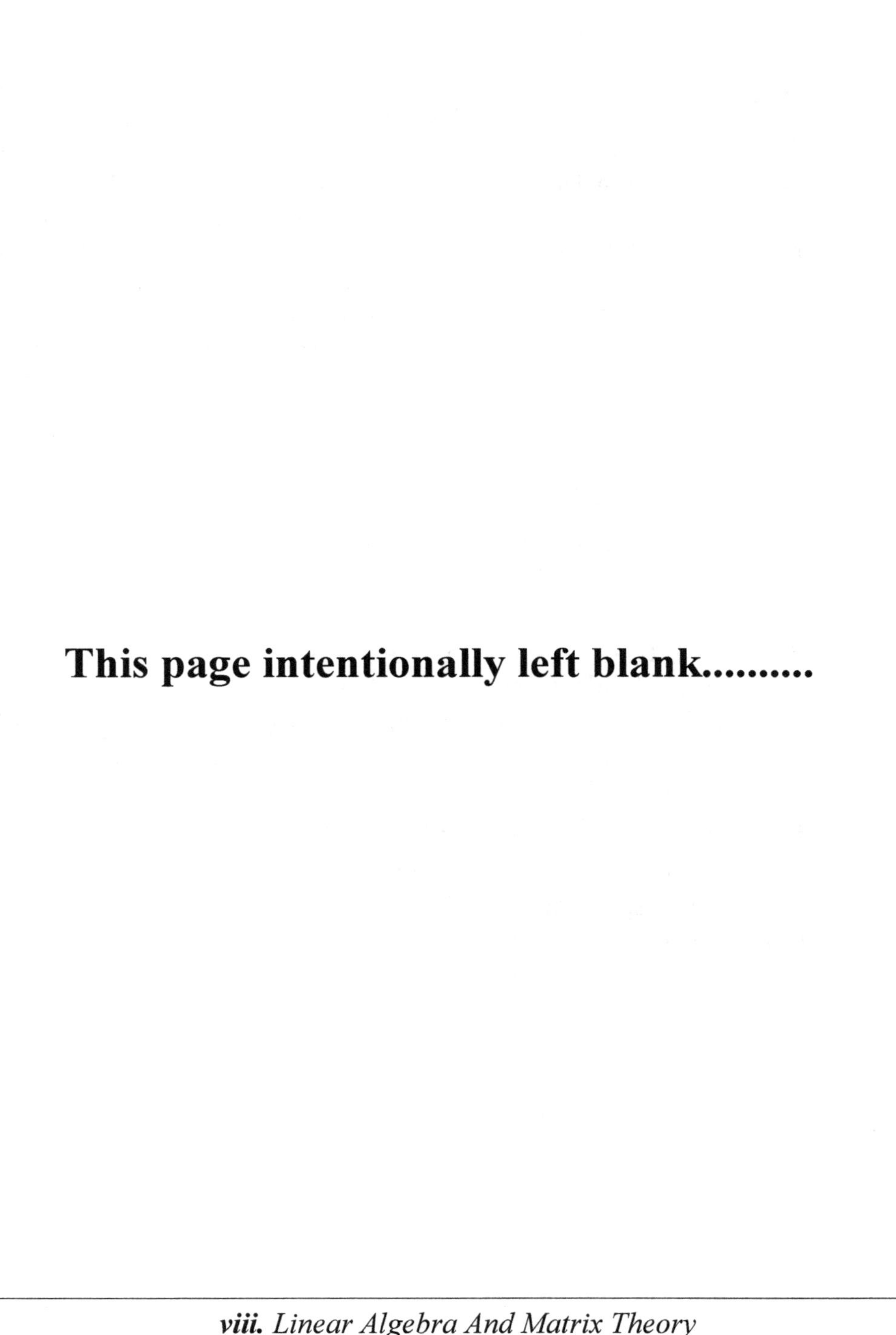

This page intentionally left blank..........

CHAPTER 1 Basic Definitions

The study of mathematics for most people begin with the study of numbers; counting numbers, integers, rational numbers and then the real numbers. Numbers represent a single dimension in the physical world; we can easily describe one-dimensional space with the use of equations and formulas. However, the physical world that we encounter on a daily basis is either two dimensional or three dimensional. When we restrict our attention to a planar surface (flat surface), we are thinking two dimensional. As soon as we think in terms of spatial coordinates (distance from a fixed location in the physical world in which we live), we are thinking three dimensional. For two, three and higher dimensional space, the concept of vectors were created to represent elements of higher dimensional spaces. In the physical world, the fourth dimension is considered to be time. For this book, the contemplation and interpretation of five or higher dimensional spaces is omitted. Such conjecture and philosophizing is left to the philosophers or science fiction writers.

As the title of this book suggests, the content of this book deals with linear transformations for vectors. Such transformations, as seen in the following chapters, are represented using matrices.

1.1 Scalar And Numerical Concepts

Definition 1.1.1 A scalar is a synonym for a number. In terms of physical properties, it is a numerical value that indicates magnitude only and does not indicate direction.

Generally, upon making reference to a numerical quantity, we think in terms of its base ten (10) representation. For instance, the number seventy nine (written 79), is interpreted as the following expression

$$79 = 9 + 7*10.$$

In this section, we investigate other bases such as binary (base 2), base 8 and hexadecimal (base 16). The discussion begins with counting numbers and integers and proceeds to real numbers.

1.1.1 Counting Numbers And Integers

The definition for counting numbers is given in Definition 1.1.2. When we are introduced to a counting system during our childhood years, we encounter these numbers. The definition for integers is given in Definition 1.1.3. The integers represent a superset of the counting numbers and allows us to define a group relative to addition for whole numbers. A graphical depiction of counting numbers and real numbers is shown in Figure 1 below. The number line, drawn to depict the counting numbers and integers, is a dashed line. The dashed line is used to indicate that the numbers are discrete (Between any two numbers there is a finite distance between them.).

Figure 1 Graphical Representation of Counting Numbers and Integers

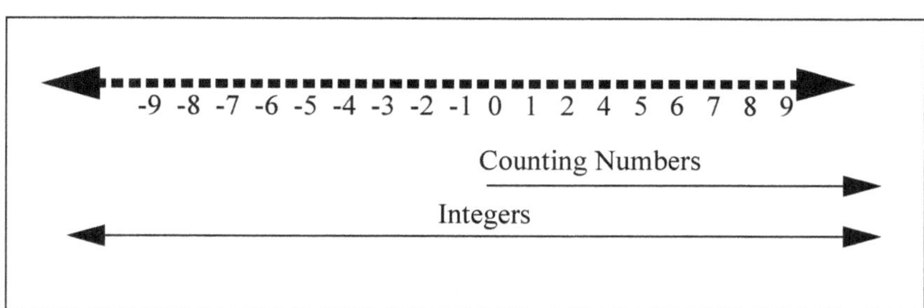

Definition 1.1.2 The counting numbers are the positive whole numbers beginning with zero (0) and increasing to positive infinity.

Definition 1.1.3 The integers are the whole numbers; the positive whole numbers, zero and the negative whole numbers.

We begin this section by defining the general methodology for converting a counting number from base 10 to an arbitrary base 'b'. But first let's assure ourselves that for any arbitrary base 'b' (b > 1), that every counting number is representable in base 'b'. To show this property, the following postulate is stated for all base 'b' systems.

Postulate 1 For any positive integer b > 0, that form a base for the set of counting numbers, the number zero (0) is representable in that base as the value 0 where

$$0 = 0 + 0 \times b^1 + 0 \times b^2 + \ldots$$

From Postulate 1, the number 0 is representable in all base 'b' representations of the counting numbers. Now let us suppose that for some non-negative value N, that N is representable in base 'b'. Then the following theorem,Theorem 1.1.1, states that N+1 is also representable in base 'b'.

Theorem 1.1.1 If a non-negative number N is representable in base 'b' as the following sum $N = \left(\sum_{i=0}^{k} n_i \times b^i\right)$, where $0 \le n_0 \le b-1, 0 \le n_1 \le b-1, \ldots, 0 \le n_k \le b-1$, then (N+1) is also representable in base b.

Proof: Given an integer N such that $N = \left(\sum_{i=0}^{k} n_i \times b^i\right)$, then we have $N+1 = \left(\sum_{i=0}^{k} n_i \times b^i\right) + 1$. To show the representation for (N+1), we prove the theorem for two cases. The first is for all coefficients n_i equivalent to b-1. The second case occurs when there exist an integer m less than or equal to k such that $n_m < (b-1)$.

<u>Case 1: $n_i = (b-1)$ **for** $i = 0, 1, \ldots, k$</u>

Since $n_i = (b-1)$ for $i = 0, 1, \ldots, k$ we can express N as the following sum $N = \left(\sum_{i=0}^{K} (b-1) \times b^i\right)$.

Using the associative and distributive properties of real numbers we have the following equation:

$$N = \sum_{i=0}^{k} b^{i+1} - \sum_{i=0}^{k} b^i = \sum_{i=1}^{k+1} b^i - \sum_{i=0}^{k} b^i.$$

Hence upon incrementing N by 1 we have

$N+1 = 1 + \sum_{i=1}^{k+1} b^i - \sum_{i=0}^{k} b^i$. By associating the number 1 with the first summation term we get $N+1 = \sum_{i=0}^{k+1} b^i - \sum_{i=0}^{k} b^i$. Upon expanding both summations, the second summation terms removes all but the largest term of the first summation term.

Therefore $N+1 = b^{k+1}$, and N+1 is represented in base b.

Case 2: $n_i = (b-1)$ for $i < m$ and $n_m < (b-1)$ where $0 \le m \le k$

By partitioning the summation for the representation of N, we have the following equation: $N = \sum_{i=0}^{m-1} n_i \times b^i + n_m \times b^m + \sum_{i=m+1}^{k} n_i \times b^i$.

Since $n_i = (b-1)$ for $i < m$ we rewrite the above equation as follows: $N = \sum_{i=0}^{m-1} (b-1) \times b^i + n_m \times b^m + \sum_{i=m+1}^{k} n_i \times b^i$. From case 1 above the first summation of this expression can be rewritten to give us the following equation:

$$N+1 = b^m - 1 + n_m \times b^m + \sum_{i=m+1}^{k} n_i \times b^i + 1$$

Since $n_m < (b-1)$ we get $n_m + 1 < b$. Hence the expression

$$N+1 = (n_m + 1) \times b^m + \sum_{i=m+1}^{k} n_i \times b^i$$ is a valid representation for

(N+1) in base b.

In conclusion, if N is representable in base b then N+1 is also representable in base b. Theorem 1.1.1 leads to

Theorem 1.1.2 Any positive integer b > 1, may form a base to represent the set of counting numbers.

Proof: Since zero (0) is representable by any positive base b, the set of counting numbers that are representable in base b is not an empty set. For any positive integer N, we can apply Theorem 1.1.1 N-times beginning with zero (0). Hence we have proven that any positive integer is representable in base b.

Since the negative integers are a mirror image of the positive numbers, Theorem 1.1.2 is easily extendable to the set of integers. Theorem 1.1.3 below states that all integers are representable within a base b where b > 1.

Theorem 1.1.3 Any positive integer b, where b > 1, forms a base that may represent the set of integers.

As promised earlier in this section, we will now convert a representation of a number in base 10 to its corresponding representation in base b (>1). From theorem 1.1.3 above, we know that for any integer "b > 1" it can serve as a basis for the integers. Hence we can express the integer "N" as a finite sum

$\sum_{i=m+1}^{k} n_i \times b^i$. To compute the terms n_i (where i = 0,....,k) of this series, we introduce the following sequential algorithm.

1. Introduce the term $N_0 = N$,

2. For any positive number X, introduce the nomenclature [X] = the smallest integer greater than or equal to X,
3. Introduce the term $N_i = [N_{i-1}/b]$ where $i > 1$,
4. Introduce the term $n_i = N_i - N_{i+1} \cdot b$ (i = 0,.....)and
5. Continue this sequential algorithm until $N_{k+1} = 0$.

This sequential algorithm results in the terms of the basis; n_i for i=0,...,k. Note that this algorithm is designed exclusively for non-negative integers. For negative integers, simply place a negative sign in front of the representation of the corresponding positive number. When the basis b is greater than ten, we will use the letters A, B, C,.... to correspond to 10, 11, 12, etcetera.

Example 1: Using the above sequential algorithm, convert the number 100 to its representation in base 16.

1. $N_1 = [100/16] = 6$ and $n_0 = 100 - 6 \cdot 16 = 4$
2. $N_2 = [6/16] = 0$ and $n_1 = 6 - 0 \cdot 16 = 6$. (The sequential approach ends, since $N_2 = 0$.)

Hence 100 (base 10) has the representation 64 in base 16.

Example 2: To convert to base 10 representation from an arbitrary basis **b**, we use the invariance properties of numbers by simply conducting the basic arithmetic operations of the numbers in base 10. Using this approach, derive the base 10 representation of the base 16 representation of the number 64

To convert the number **AA** base 16 to its corresponding base 10 representation, we simply expand the following: N = **(10)***16+**(10)** = 160 + 10 = 170 (Note A=10 in base 16.). Hence the base 10 representation of the number is 170.

Exercises

For the following base 10 representation, convert the numbers to the representation in the base specified.

1.	48 to base 2	2.	1000 to base 12
3.	96 to base 3	4.	11 to base 5
5.	512 to base 16	6.	51 to base 11
7.	143 to base 4	8.	91 to base 8
9.	48 to base 24	10.	151 to base 9

For the following base b representations, convert the numbers to the corresponding base 10 representation.

11.	b = 9, N = 84	12.	b = 12, N = BB
13.	b = 7, N = 36	14.	b = 2, N = 10001001
15.	b = 16, N = A11	16.	b = 4, N = 302
17.	b = 3, N = 2201	18.	b = 11, N = A11
19.	b = 8, N = 167	20.	b = 5, N = 4403

1.1.2 Rational Numbers

When the physical world around us consists only of whole items, the set of integers is adequate for measuring size. But in many instances, parts of the whole must be considered while taking measurements. Consider the example of a single pie that is divided into five slices. If two of the five slices are sold, how do we measure the quantity of pie that is sold. In this instance we must extend the set of integers to fractions or what we define as rational numbers.

The above example illustrates the need for an extension of the integers. Using the following, Definition 1.1.4, we extend the integers to the rational numbers.

Definition 1.1.4 A number **r** is defined to be a rational number if it can be expressed as the ratio of two integers **m/n** where **n** is not equal to zero.

With this definition we begin by illustrating the rationals along a number line as shown in Figure 2 below. Note, that we use a solid line to represent the rationals. The solid line indicates that the rationals are dense (There is an infinite number of rational numbers within an interval.) within any given interval of rationals. This leads to Theorem 1.1.1 which indicates that every interval of rational numbers is dense.

Figure 2 Graphical Representation Of Rational Numbers

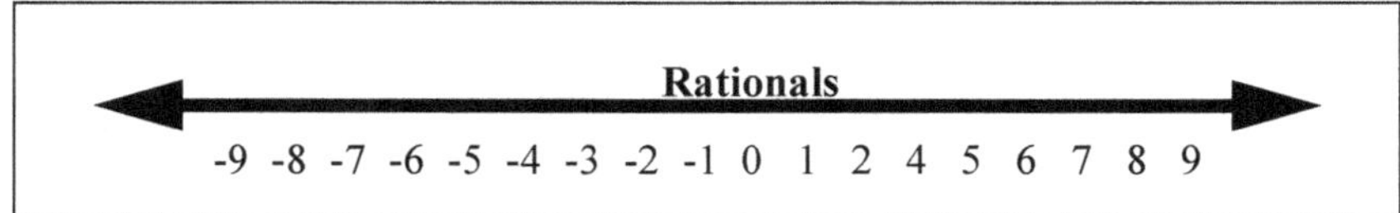

Theorem 1.1.1 For any two rational numbers $r_1 = m_1/n_1$ and $r_2 = m_2/n_2$ there is another rational number that lies between them. Hence for each pair of rational numbers there are an infinite number of rational numbers lying between them.

Proof: To prove this theorem, show that the rational number $r = (m_1 + m_2)/(n_1 + n_2)$ satisfy the inequality $r_1 < r < r_2$. This proof is left as an exercise for the reader.

(Representing A Number Less Than One (1) In Base b)

Notation: If the number R is a non-negative number, then the expression $[R]$ represents the largest positive integer that is less than or equal to R.

Consider the non-negative number r that is less than 1. Then if we multiply it by b and subtract $[b \times R]$, we get the following expression:

$$(R_1 = b \times R - a_1) \text{ , where } a_1 = [b \times R] \text{ .}$$

Since a_1 is the largest positive integer less than $b \times R$, it is also less than b. Furthermore we get that R_1 is a non-negative number that is less than 1. Hence we can continue this process for a finite number of times to construct the following sequence of non-negative numbers less than 1.

$$R_1 = b \times R - a_1 \quad \text{where} \quad a_1 = [b \times R] \text{ ,}$$

$$R_2 = b \times R_1 - a_2 = b^2 \times R - (a_1 \times b - a_2) \quad \text{where} \quad a_2 = [b \times R_1] \text{ ,}$$

$$R_3 = b \times R_2 - a_3 = b^3 \times R - (a_1 \times b^2 - a_2 \times b - a_3) \text{ where } a_3 = [b \times R_2] \text{ ,}$$

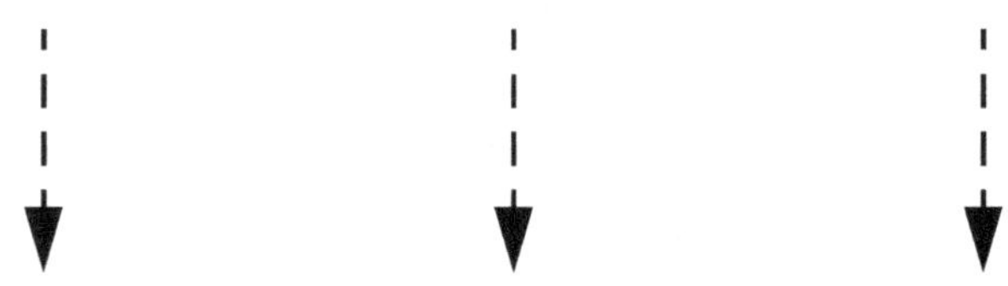

$$R_k = b \times R_{k-1} - a_k = b^k \times R - \sum_{i=1}^{k} a_i \times b^{k-i} \quad \text{where} \quad a_k = [b \times R_{k-1}].$$

Up to this point we have expressed each R_i as a linear expression of R. Using some basic arithmetic we can express R in terms of R_k. The resulting expression is the following: $R = \left(\sum_{i=1}^{k} a_i \times b^{k-i}\right) / b^k + (R_k / b^k)$ or

$$R = \left(\sum_{i=1}^{k} a_i / b^i\right) + (R_k / b^k).$$

But remember for each term of the sequence R_k, we have the following inequality $0 \le R_k < 1$. With this inequality we can squeeze R between two series as in the following: $\left(\sum_{i=1}^{k} a_i / b^i\right) \le R < \left(\sum_{i=1}^{k} a_i / b^i\right) + (1 / b^k)$. As **k** becomes infinitely large we can express R as the limit of a series as **k** approaches infinity. In summation notation, this becomes the following expression:

$$R = \left(\sum_{i=1}^{\infty} a_i / b^i\right).$$

I leave the burden of proving that this series is a convergent series as an exercise for the reader. With this expression we now have the following theorem.

Theorem 1.1.2 For all non-negative numbers R, such that $R < 1$, there is an infinite sequence of non-negative integers $a_i < b$ such that $R = \left(\sum_{i=1}^{\infty} a_i / b^i\right)$. Hence for all non-negative integers less than 1, we can construct a representation in base **b**, using the negative powers of **b**.

(Properties Of Rational Numbers)

When attempting to identify a rational number that is represented in base **b**, it is not always obvious that the number can be expressed as the ratio of two integers. Consider the number 1.2345 (base 10). By notation, it is actually the following summation: $1.2345 = 1 + 2\times10^{-1} + 3\times10^{-2} + 4\times10^{-3} + 5\times10^{-4}$. We can therefore express this number as the following mixed number 1.2345 = 1 + 2345/10000. By combining the whole number and the fraction we have 1.2345 = 12345/ 10000. Hence, it is expressed as a ratio of two integers. We can extend this logic to all bases **b** and state that if a number can be expressed as a finite sum of power terms in base **b**, then it is a rational number.

So we have illustrated that a number that can be expressed as a summation of a finite number of terms of powers of **b** then it is a rational number. Now we question if there are numbers that are expressed as an infinite series of powers of **b** that are rational. The answer to this question is “yes”. Theorem 1.1.3 shows that if a number is an infinite series in powers of base **b** and it is repeating then it is a rational number.

Note, using the definition of a repeating number in base b, all series having only a finite number of powers are considered repeating.

Theorem 1.1.3 Let R be a number that is expressed as the sum of a whole number and a fraction and the fraction is the sum of an infinite number of powers of **b**. Then we can write $R = N + r$, where N is an integer, and $r = \left(\sum_{i=1}^{\infty} a_i / b^i\right)$. Using the notation introduced in the proof of Theorem 1.1.2, we state that r is repetitive in base b if there are two distinct positive integers K and L such that $r_K = r_L$. If r is repetitive in base **b**, then it is a rational number.

Proof:

Let the rational number R be expressed as the sum of a whole number and a fraction that is less than one (1). The expression for R is $R = N + r$, where N is an integer, and $r = \left(\sum_{i=1}^{\infty} a_i / b^i\right)$. If r is repetitive in base b then there are two distinct integers K and L such that $r_K = r_L$ where

$$r_K = b \times r_{K-1} - a_K = b^K \times r - \sum_{i=1}^{K} a_i \times b^{K-i} \;;\; a_K = [b \times r_{K-1}] \text{ and}$$

$$r_L = b \times r_{L-1} - a_L = b^L \times r - \sum_{i=1}^{L} a_i \times b^{L-} ;\; a_L = [b \times r_{L-1}] \,.$$

With this equality we now have a solvable linear equation for the value of r;

$b^K \times r - \sum_{i=1}^{K} a_i \times b^{K-i} = b^L \times r - \sum_{i=1}^{L} a_i \times b^{L-i}$. Since K and L are distinct integers we have the following solution for this equation: $r = M/(b^L - b^K)$ and M is an integer given by the following equation:

$$M = \sum_{i=1}^{L} a_i \times b^{L-i} - \sum_{i=1}^{K} a_i \times b^{K-i}.$$

Hence we expressed r in terms of the ratio of two integers. Therefore

we have $R = N + M/(b^L - b^K) = (N \times (b^L - b^K) + M)/(b^L - b^K)$. Therefore we can conclude that R is a rational number.

For Theorem 1.1.3, we have established two possible instances of rational numbers. The first instance is a number with a fractional part that can be represented in base **b** using a finite series of negative powers of **b**. The second instance is a number with a fractional part that is a repetitive series in the negative powers of base **b**. Since the finite series is a special instance of the repetitive series, this begs the question "what is a sufficient condition in terms of the series of base b powers for a number to be rational?". The sufficiency of the repetitive series is shown in Theorem 1.1.4 below.

Theorem 1.1.4 A number R is a rational number if and only if its fractional part can be expressed as a repetitive series of the negative powers of an arbitrary base **b**.

Proof: The "if" clause of this theorem is readily proved by the use of Theorem 1.1.3. The "only if" clause of this theorem is left as an exercise for the reader.

The conversion of a rational number when the fractional portion is present (not an integer) requires two algorithms. First we represent the rational number as the sum of an integer and a rational number with value less than 1. Again we will apply the algorithm to non-negative numbers since the negative number conversion is simply the corresponding positive number conversion preceded by a negative sign. To represent the rational number R as the sum of an integer and the fractional part (rational number less than 1) R = N + r; we use the formulas N = [R] and r = R - [R]. The integer portion of the rational number is then obtained using the sequential algorithm shown in section 1.1.1. For the fractional portion, the following algorithm is used:

1. Introduce the notation $r_0 = r$,
2. Introduce the term $m_i = [b \cdot r_{i-1}]$
3. Define the ith term $r_i = b \cdot r_{i-1} - m_i$
4. Continue this algorithm until either $r_k = 0$ or until a sequence of numbers begin to repeat.

In this algorithm, the m_i terms are used to represent the fractional portion **r** of the rational number as follows: $r = \sum_{i=0}^{\infty} m_i/b^i$. For convenience we introduce the notation R =.$m_1m_2m_3$...(repeating sequence of numbers) when a repeating sequence of numbers is in the representation. An example in the representation of the number 1/3 in decimal (base 10) notation; 1/3 = 0.(3) (equals 0.33333.....).

Example 3: For this example, we will convert the decimal number 0.(3) to its base 9 representation. Since it is a positive number less than one, we are only concerned with the representation of the fractional portion of the number. Using the approach described above we have the following steps:

1. r_0 = 0.33333....
2. m_1 = [9*0.333] = (not determined yet). The integer portion of this multiplication is not yet clear, since we have not yet defined the product of a number with an infinite repeating terms. To over-

come this obstacle, we use the properties of multiplying by powers of 10. In this instance 10*(0.333….) = (9 + 1)*(0.333….) = (3.333….). Using the distribution properties of real numbers we get 9*(0.333…) + (0.333…) = 3 + (0.333…). The term (0.333…) occurs on both sides of the equation hence we have 9*(0.333…) = 3. Hence we write $m_1 = 3$.

3. $r_1 = 9*r_0 - m_1 = 9*(0.333...) - 3 = 0$. (We stop at this point.)

Hence the representation of 0.333… in base 9 is 0.1.

Example 4: This example highlights the fact that a repeating fraction is a rational number (It can be expressed as a ratio of two integers.). Determine the smallest integers whose ratio gives the decimal number 0.(7). The approach for finding these two integers is shown in the following steps.

1. Label original number as R = 0.(7).
2. Multiply original number by 10 to get 10*R = 7.(7).
3. Substitute for the value of 0.(7) to yield 10*R = 7 + R.
4. Combining like terms we get 9*R = 7 or R = 7/9.

Hence this repeating number is a rational number and 7/9 = 0.(7). Since these integers have no factors in common, they are the smallest integers whose ratio is equal to this number.

Exercises

Using the algorithm described above, convert the following base 10 numbers to the base specified.

1. 1.2 to base 3
2. 12.3 to base 16
3. 25.7 to base 2
4. 101.25 to base 4
5. 84.(3) to base 12
6. 0.333(3) to base 3
7. 0.222 to base 5
8. 23.44 to base 7
9. 148.42 to base 15
10. 201.2121 to base 14

Express the following decimal numbers as ratios of integers. Determine the smallest pair of integers with ratio resulting in the specified decimal number.

11. 0.512
12. 0.(3)
13. 0.(9)14
14. 1.55
15. 2.1(4)16
16. 0.(12)
17. 1.(30)18
18. 1.(03)
19. 0.12(201)20
20. 2.(112)

1.1.3 Irrational Numbers

The previous section discussed rational numbers, which are numbers that can be expressed as ratios of two integers. What about those numbers that cannot be expressed as a ratio of two integers? Can such numbers exist? This section gives examples of such numbers. Such numbers are called irrational numbers and where realized by the Greeks more than 2000 years ago.

The first example of an irrational number is the square root of two (2). We will discuss the square root function in more detail in the following section. To avoid discussing the square root function at this point we examine the square of the number that is equivalent to 2. Theorem 1.1.4 proves that there is no rational number such its square is equal to 2. Hence, the square root of 2 is not a rational number. It is an irrational number.

Theorem 1.1.4 There does not exist a rational number $R = m/n$ such that $2 = R^2$.

Proof: Since we are only concerned with the ratio of the number m/n, we can stipulate that m and n have no common factors. In other words we can assume that m and n are the smallest integers (integers closest to zero) satisfying the relationship $2 = m^2/n^2$. This relationship can be rewritten as $2 \times n^2 = m^2$.

Since m^2 is an even number, m must also be an even number. That is, there exist an integer a such that $m = 2 \times a$. Substituting this expression we have $2 \times n^2 = 4 \times a^2$ or $n^2 = 2 \times a^2$.

Since n^2 is an even number, n must also be an even number. That is, there exist an integer b such that $n = 2 \times b$.

This implies that m and n are not the smallest integers such that the square of their ratios is equal to 2. This is a contradiction, hence we cannot have a ratio satisfying this relationship.

In this section, we proved that $\sqrt{2}$ is an irrational number. Now that we know that it is an irrational number, we need a method for computing it. Specifically, we need to compute the square root of a number. To compute the square root of a number, we use a popular sequential algorithm. The algorithm, for computing the square root of **x**, begins with an initial guess x_0 and then we derive the elements $x_{k+1} = (x/x_k + x_k)/2$ of the sequence (where $k \geq 0$). For positive numbers less than 100, this sequential algorithm converges within 15 decimal places (base 10) of the true value after 7 iterations.

Example 5: Using the algorithm described above, show the first 4 iterations of the sequence described for computing the square root of two (2). Show at least 10 decimal places to illustrate the non repeating nature of these numbers.

1. For the square root of 2, lets begin with an initial guess of 1.5. The initial four elements of this sequence is given in steps 2 through 5.
2. $x_1 = (2/1.5 + 1.5)/2 = 1.4166667$,
3. $x_2 = (2/1.4166667 + 1.4166667)/2 = 1.4142157$
4. $x_3 = (2/1.4142157 + 1.4142157)/2 = 1.4142136$
5. $x_4 = (2/1.4142136 + 1.4142136)/2 = 1.4142136$
6. Hence, the square root of 2 is approximately 1.4142135624 (with a precision of 9 decimal places).

Example 6: Later in this chapter, we will discuss trigonometric concepts and the well known irrational physical constant π (pronounced pi). Like the previous irrational number, $\sqrt{2}$, π can be approximated using elements of a sequence. The sequence that we will use in this example is defined as

$Y_{k+1} = Y_k / C_{k+1}$ where $C_{k+1} = \sqrt{(1 + C_k)/2}$ and

$\{Y_0 = 2, C_0 = 0\}$. Using this sequence, compute the initial six elements of the sequence (Compute each element with 6 decimal places.).

1. $C_1 = 0.707107$ and $Y_1 = 2.828427$
2. $C_2 = 0.923880$ and $Y_2 = 3.061467$
3. $C_3 = 0.980785$ and $Y_3 = 3.121445$
4. $C_4 = 0.995185$ and $Y_4 = 3.136548$
5. $C_5 = 0.998795$ and $Y_5 = 3.140331$
6. $C_6 = 0.999699$ and $Y_6 = 3.141277$
7. Hence π can be approximated as 3.141277. A more precise estimate is obtained by computing a higher order element of the sequence.

Exercises

These exercises illustrate the computation of irrational numbers as was shown in the examples above.

1. Compute value of $\sqrt{3}$ with 8 significant decimal places.

2. Compute value of $\sqrt{7}$ with 8 significant decimal places.
3. Compute value of π with 10 significant decimal places.
4. Compute value of $\sqrt{19}$ with 12 significant decimal places.
5. Compute value of $\sqrt{17}$ with 12 significant decimal places.

1.2 Sequences, Limits, Functions And Continuous Functions

An example that was introduced in section 1.1 was that of converting the decimal number 0.9999(9) as a ratio of positive integers. We showed that this number represents the number 1. But how can a fraction (a number less than 1) be a whole number. Is this a contradiction or is something wrong with the logic of converting decimal numbers to ratios of integers. The answer lies in the notion of limits. If we think of each of the finite representations $(r_1 = 0.9), (r_2 = 0.99)$ and so forth, then we have $1.0 = \lim_{n \to \infty} r_n$. The concept of limits is discussed below.

For this section, we discuss some of the basic concepts and definitions of mathematics. These concepts will greatly increase our ability to derive some basic theorems that are important to the underlying mathematics of this material.

Definition 1.2.1 - A mapping is a transformation of elements contained in one space to that of another space. A mapping may be one-to-many, many-to-one or one-to-one.

Definition 1.2.2 - A function is a mapping that is many-to-one or one-to-one.

Definitions 1.2.3 - The domain of a function is the set of independent values over which it is defined.

Definition 1.2.4 - The range of a function is the space containing the mapped values of the function.

Definition 1.2.5 - A function is called a discrete function if it maps the set of integers or a subset integers to a set of values. When the domain of the discrete real function is the set of integers, it is called a sequence of real numbers.

Definition 1.2.6 - A function is a called a real function if its domain and range are subsets of real numbers.

Definition 1.2.7 - A real function is continuous about a point in its domain if for any selected (interval containing real numbers) about its corresponding value in the range, we can specify a neighborhood in the domain about the domain point such that the function maps all values of this domain neighborhood into the selected neighborhood of the range.

In mathematical relationship terminology, we define a real function continuous about a point x_0 in the domain of the function if for any d > 0, we can find an h > 0 such that for x in the interval $x_0 - h < x < x_0 + d$, have $f(x_0) - d < f(x) < f(x_0) + d$.

When dealing with continuous functions, it is helpful to be able to verify the continuity or non-continuity of composites of continuous functions. For example, if both functions f(x) and g(x) are continuous, can we state that f(x) + g(x) is continuous? It turns out that the sum of two continuous functions is itself a continuous function. For the product of two continuous functions, we will show, in example 1 below, that the product is a continuous function.

Definition 1.2.8 - The limit of a sequence of real numbers exists if there is a real value x_0 such for any selected neighborhood about the value x_0, we can find a finite positive integer N_0 that for all integer values greater than N_0, the sequence of real values is mapped into the neighborhood about the value x_0.

In mathematical relationship terminology, we define a sequence $\{ r_N \}$ as converging to r_0 if for d > 0 we can find a large integer N_0 such that for all N > N_0 we have $r_0 - d < r_N < r_0 + d$.

In section 1.1, we converted the decimal representation 0.999(9) to its fractional counterpart and that it is a representation of the whole number 1. But, how can a fraction be a whole number? Is there something wrong with the logic of the algorithms? The answer to this question is **NO**. The representation 0.999(9) is a sequence of numbers. If we consider the nth term of the number to be the decimal number with 'n' nines (9) behind the decimal point, then the limit of sequence is equivalent to 1. This concept leads to the following corollary for representing all numbers including irrational numbers as a limit of a sequence of rational numbers.

Corollary 1.2.1 - Any real number can be expressed as a limit of a sequence of rational numbers.

Proof: The proof of this theorem is an application of representing any number in a given base **b**. Recall that any number "**r**" can be represented in a base using an infinite sequence of $\{a_i\}$ such that

$r = W + \sum_{i=1}^{\infty} a_i/b^i$. If we define $m_N = W + \sum_{i=1}^{N} a_i/b^i$, then the $\{rn_N\}$ converges to the value r.

Hence each real number can be expressed as a limit of a sequence of rational numbers.

Example 7: Given two continuous functions, f(x) and g(x), show that the function $f(x) \cdot g(x)$ is a continuous function.

1. Since both functions f(x) and g(x) are continuous we have, for any real value x_0 of the domain, each the following:

For any real number $\delta > 0$, there is a number $h_1 > 0$ such that for all values contained in the interval $x_0 - h_1 < x < x_0 + h_1$, we have

$$f(x_0) - \delta < f(x) < f(x_0) + \delta \cdot$$

For any real number $\delta > 0$, there is a number $h_2 > 0$ such that for all values contained in the interval $x_0 - h_2 < x < x_0 + h_2$, we have

$$g(x_0) - \delta < g(x) < g(x_0) + \delta \cdot$$

2. For the real number $\delta > 0$, we can apply the inequalities in step 1 to obtain

$$|f(x) - f(x_0)| \cdot |g(x)| + |g(x) - g(x_0)| \cdot |f(x_0)| < h_1 \cdot |g(x)| + h_2 \cdot |f(x_0)| \cdot$$

3. Combining the terms on the left side of the inequality in step 2, we obtain the following inequality:

$$|f(x) \cdot g(x) - f(x_0) \cdot g(x_0)| < h_1 \cdot |g(x)| + h_2 \cdot |f(x_0)| \cdot$$

4. From the inequalities established in step 1, we obtain

$$|g(x)| < |g(x_0) + \delta| < |g(x_0)| + \delta \cdot$$

5. Combining inequalities in steps 3 and 4 and defining

$h = h_1 \cdot (|g(x_0)| + \delta) + h_2 \cdot |f(x_0)|$, we have

$|f(x) \cdot g(x) - f(x_0) \cdot g(x_0)| < h$. Therefore, for any real value x_0 of the domain and for each $\delta > 0$, there is a number $h > 0$ such that $|f(x) \cdot g(x) - f(x_0) \cdot g(x_0)| < h$.

6. From the definition of a continuous function over a domain and the results shown in step 5, we have shown that the product of a continuous functions is again a continuous function.

Exercises

(Series, Sequences and Limits)

1. Under what conditions will the sequence p^N converge as N approaches infinity. What is (are) the convergent value(s).

2. Prove that the series $\sum_{k=0}^{N} p^k$ is equivalent to $\frac{1-p^{N+1}}{1-p}$.

3. Prove that the series $\sum_{k=1}^{N} k$ is equivalent to $\frac{N \cdot (N+1)}{2}$.

4. Prove that the series $\sum_{k=1}^{N} k^2$ is equivalent to $\frac{N \cdot (N+1) \cdot (2 \cdot N+1)}{6}$.

5. Compute the following limit $\lim_{x \to \infty} \frac{4 \cdot x^3 + x^2 - 1}{7 \cdot x^3 - 1}$.

(Continuous Functions)

6. Using a similar approach as shown in the example above, prove that for any two continuous functions $f(x)$ *and* $g(x)$, the quotient $f(x) / g(x)$ is a continuous function over the domain of numbers for which it is defined.

7. Prove that $f(x) = x^2$ is a continuous function.

8. Prove that $f(x) = 1/x$ is a continuous function.

9. Prove that $f(x) = 1/x^2$ is a continuous function for $x \neq 0$.

10. Prove that $f(x) = |x|$ is a continuous function.

1.3 Linear Functions

A function f(x) is a linear function over the domain of real numbers if the following relationship holds for all real values x and y:

$f(x+y) = f(x) + f(y) - b$. The value of 'b' is constant for all values of x and y. With this definition, for the linear function, we derive the following properties of linear functions by proving corollaries 1.3.1 through 1.3.2 and theorem 1.3.1.

Corollary 1.3.1 If f(x) is a linear function such that

$f(x+y) = f(x) + f(y) - b$ for all real values of x and y, then $f(0) = b$.

Proof: By selecting y = 0 in the definition of a linear function, the equation in the definition is rewritten as $f(x+0) = f(x) + f(0) - b$. Using the identity property of zero, we replace the left side of this equation to yield $f(x) = f(x) + f(0) - b$. Since this is true for all real values of 'x', we can subtract $f(x)$ from both sides of the equation. This yields the following results: $f(0) = b$. Hence, the statement of the corollary is proven to be true.

With corollary 1.3.1, we can rewrite the definition as $f(x+y) = f(x)+f(y)-f(0)$.Now, let's investigate the properties of $f(r \cdot x)$, where 'r' is a real number. From corollary 1.3.2 below, we derive $f(r \cdot x)$ as a function of 'r' and $f(x)$.

Corollary 1.3.2 If $f(x)$ is a linear function, then for any real number "r" $f(r \cdot x) = r \cdot f(x) - (r-1) \cdot f(0)$.

Proof: The proof of this corollary is split into four parts; **r** as an integer, **r** as rational number, **r** as an irrational number and **r** as a negative number.

(A. When r Is A Positive Integer) When **r** is an integer, we have $f(r \cdot x) = f((r-1) \cdot x + x)$. By definition of linear function this equation can be rewritten as follows:

$$f(r \cdot x) = f((r-1) \cdot x) + f(x) - f(0)$$

. If we reapply the definition of a linear function to the term $f((r-1) \cdot x)$ and the subsequent terms for (r-1) times, we get the following: $(A)........f(r \cdot x) = r \cdot f(x) - (r-1) \cdot f(0)$.

(B. When r Is A Positive Rational Number) In this instance 'r' is a rational number and by definition of a rational number, we can represent it as the ratio of two integers. We will assume 'r' can be represented as the following ratio: $r = m/n$. With 'm' being an integer, the application of part (A) of this proof yields the following:

$$f((m/n) \cdot x) = m \cdot f(x/n) - (m-1) \cdot f(0) .$$

Now all that is left is to derive the relationship of $f(x/n)$ as it relates to $f(x)$. If we use the identity $x = n/(x/n)$, we have

$$f(x) = n \cdot f(x/n) - (n-1) \cdot f(0)$$

. Solving for $f(x/n)$ in this equations, we are able to rewrite it as follows:

$$f(x/n) = f(x)/n + (n-1) \cdot f(0)/n.$$

Through substitution for $f(x/n)$, we can rewrite the equation for $f(((m/n) \cdot x))$ as follows:

$$f((m/n) \cdot x) = (m/n) \cdot f(x) + m \cdot (n-1) \cdot f(0)/n - (m-1) \cdot f(0)$$

.

Combining the coefficients for the $f(0)$ terms on the right hand side of the equation, we reduce the above relationship to the following:

$$f((m/n) \cdot x) = (m/n) \cdot f(x) - (m/n - 1) \cdot f(0) .$$

Therefore, if 'r' is a positive rational number we have

$$(B)\ldots\ldots\ldots f(r \cdot x) = r \cdot f(x) - (r-1) \cdot f(0) \;.$$

(C. When 'r' is A Positive Irrational Number) In considering the case in which 'r' is a positive irrational number we define the linear function for the irrational coefficient to be the following:

$f(r \cdot x) = \lim_{N \to \infty} f(r_N \cdot x)$, where the sequence $\{r_N\}$ is an infinite sequence of positive rational numbers such that $r = \lim_{N \to \infty} r_N$.

Since for any finite value of 'N', r_N is a positive rational number, we can apply part **(B)** of the proof of this corollary and rewrite the following relationship for $f(r \cdot x)$ as the following:

$f(r \cdot x) = \lim\limits_{N \to \infty} (r_N \cdot f(x) - (r_N - 1) \cdot f(0))$. Since the limit of a sum is the sum of a limit, we replace the right side of this equation as follows: $f(r \cdot x) = \lim\limits_{N \to \infty} (r_N) \cdot f(x) - \lim\limits_{N \to \infty} (r_N - 1) \cdot f(0)$.

Since $r = \lim\limits_{N \to \infty} r_N$, we have shown the corollary is true for irrational positive numbers. Hence we have the following for positive irrational numbers: $(C)........f(r \cdot x) = r \cdot f(x) - (r-1) \cdot f(0)$.

(D. When r Is A Negative Real Number) Let **r** be a negative real number. Then (-r) is a positive real number and

$(-r) \cdot x + r \cdot x = 0$. Hence, using this identity, we have

$f((-r) \cdot x + r \cdot x) = f(0)$. Now by applying the definition for a linear function we have the following relationship:

$f((-r) \cdot x) + f(r \cdot x) - f(0) = f(0)$. Since (-r) is a positive real number, we can apply (A), (B) and (C) parts of the proof of this corollary to obtain the following relationship:

$(-r) \cdot f(x) - ((-r) - 1) \cdot f(0) + f(r \cdot x) - f(0) = f(0)$. Collecting all coefficients of $f(0)$ and constructing a relationship for $f(r \cdot x)$, we have the following relationship:

$(D)........f(r \cdot x) = r \cdot f(x) - (r-1) \cdot f(0)$ when **r** is a negative real number.

Combining parts (A), (B), (C) and (D) we have verified the statement of this corollary.

As we wrap up the discussion of linear functions, it would be convenient to construct a form of the linear function that would be easy to apply when evaluating the function for various real values. Applying corollary 1.3.2, we can derive theorem 1.3.1 below.

Theorem 1.3.3 If $f(x)$ is a linear function, then the general form of the function is $f(x) = a \cdot x + b$, where both 'a' and 'b' are real numbers such that the intercept $b = f(0)$ and slope $a = f(1) - f(0)$.

Proof: Using the multiplicative identity property of real numbers and applying corollary 1.3.2, we can construct the following relationship: $f(x) = f(x \cdot 1) = x \cdot f(1) - (x-1) \cdot f(0)$. From the distributive and associative properties of real numbers, we can rearrange the terms in the equation to give the following results: $f(x) = x \cdot (f(1) - f(0)) + f(0)$. Hence the statement of the theorem is true.

Example 1: For the linear equation $f(x)$ such that $f(0) = 0.5$ and $f(1) = 2.0$, construct the graph for the function between the values -5 and +5.

Applying theorem 1.3.3 for linear functions we produce the working formula for this function is

$f(x) = x \cdot (2 - 0.5) + 0.5 = 1.5 \cdot x + 0.5$. However, to construct a graph of a linear function, it is sufficient to evaluate the function at two distinct points and use a straight edge to complete the curve. Since we have the values at x = 0 and x = 1, we can use these points.

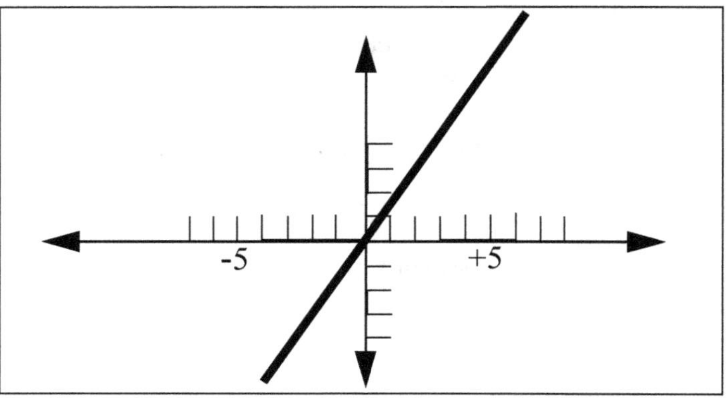

Example 2: Given a linear function $f(x)$ such that $f(0) = -2$ and $f(1) = 2$, evaluate the function at x = 10.

To evaluate this linear function at x = 10, it is easiest to use the results of corollary 1.3.2. Applying corollary 1.3.2, we get $f(10) = f(10 \cdot 1) = 10 \cdot 2 - (9) \cdot (-2)$. Hence we have $f(10) = 38$.

Exercises

For the exercises 1 through 5 below, the functions designated as $f(x)$ are linear functions. Compute the slope (a) and the intercept (b) of the function.

1. $f(1) = 7$ and $f(0) = 0$
2. $f(5) = 6$ and $f(0) = 1$

3. $f(1) = 6$ and $f(-1) = -4$

4. $f(0) = 3$ and $f(10) = 3$

5. $f(2) = 8$ and $f(1) = 6$

(Exercises - General Concepts)

6. Using theorem 1.3.3 show that the slope (a) of the linear function $f(x)$ is $a = \dfrac{f(x_2) - f(x_1)}{x_2 - x_1}$ where x_1 and x_2 are two distinct values in the domain of $f(x)$.

7. Applying theorem 1.3.3 to the linear function $f(x)$ and the two distinct points x_1 and x_2 of its domain, show that the intercept of the function is given by $b = f(x_1) - a \cdot x_1$.

8. For the linear function $f(x)$ and two distinct points x_1 and x_2 of its domain, show that for any arbitrary point x of the domain, we get the relationship $\dfrac{f(x) - f(x_1)}{x - x_1} = \dfrac{f(x_2) - f(x_1)}{x_2 - x_1}$.

9. For the linear function with slope $a = 5$ and $b = -1$, construct the graph of the function between domain values 0 and 10.

10. For the linear function with $f(2) = 8$ and $f(-1) = 2$, construct the graph of the function between domain values -4 and 4.

1.4 Exponential Function

In this section we introduce the exponential family of functions. Like the linear family, a specific relationship is used to identify the defining characteristics of this family of functions. When a function satisfy the following relationship $f(x+y) = f(x) \cdot f(y)$; for all real values of x and y, it is defined to be an exponential function. If we can find two distinct points in the domain of the function such that the function maps each point into distinct values in the range of the function, the exponential function is said to be non-degenerate exponential function.

If we consider the function $f(x) = 0$ for all real values of x, then clearly $f(x+y) = f(x) \cdot f(y)$ since $0 = 0 \cdot 0$. Hence by definition of an exponential function, this is an exponential function. However, we can not find two distinct points in the domain of real numbers that maps into a number other than zero. Hence it is a degenerate exponential function. What is another instance of a degenerate exponential function?

This section derives the properties of the exponential family of functions using the definition above. Similar to the preceding section (Linear Functions), these properties will lead to a standard formula (mathematical expression) for the exponential function. In keeping with the previous trend of establishing corollaries to present the properties, we will prove several corollaries that will be helpful in establishing some important properties of the exponential functions that are used throughout other disciplines of mathematics.

Corollary 1.4.1 If the function $f(x)$ is a nonzero exponential function then $f(0) = 1$ is always true.

Proof: Since the function $f(x)$ is an exponential function it must satisfy the following condition for all real values of **x** and **y**:

$f(x+y) = f(x) \cdot f(y)$. If we let the value of **y** be zero, we get $f(x) = f(x) \cdot f(0)$. Since this is true for all values of **x** in the domain of real numbers and this function is a nonzero exponential function then there is a real value x_0 such that $f(x_0) \neq 0$.

Hence we must have $f(0) = 1$ for all nonzero exponential functions.

Corollary 1.4.1 fixes the corresponding range value for domain value of zero (0) to be one (1). If you are curious, you a probably asking yourself; 'what about the other values of in the range of the exponential function. The next corollary defines the range of the exponential function.

Corollary 1.4.2 If the function $f(x)$ is a nonzero exponential function, then for all real values of **x** the relationship $f(x) > 0$ is true.

Proof: By applying the definition of an exponential function for the identity $x = x/2 + x/2$, we have the relationship $f(x) = f(x/2) \cdot f(x/2)$ or we are able to write $f(x) = (f(x/2))^2$. Therefore each range value can be expressed as the square of a real number. This implies that $f(x) \geq 0$ for all real values of **x**.

To show that the inequality is strictly greater than zero, we will consider the x_0 domain value such that $f(x_0) > 0$. This is possible since $f(x)$ is a nonzero exponential function. Using the identity $x_0 = (x_0 - x) + x$, we are able to write the relationship $f(x_0) = f(x_0 - x) \cdot f(x)$. Since the right side of the equation is greater than zero, we must have that each term on the left side of the equation must be greater than 0 (both are not negative by first part of this proof). Hence we must conclude that $f(x) > 0$ for all real values of 'x'.

With corollary 1.4.2, we can state that the range of any exponential function is the set of positive real numbers. As we derive other properties of this family of functions, this property will prove very useful. Another useful property of the relationship of the function of positive domain values to the function of the negative domain values. This property is illustrated in the following corollary.

Corollary 1.4.3 For every 'x' value in the domain of the nonzero exponential function $f(x)$, the following relationship $f(-x) = \frac{1}{f(x)}$ is true.

Proof: Using the identity $x + (-x) = 0$, applying the definition of the exponential function and applying corollary 1.4.1, the following relationships are true:

1. $f(x + (-x)) = f(0)$

2. $f(x) \cdot f(-x) = 1$.

Applying corollary 1.4.2, the values $f(x)$ and $f(-x)$ are each greater than zero and hence $f(-x) = \frac{1}{f(x)}$ is true.

The above corollary, corollary 1.4.3, presents a natural partition for the domain of the exponential function since its shows that all functions of the negative values of the domain can be expressed as a relationship of the function of the corresponding positive value in the domain. Hence we partition the domain into negative values and non-negative values (We include the zero value with the positive values so that it is not lonely.). In the remainder of this section, all properties will be illustrated for each of these sub-domains. First we will prove the property true for the non-negative sub-domain and then we will use corollary 1.4.3 to infer the properties for the negative sub-domain.

A good starting point for applying this concept is the examination of the set of exponential functions that are strictly increasing over the domain of non-negative real numbers. I realize that the assumption of the existing of a strictly increasing exponential function over the domain of non-negative real numbers is a big leap of faith at this stage, but let's be daring. For such a function we derive corollary 1.4.4 below.

Corollary 1.4.4 If $f(x)$ is a strictly increasing function over the domain of non-negative real numbers, then it is a strictly increasing function over the domain of negative real numbers. Furthermore, it is strictly increasing over the domain of all real numbers.

Proof: To begin this proof, we select two arbitrary negative real numbers from the domain of negative real numbers. We will label these numbers x_1 and x_2 . Also we will assume that we have the following relationship between the numbers: $x_1 < x_2$. In terms of the positive counterparts of these numbers we have $(-x_2) < (-x_1)$. Applying the strictly increasing property of the non-negative real domain and corollary 1.4.3 we get the following properties: **1.** $f(-x_2) < f(-x_1)$

2. $\frac{1}{f(x_2)} < \frac{1}{f(x_1)}$

3. $f(x_1) < f(x_2)$ (Consequence of corollary 1.4.2).

If $f(x)$ is strictly increasing in the domain of non-negative real numbers we must have $f(x) > 1$ for all positive values of 'x' since $f(0) = 1$ (corollary 1.4.1). With the domain of positive real numbers mapping into the range of real values greater than 1 (one), the domain of negative numbers must map into values less than 1. This is a implied from corollary 1.4.3. Hence all negative real values map into values that are less than the range values mapped by the non-negative real values of domain. Therefore, the exponential function is strictly increasing over the domain of all real numbers when it is strictly increasing over the domain of non-negative real numbers.

Using similar steps when $f(x)$ is a strictly decreasing function over the domain of non-negative real numbers, we can verify corollary 1.4.5 below.

Corollary 1.4.5 If $f(x)$ is a strictly decreasing function over the domain of non-negative real numbers, then it is a strictly decreasing function over the domain of negative real numbers. Furthermore, it is strictly decreasing over the domain of all real numbers.

Proof: The proof for this corollary is similar to the proof for corollary 1.4.4 above and will be left as an exercise for the reader.

The above arguments for showing an increasing or a decreasing exponential function is good, but it does ask for a leap of faith from the user. So you may ask "Under what conditions can you be certain that a non-degenerate exponential function may be strictly increasing or strictly decreasing over the domain of non-negative real numbers." It turns out that there is a simple condition for establishing "strictly increasing" or "strictly decreasing" over the domain of non-negative real values. To establish this criterion, we prove corollary 1.4.6 below.

Corollary 1.4.6 If there is a value x_0 in the domain of non-negative real numbers such that $f(x_0) > 1$, for the non-degenerate exponential function $f(x)$, then the function $f(x)$ is strictly increasing over the domain of real numbers.

Proof: We will prove this corollary for the domain of non-negative real numbers, then we will extend the assertion of strictly increasing to the domain of all real numbers by applying corollary 1.4.4. In the statement of this corollary, we have the existence of a non-negative number x_0 such that $f(x_0) > 1$. If we have a positive integer 'N', then the following identity holds: $x_0 = N \cdot (x_0 / N)$. By applying the definition of exponential function to the quantity $f(N \cdot (x_0 / N))$ we establish the relationship

$f(x_0) = (f(x_0 / N))^N$.

(A.) Since $f(x_0) > 1$, we must have that

$f(x_0) = (f(x_0 / N))^N > 1$. With N-products of a single number being greater than one (1) we must have the individual number being greater than one (1). Hence $f(x_0 / N) > 1$ and all K-products, $(f(x_0 / N))^K$ (with K > 0), are also greater than zero (0). But recall that $f((K / N) \cdot x_0) = (f(x_0 / N))^K$ and hence $f((K / N) \cdot x_0) > 1$ for all K > 0. With this inequality, we can now prove the statement of this corollary.

To prove the statement of this corollary, it is sufficient to prove the relationship of $f(r \cdot x)$ to $f(x)$ where 'r' is a real number and 'x' is any arbitrary non-negative domain value.

(B. 'r' Is A Positive Rational Number) When 'r' is a positive rational number, then there are two positive integers 'M' and 'N' such that $r = M / N$. If $r < 1$, we can write the identity $1 = M / N + (N - M) / N$ and we have $(N - M) > 0$. With this identify and the definition of an exponential function the relationship $f(x_0) = f((M / N) \cdot x_0) . f(((N - M) / N) \cdot x_0)$ is true. Since the left side of this equation is greater than one and both terms of the product on the right side of the equation is greater than one, we must have that each of the terms on the right is less than the expression on the right. Therefore, for $r < 1$, we have $f(x_0) > f(r \cdot x_0)$.

When 'r' is a positive rational number such that $r > 1$, then we can represent 'r' as the ratio of two positive integers 'M' and 'N' such that $M > N$. and therefore we have the identity

$M/N = 1 + (N-M)/N$, where the inequality $(N-M) > 0$ is true. From this identity we have the relationship

$f((M/N) \cdot x_0) = f(x_0) . f(((N-M)/N) \cdot x_0)$. From statement **(A)** above, the quantities $f((M/N) \cdot x_0)$ and $f(((N-M)/N) \cdot x_0)$ are greater than one which implies that $f(r \cdot x_0) > f(x_0)$ for $r > 1$.

In general terms (Using similar logic as seen above.), if there are two positive rational numbers, r_1 and r_2 , such that $r_1 < r_2$. Then we must have $f(r_1 \cdot x_0) < f(r_2 \cdot x_0)$. This shows that for the domain or real positive values 'x' that can be expressed as the product of a ration number times x_0 , the non-degenerate exponential function $f(x)$ is strictly increasing when $f(x_0) > 1$.

(C. 'r' Is A Positive Irrational Number) To extend this corollary for all positive real numbers it is sufficient to show that $f(x)$ is strictly increasing for all $x = r \cdot x_0$, where 'r' is a positive irrational number. To define the function $f(r \cdot x_0)$ when 'r' is an irrational number we use the following limit: $f(r \cdot x_0) = \lim_{N \to \infty} f(r_N \cdot x_0)$; where the sequence $\{r_N\}$ is such that $r = \lim_{N \to \infty} r_N$. With this definition for the function $f(r \cdot x_0)$, for 'r' an irrational number, the first task that we will undertake is to verify that it is greater than one for positive irrational values of 'r'.

Using the definition of limits; for any $\delta > 0$ there is an N_0 such that for all $N > N_0$ we must have

$f(r_N \cdot x_0) - \delta < f(r \cdot x_0) < f(r_N \cdot x_0) + \delta$. From part (B) of

this proof, we showed that for r_N a positive rational number we have $f(r_N \cdot x_0) > 1$. Therefore we have the following inequality: $1 - \delta < f(r \cdot x_0)$. Since $\delta > 0$, is arbitrarily close to zero, we must have $1 < f(r \cdot x_0)$. From part **(B)**, this inequality is also true when 'r' is a rational number and hence it is true for all positive real numbers.

Now, let's get back to the original statement of this corollary. If we select a positive real valued number r_1 such that $r_1 < r_2$, then we can write the following relationship:

$$f(r_2 \cdot x_0) - f(r_1 \cdot x_0) = f(r_1 \cdot x_0) \cdot (f((r_2 - r_1) \cdot x_0) - 1) .$$

Since the relation $(r_2 - r_1) \cdot x_0 > 0$ is true we have that the right side of the relation above is positive and hence

$f(r_2 \cdot x_0) - f(r_1 \cdot x_0)$ is positive. This implies that $f(r \cdot x_0)$ is strictly increasing as 'r' increases in the domain of positive irrational numbers. Combining this with the results of part **(B)**, we conclude that $f(x)$ is strictly increasing over the domain of positive real numbers.

(D. Behavior of $f(x)$ Over Domain Of Negative Numbers) To prove that this function is strictly increasing over the set of negative numbers, we select two negative numbers x_1 and x_2 such that $x_1 < x_2$. Then from part **(C)** we have $f(-x_2) < f(-x_1)$ since $0 < -x_2 < -x_1$. Applying corollary 1.4.3 we can rewrite the inequality as $\frac{1}{f(x_2)} < \frac{1}{f(x_1)}$ and hence we have $f(x_1) < f(x_2)$. Since x_1 and x_2 are arbitrary negative values, we conclude that $f(x)$ is strictly increasing over the domain of negative values.

(E. $f(x)$ is strictly increasing over the domain of real numbers) The proof that $f(x)$ is strictly increasing follows from parts **(B)**, **(C)**, **(D)** and the fact that for negative values the inequality $f(x) < 1$ is true and for positive values the inequality $f(x) > 1$ is true.

Now that we have established conditions under which a non-degenerate exponential function is strictly increasing, we ask ourselves "under what conditions is the function strictly decreasing?" If you guessed the existence of a positive value **x** such that $f(x) < 1$, then you are correct. This condition is stated formally in corollary 1.4.7 below.

Corollary 1.4.7 If there is a value x_0 in the domain of non-negative real numbers such that $f(x_0) < 1$, for the non-degenerate exponential function $f(x)$, then the function $f(x)$ is strictly decreasing over the domain of real numbers.

Proof: This proof is similar to the proof for corollary 1.4.6.

Another property of this non-degenerate exponential function is that it is a continuous function over the domain of real numbers. I have chosen not to show a rigorous proof for this property, I leave it as an exercise for the reader. A hint for such a proof is to evaluate $f(x \cdot (1 + 1/N))$ and apply corollaries 1.4.6 and 1.4.7 for each small interval.

It is not unusual, in the study of mathematics, to use limits of a function to estimate the behavior of the function at specific points in the domain. One case, as shown in the two preceding corollaries, is the evaluation of a continuous function when the domain's point of interest is an irrational number. Since an irrational number can be expressed as the limit of an infinite sequence of rational numbers, we define the function of an irrational number as the limit of the function over a sequence of rational numbers that has the irrational number of interest as its limit.

For the non-degenerate exponential function, another subset of interest is a small interval of real numbers about the point zero in the domain. Can we find a linear function, $a \cdot x + b$, that approximates the exponential function as the value in the domain approaches zero? Looking ahead, it turns out that for values close to zero we can define a linear function that approximates the exponential function in the small interval. When the value of the domain point is zero, the exponential function maps into the real value one (1). Using **theorem 9**, the linear function for approximating any non-degenerate exponential function must be of the form $a \cdot x + 1$ since $f(0) = 1$ for all non-degenerate exponential functions.

Using the approximation $f(x) \cong a \cdot x + 1$, an estimate for the constant 'a' is $\frac{f(x) - 1}{x}$, where 'x' is a real value contained in a small neighborhood about zero. But, there are an infinite number of values in any given interval about zero. This is true no matter how small the interval of real numbers. To make this a unique estimate, let's use the domain value zero. But the function $\frac{f(x) - 1}{x}$ has 'x' in the denominator and division by zero is undefined. This statement is only true if we cannot define a finite limit of this function as 'x' approaches zero. If there is a finite limit that exists, then define the value of this function to be $\lim_{x \to \infty} \frac{f(x) - 1}{x}$. Some properties of this function that will prove useful in deriving the estimate are stated in the corollaries 1.4.8, 1.4.9 and 1.4.10 below.

Corollary 1.4.8 If the function $f(x)$, is a non-degenerate exponential function, then $\frac{f(x) - 1}{x}$ is a continuous function.

Proof: Since this function is the ratio of two continuous functions, it is a continuous function for those real values for which it is defined.

Other properties of this ratio, is that of strictly increasing or strictly decreasing. The corollaries 1.4.9 and 1.4.10 describe the conditions under which the function $\frac{f(x) - 1}{x}$ is either strictly increasing or strictly decreasing.

Corollary 1.4.9 If the function $f(x)$, is a strictly increasing exponential function, then $\frac{f(x)-1}{x}$ is a strictly increasing function.

Proof: This proof is first accomplished for positive values of the domain. The negative values of the domain will be included at the conclusion of this proof.

Let x_0 be a positive value and let N be a positive integer. Applying the definition of the exponential function "N" times we have that relationship $\frac{f(x_0)-1}{x_0} = \frac{f^N(x_0/N)-1}{x_0}$. Remember that for a sequential sum of integer powers of a real number, we have the following relationship $\frac{p^N-1}{p-1} = 1+p+\ldots+p^{N-1}$. Applying this relationship to this function we have

$$\frac{f(x_0)-1}{x_0} = \frac{f(x_0/N)-1}{(x_0/N)} \cdot \frac{\left(1+f(x_0/N)+\ldots+f(x_0/N)^{N-1}\right)}{N}.$$

With $x_0 > 0$ and $f(x)$ strictly increasing, we get $f(x_0) > 1$ and hence $\frac{\left(1+f(x_0/N)+\ldots+f(x_0/N)^{N-1}\right)}{N} > \frac{N}{N} = 1$. Therefore the following inequality $\frac{f(x_0)-1}{x_0} > \frac{f(x_0/N)-1}{(x_0/N)}$ is true.

Using similar power representations for the quantities

$$\frac{f(\frac{M+1}{N}\cdot x_0)-1}{\left(\frac{M+1}{N}\cdot x_0\right)} \quad \text{and} \quad \frac{f(\frac{M}{N}\cdot x_0)-1}{\left(\frac{M}{N}\cdot x_0\right)}$$

, we have the following equalities any positive integer 'M':

$$\frac{f(\frac{M+1}{N}\cdot x_0)-1}{\left(\frac{M+1}{N}\cdot x_0\right)} = \frac{f(x_0/N)-1}{(x_0/N)}\cdot\frac{(1+f(x_0/N)+\ldots+f(x_0/N)^M)}{M+1}$$

and

$$\frac{f(\frac{M}{N}\cdot x_0)-1}{\left(\frac{M}{N}\cdot x_0\right)} = \frac{f(x_0/N)-1}{(x_0/N)}\cdot\frac{(1+f(x_0/N)+\ldots+f(x_0/N)^{M-1})}{M}.$$

By defining 'd' to be the difference between the power series in

each term: $d = \frac{(1+\ldots+f(x_0/N)^M)}{M+1} - \frac{(1+\ldots+f(x_0/N)^{M-1})}{M}$,

we can write the following relationship:

$$\frac{f(\frac{M+1}{N}\cdot x_0)-1}{\left(\frac{M+1}{N}\cdot x_0\right)} - \frac{f(\frac{M}{N}\cdot x_0)-1}{\left(\frac{M}{N}\cdot x_0\right)} = \frac{f(x_0/N)-1}{(x_0/N)}\cdot d$$

. If we can establish that 'd' is positive (or negative), we can establish a relationship for the left side of the above equation. We begin by combining the like terms in the defining equation for 'd'. This step yields the relationship $$d = \frac{\sum_{k=0}^{M-1}\left(f(\frac{M}{N}\cdot x_0) - f(\frac{k}{N}\cdot x_0)\right)}{M\cdot(M+1)}.$$

With $f(x)$ being a strictly increasing function and **k** less than **M**, we have $f(\frac{M}{N}\cdot x_0) > f(\frac{k}{N}\cdot x_0)$. Hence each term of the summation is positive and therefore 'd' is positive. Now that we have established the inequality $\frac{f(\frac{M+1}{N}\cdot x_0) - 1}{\left(\frac{M+1}{N}\cdot x_0\right)} > \frac{f(\frac{M}{N}\cdot x_0) - 1}{\left(\frac{M}{N}\cdot x_0\right)}$ for all positive integers **M**, we can apply it sequentially **N** times and conclude that $\frac{f(x_0) - 1}{x_0} > \frac{f(\frac{M}{N}\cdot x_0) - 1}{\left(\frac{M}{N}\cdot x_0\right)}$ for all 'M' such that $0 < M < N$.

Letting $\frac{M}{N}$ represent any rational number **r**, then we can conclude that for any rational number **r** such that $r < 1$:

$\frac{f(x_0) - 1}{x_0} > \frac{f(r\cdot x_0) - 1}{(r\cdot x_0)}$ .Using the fact that a positive irrational number can be expressed as the limit of a sequence of positive rational numbers we extend the above inequality to all positive real numbers **r** such that $r < 1$. Hence, we conclude that this function is strictly increasing over the domain of all positive real numbers. To verify that this function is strictly decreasing over the domain of negative numbers, we use a similar approach as used for positive values. (I leave this as an exercise for the reader.) Hence we conclude that this function is strictly increasing when the conditions of the corollary are met.

Since the function $\frac{f(x)-1}{x}$ is strictly increasing over both positive and negative numbers and it is a continuous function, then it must be strictly increasing over the domain of all real numbers.

Just as you might expect, if the function $f(x)$ is a strictly decreasing exponential function, then the function $\frac{f(x)-1}{x}$ is strictly decreasing. This is formally stated in corollary 1.4.10 below.

Corollary 1.4.10 If the function $f(x)$, is a strictly decreasing exponential function, then $\frac{f(x)-1}{x}$ is a strictly decreasing function.

Proof: (The proof is left as an exercise for the reader.)

At the beginning of the discussion for the properties of the function $\frac{f(x)-1}{x}$, we stated an interest in computing $\lim_{x \to 0} \frac{f(x)-1}{x}$. The following theorem gives us conditions for the convergence of the function of the function as 'x' approaches zero.

Theorem 1.4.11 If $f(x)$ is a strictly increasing or strictly decreasing exponential function and "a" is the largest real number such that $f((x) \geq 1 + a \cdot x)$ for all real numbers 'x', then we must have

$$\lim_{x \to 0} \frac{f(x) - 1}{x} = a \cdot$$

Proof: From the statement of this theorem we have $f((x) \geq 1 + a \cdot x)$. Hence we have the inequalities:

A. $\frac{f(x) - 1}{x} \geq a$ for $x > 0$ and

B. $\frac{f(x) - 1}{x} \leq a$ for $x < 0$.

Using the continuous property of $\frac{f(x) - 1}{x}$, we have, for any

$\delta > 0$, there is an $h > 0$ such that for any 'x' for which

$-h < x < h$ then $-\delta < \frac{f(x) - 1}{x} - \frac{f(-x) - 1}{(-x)} < \delta$ or

$\frac{f(-x) - 1}{(-x)} - \delta < \frac{f(x) - 1}{x} < \frac{f(-x) - 1}{(-x)} + \delta$. Assuming 'x' positive, we

apply inequality **B** above to get $\frac{f(-x) - 1}{(-x)} \leq a$ and therefore

$\frac{f(-x) - 1}{(-x)} - \delta < \frac{f(x) - 1}{x} < \delta + a$. But from inequality **A** above we have that $a \le \frac{f(x) - 1}{x} < \delta + a$. This implies that for each $\delta > 0$, there is an $h > 0$ such that for any 'x' for which $0 < x < h$ then $\left|\frac{f(x) - 1}{x} - a\right| < \delta$. We chose 'x' as a positive value in this proof. If we choose x to be negative, the same logic applies and the same results is obtained for $-h < x < 0$. Now we can state the above for any 'x' for which $-h < x < h$. Therefore, by definition of limit we have $\lim_{x \to 0} \frac{f(x) - 1}{x} = a$. Hence the theorem is proven true for all values of 'x'.

The general form of the exponential function is derived from the basic definition of the exponential function. If we let $f(1) = B$, then we call 'B' the base of the exponential function. Applying the definition of the exponential function 'N' times we get $f(N) = B^N$. For any rational number $r = M/N$, we apply the definition of the exponential to produce $f(M/N)^N = f(M) = B^M$. By definition of Nth root, we can rewrite the previous relationship as $f(M/N) = (B^M)^{1/N} = B^{M/N}$ and hence $f(r) = B^r$ for all rational numbers 'r'. Using the limit of a sequence of rational numbers to represent an irrational number we define $f(r) = B^r = \lim_{n \to \infty} B^{r_n}$ where $r = \lim_{n \to \infty} r_n$. Therefore we conclude that the general form of the exponential function is B^x for all real values of 'x'. Furthermore, we have that it is strictly decreasing when $B < 1$ and strictly increasing when $B > 1$.

We conclude the discussion of the exponential with the following theorem. The proof of the theorem is left as an exercise for the reader.

Theorem 1.4.12 If 'a' is a real number such that $\lim_{x \to 0} \frac{f(x) - 1}{x} = a$, then

we have $\lim_{N \to 0} (1 + a \cdot x / N)^N = B^x$.

Proof: (The proof of this theorem is shown in the following example.)

In the special case of above theorem, when a = 1, we have

$\lim_{N \to 0} (1 + x / N)^N = B^x$. Furthermore, if we evaluate the limit at x = 1, we can establish the value of 'B' in this special case. In this instance, we get the Naperian constant $e \cong 2.718...$ or $\lim_{N \to 0} (1 + 1 / N)^N = e$.

Example

The proof of theorem 1.4.12 showcases the concepts of convergence of functions and continuous functions. Using the concepts introduced in this section, we will now verify the statement of theorem 1.4.12 in the statements below.

1. By definition of $\lim_{y \to 0} \frac{f(y) - 1}{y} = a$, we have for any arbitrary number $\delta_1 > 0$ there is an $h_1 > 0$ such that for all y contained in the interval $-h_1 < y < h_1$ we have

   $-\delta_1 < \frac{f(y) - 1}{y} - a < \delta_1$.

2. By making the substitution $y = x/N$ where x is a constant arbitrary value, then y approaches zero as N becomes larger (approaches infinity). The limit in step 1 implies, for $N > h_1/|x|$, we get the inequality $-\delta_1 < \frac{f(x/N) - 1}{(x/N)} - a < \delta_1$.

3. Using basic arithmetic operations, we convert inequalities in step 2 to the inequalities

   $(1 + a \cdot x/N) - \delta_1 \cdot x/N < f(x/N) < (1 + a \cdot x/N) + \delta_1 \cdot x/N$.

4. By the definition of the constant **a**, we have $(1 + a \cdot x/N) \le f(x/N)$. If we raise both sides of the equation to the power of **N**, we get the inequality $(1 + a \cdot x/N)^N \le f(x)$, for all values of **x** using the properties of exponential functions.

5. Raising the right inequality in step 3 to the power of **N**, we get the inequality $f(x) < (1 + a \cdot x / N + \delta_1 \cdot x / N)^N$ or

$f(x) < (1 + (a + \delta_1) \cdot x / N)^N$. For each value of **N**, the function $(1 + z \cdot x / N)^N$ is continuous over the domain of **z** and hence for any arbitrary number $\delta > 0$ there is an $h > 0$ such that for all z contained in the interval $a - h < z < a + h$ we have

$(1 + a \cdot x / N)^N - \delta < (1 + z \cdot x / N)^N < (1 + a \cdot x / N)^N + \delta$.

6. With the inequalities in steps 4 and 5 we get for any arbitrary number $\delta > 0$ there is an N_1 such that for all

$N > N_1(h_1 / |x|)$, we have

$(1 + a \cdot x / N)^N - \delta < f(x) < (1 + a \cdot x / N)^N + \delta$ or

$f(x) - \delta < (1 + a \cdot x / N)^N < f(x) + \delta$ and hence

$\lim_{N \to 0} (1 + a \cdot x / N)^N = f(x) = B^x$.

Exercises

Using the techniques shown in this section, work through the logic of the following problems.

1. Prove the statement of corollary 1.4.5 that the exponential function is strictly decreasing over the domain of all real numbers when it is strictly decreasing over the domain of non-negative real numbers.
2. Prove the statement of corollary 1.4.7.

3. Prove the statement of corollary 1.4.10 for the non-degenerate exponential function $f(x)$.

4. Using the sequence $(1 + a \cdot x / N)^N$, determine the approximate value of "B" when the value of "a" is one (1).
5. Let the value of "B" have the value of ten (10), determine the approximate value of "a".

1.5 Logarithm Function

The previous section showed that the non-constant exponential function is either strictly increasing or strictly decreasing. For such functions, the mapping is one-to-one from the domain to the range. For one-to-one functions, the concept of an inverse function is given by the following definition.

Definition 1.5.1 If the function f(x) is a one-to-one function, then g(x) is an inverse function of f(x) if for all x in the domain we have $g(f(x)) = x$. Furthermore, the range of f(x) is equivalent to the domain of g(x).

Definition 1.5.2 The inverse function of a non-constant exponential function is called a logarithm function. If 'B' is the base of the exponential function then we use the notation $log(x)_B$.

The first property that is examined for the $log(x)_B$ function is the identification of its inverse function. As you would expect, its inverse is the exponential function.

Corollary 1.5.3 The inverse function of the $log(x)_B$ function is B^x.

Proof: By definition of inverse function, we have $log(B^x)_B = x$. Letting $y = B^x$, we have $log(y)_B = x$. Now we compute $B^x = B^{log(y)_B}$ or we may restate this as $y = B^{log(y)_B}$. By definition of inverse function, B^x is an inverse function of $log(x)_B$.

The next property of logarithms deals with computing the logarithm of the products of positive real numbers.

Corollary 1.5.4 The logarithm of the product of two real numbers in the domain of positive real numbers is the sum of the logarithms of each real number. The mathematical formula for this property is

$$log(x \cdot y)_B = log(x)_B + log(y)_B .$$

Proof: Using corollary 1.5.1, we have $x = B^{log(x)_B}$ and $y = B^{log(y)_B}$. If we compute the product of 'x' times 'y' we get the following equation $x \cdot y = B^{log(x)_B} . B^{log(y)_B}$. Recall from the definition of exponential functions that the function of a sum is equivalent to the product of each function. This gives the relationship $x \cdot y = B^{log(x)_B + log(y)_B}$. Since $log(x)_B$ is the inverse function of B^x, when we take the logarithm of each side of this equation we get $log(x \cdot y)_B = log(x)_B + log(y)_B$. Hence the statement of the corollary is true.

Now we examine the logarithm of the ratio of two positive numbers.

Corollary 1.5.5 The logarithm of the ratio of two real numbers in the domain of positive real numbers is the difference of the logarithms of each real number. The mathematical formula for this property is

$$log(x/y)_B = log(x)_B - log(y)_B \ .$$

Proof: Using the identify $x = (x/y) \cdot y$, and applying corollary 1.5.2 we have $log(x)_B = log(x/y)_B + log(y)_B$. Subtracting the $log(y)_B$ term from each side, the corollary statement is proven true.

As with the exponential function, we end this section with a limit that proves very useful in the study of logarithms in mathematics.

Theorem 1.5.6 Let B^x be an exponential function with **a**; the largest real number such that $B^x \geq (1 + a \cdot x)$ for all real numbers **x**. Then for the logarithm function, $log(x)_B$, we have

$$\lim_{h \to 0} \frac{log(1 + a \cdot h \cdot x)_B}{h} = x \ .$$

Proof: The proof of this theorem is left to the reader as an exercise. (Apply theorem 1.4.1.)

Exercises

To extend the properties of the family of logarithm functions, solve the following problems.

1. Given the exponential function B^x with an associated value **a**, the largest real number such that $B^x \geq (1 + a \cdot x)$ for all real numbers **x**. Prove for the logarithm function, $log(x)_B$, we have $\lim_{h \to 0} \frac{log(1 + a \cdot h \cdot x)_B}{h} = x$.

2. Let **x** be a positive real number and **c** a real number where $y = x^c$; show that $log(y)_B = c \cdot log(y)_B$.

3. If **A** and **B** are two positive numbers prove that $log(y)_B = log(y)_A / log(B)_A$.

1.6 Trigonometric Functions

Using the concept of similar triangles and the relationship of the sides of the right triangles to each other, the concept of trigonometric relationships are formed. As we progress in the discussion of matrices, the concept of trigonometric identities will become more important. They are useful in translated some matrix operations into physical applications. To begin our discussion, we examine some basic concepts about triangles. These basic concept are traditionally introduced in the study of geometry. So think of this section as a brief review of geometric concepts.

Definition 1.6.1 Two distinct triangles are similar triangles if their corresponding angles are equal in measurement.

Using definition 1.6.1, the relationship of corresponding segments opposite corresponding angles is derived. This relationship is stated in theorem 1.6.1 below.

Theorem 1.6.1 Let the two distinct triangles ABC and abc be similar triangles. If the length of the segments of the first triangle are A, B, C and the length of the corresponding segments of the second triangle are a, b, c; then the following relationship is true: $A/a = B/b = C/c$.

Proof: The underlying principle of this proof is that the area of a specific triangle is constant regardless of how we compute the area of the triangle. The proof of this theorem will first be accomplished for a right triangle (a triangle with one right angle (90 degree angle). Then we will extend the proof to all triangles.

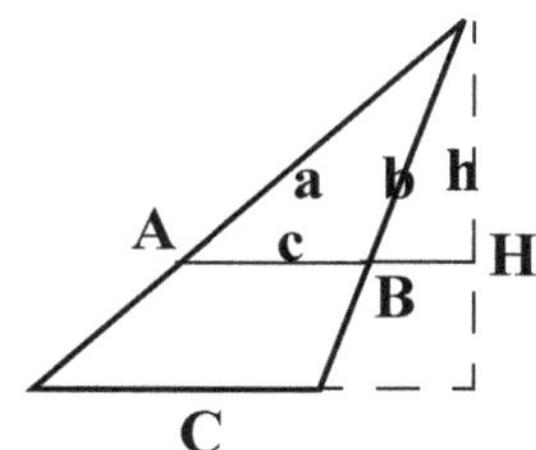

To begin, let's construct two distinct triangles. The smaller triangle will be enclosed by the larger triangle. We will denote the length of the sides of the larger triangle as **A**, **B** and **C**. We will denote the length of the corresponding sides of the smaller triangle as **a**, **b** and **c**. The area of the larger triangle is computed to be $\frac{1}{2} \cdot H \cdot C$. Since the larger triangle is made up of two mutually exclusive shapes (one triangle and a trapezoid) its area is equivalent to the sum of the individual areas that comprise the large area. The sum of the areas of these mutually exclusive shapes is $\frac{1}{2} \cdot h \cdot c + \frac{1}{2} \cdot (H - h) \cdot (C + c)$ .Through basic arithmetic operations such as distributive property, commutative property and associative property of numbers, this cumulative area is rewritten as $\frac{1}{2} \cdot H \cdot C + \frac{1}{2} \cdot H \cdot c - \frac{1}{2} \cdot h \cdot C$. Since the area of the large triangle is an invariant quantity, these two approaches yield equivalent values. Hence we have the equality $\frac{1}{2} \cdot H \cdot C = \frac{1}{2} \cdot H \cdot C + \frac{1}{2} \cdot H \cdot c - \frac{1}{2} \cdot h \cdot C$. Combining like terms, we get $\frac{1}{2} \cdot H \cdot c = \frac{1}{2} \cdot h \cdot C$ or $\frac{H}{h} = \frac{C}{c}$. This equality can be applied to each of the sides of the triangle, whether it's opposite an acute angle, a right angle or an obtuse angle. In each case, the 'H' and 'h' represent corresponding heights perpendicular to the segment of interest.

More generally, we can multiply this equation by $\frac{C}{c}$ to yield

$$\frac{C^2}{c^2} = \frac{Area(ABC)}{Area(abc)} \quad \text{or} \quad \frac{C}{c} = \frac{\sqrt{Area(ABC)}}{\sqrt{Area(abc)}} .$$

The ratio of the areas is a constant value, regardless of which segment we are using as the base. Hence we have the equality

$A/a = B/b = C/c$.

Using theorem 1.6.1 for pairs of equalities, we can establish that the ratio of corresponding sides is constant for similar triangles. Mathematically we write $A/C = a/c$, $B/C = b/c$ and $A/B = a/b$. Having established the constancy of the ratio of the corresponding segment lengths for similar triangles, we are now able to define the family of trigonometric functions.

Definition 1.6.2 For the family of triangles having at least one angle of measure θ, the generalized sine function of the angle α also contained in the triangle, is defined as the ratio of the length of the segment opposite the angle α, over the length of the segment opposite the angle θ. Using this diagram, we express this generalized sine function as $sine_{\theta}(\alpha) = \frac{A}{C}$.

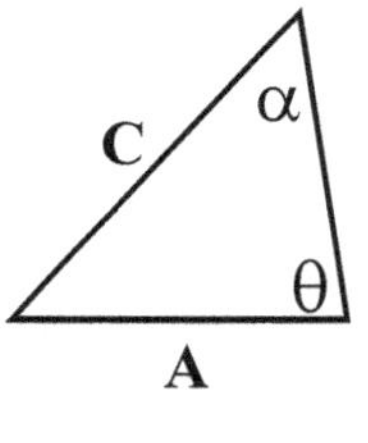

Definition 1.6.3 For the family of triangles having at least one angle of measure θ, the generalized cosine function of the angle α also contained in the triangle, is defined as the ratio of the length of the segment adjacent the angle α, over the length of the segment opposite the angle θ. Using this diagram, we express this generalized cosine function as $cosine_{\theta}(\alpha) = \frac{B}{C}$.

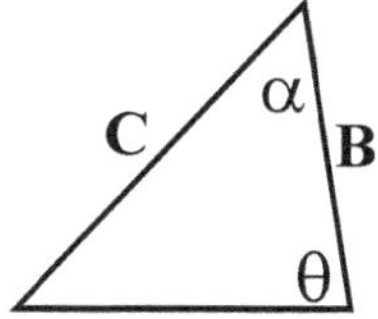

For both definitions, theorem 1.6.1 insures that for each family, specific value of θ, the $sine_{\theta}(\alpha)$ and $cosine_{\theta}(\alpha)$ functions map the angle α into a single real value. Traditionally, the trigonometric function is defined for $\theta = 90°$ and the subscript θ is removed from the function name. For example the function $sine(\alpha)$ is synonymous with the function $sine_{90°}(\alpha)$. This will be the notation throughout the remaining chapters of this book for all trigonometric functions. As tempting as it may be to probe into the properties of the generalized trigonometric functions, we will restrict our attention to the traditional trigonometric functions. However, we will show the relationship of the generalized trigonometric identities to the traditional trigonometric identities.

Using similar triangles, we can prove one of most famous theorems of geometry; Pathagorean Theorem. This will also lead to a relationship of sine() and cosine() functions of an angle.

Theorem 1.6.2 **(Pathagorean Theorem)** Let the triangle ABC be a right triangle such that 'C' (hypotenuse) represents the length of the side opposite the right angle and 'A' and 'B' (legs) represent the lengths of the sides adjacent to the right angle. Then the following relationship is true for the lengths of the sides of the right triangle: $C^2 = A^2 + B^2$. In terms of the other two remaining angles of the triangle this relationship yields

$1 = (sine(\alpha))^2 + (cosine(\alpha))^2$; where 'α' is an angle contained in the right triangle.

Proof: To prove this theorem, consider the right triangle shown in the figure to the right. By drawing a line segment from the vertex at the right angle of the triangle that is perpendicular to the line segment opposite the right angle we can construct three similar triangles.

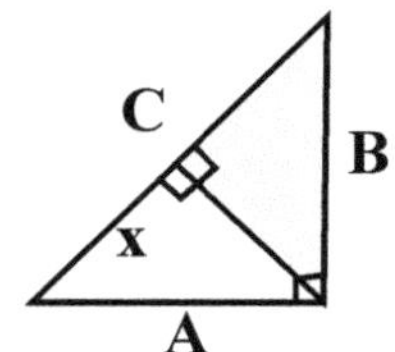

The shaded triangle enclosed in the larger triangle is similar to the larger triangle and therefore, we can apply theorem 1.6.1 for the corresponding segments. This gives us the following equality: $\frac{B}{C} = \frac{(C-x)}{B}$. The unshaded triangle is also similar to the larger triangle that encloses it. Applying theorem 1.6.1 again gives the relationship: $\frac{A}{C} = \frac{x}{A}$. Upon multiplying the first equation by

$B \cdot C$ and the second equation by $A \cdot C$, we get the:

$B^2 = (C^2 - C \cdot x)$ and $A^2 = C \cdot x$.When we combine the two equations we get $A^2 + B^2 = C^2$. Hence we have proven the first statement of the theorem. For the second statement of the theorem, we divide this results by C^2 to get $\frac{A^2}{C^2} + \frac{B^2}{C^2} = 1$. If we assume

the angle opposite segment of length 'A' to be α, then we have the equation $1 = (sine(\alpha))^2 + (cosine(\alpha))^2$ and the theorem is proven true.

Other properties of the traditional trigonometric functions are given in two additional well known theorems of geometry. They are "Law of Sines Theorem" and the "Law of Cosines Theorem". The proof of these theorems are left as exercises for the reader.

Theorem 1.6.3 **(Law Of Sines)** Let the angles of a triangle be α, β, θ and the corresponding lengths of the segments opposite these angles be 'A', 'B', 'C'. Then the following equalities are true $\frac{sin\alpha}{A} = \frac{sin\beta}{B} = \frac{sin\theta}{C}$.

Theorem 1.6.4 **(Law Of Cosines)** Let the angles of a triangle be α, β, θ and the corresponding lengths of the segments opposite these angles be 'A', 'B', 'C'. Then the following equation is true $A^2 + B^2 - 2 \cdot A \cdot B \cdot cos\theta = C^2$.

Without going into too much detail concerning the generalized trigonometric functions, we can express the generalized sine and cosine function in terms of the traditional sine and cosine function. (Note: From this point forward we will use the notation sin() to represent the sine function and cos() to represent the cosine function.)

(Reference Angle $\theta < 90°$ **)** When the reference angle θ is acute, the generalized sine and cosine is computed as follows: $sin_{\theta}(\alpha) = \frac{sin(\alpha)}{sin(\theta)}$ and

$$cos_{\theta}(\alpha) = cos(\alpha) + sin(\alpha) \cdot \frac{cos(\theta)}{sin(\theta)}.$$

(Reference Angle $\theta > 90°$ **)** When the reference angle θ is obtuse, the generalized sine and cosine is computed as follows: $sin_{\theta}(\alpha) = \dfrac{sin(\alpha)}{sin(180° - \theta)}$ and

$$cos_{\theta}(\alpha) = cos(\alpha) - sin(\alpha) \cdot \frac{cos(180° - \theta)}{sin(180° - \theta)}.$$

(Generalized Law Of Sines) Let the angles of a triangle be α, β, θ and the corresponding lengths of the segments opposite these angles be 'A', 'B', 'C'. Then the following equalities are true $\dfrac{sin_{\theta}(\alpha)}{A} = \dfrac{sin_{\theta}(\beta)}{B} = \dfrac{1}{C}$.

(Generalized Law Of Cosines) Let the angles of a triangle be α, β, θ and the corresponding lengths of the segments opposite these angles be 'A', 'B', 'C'. Then the following equation is true

$$(sin_{\theta}(\alpha))^2 + (cos_{\theta}(\alpha))^2 - 2 \cdot sin_{\theta}(\alpha) \cdot cos_{\theta}(\alpha) \cdot cos\theta = 1.$$

Additional properties of the traditional trigonometric functions will be discussed in later chapters as we discuss applications associated with matrices.

Exercises

To extend the concept of the generalized sine and cosine functions, derive the relationships in the following exercises. The user will find the following figure helpful in proving the relationships that follows.

FIGURE 2. **Relationship of Generalized Trigonometric Functions to Standard**

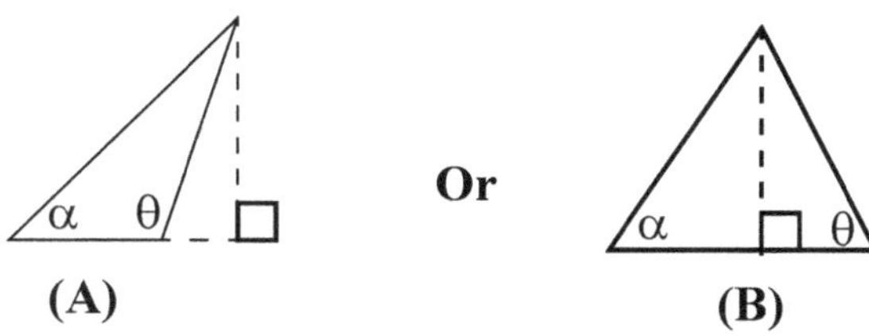

1. Using the triangle labeled with **(A)** in figure 2, prove the following relationship for the generalized sine function for the obtuse angle θ: $sin_\theta(\alpha) = \dfrac{sin(\alpha)}{sin(180° - \theta)}$.

2. Using the triangle labeled with **(B)** in figure 2, prove the following relationship for the generalized sine function for the acute angle θ: $sin_\theta(\alpha) = \dfrac{sin(\alpha)}{sin(\theta)}$.

3. Using the triangle labeled with **(A)** in figure 2, prove the following relationship for the generalized cosine function for the obtuse angle θ:

$$cos_\theta(\alpha) = cos(\alpha) - cos(180° - \theta) \cdot sin_\theta(\alpha) .$$

4. Using the triangle labeled with **(B)** in figure 2, prove the following relationship for the generalized cosine function for the acute angle θ:

$$cos_\theta(\alpha) = cos(\alpha) + cos(\theta) \cdot sin_\theta(\alpha) .$$

(Computing Traditional Sine and Cosine Functions)

Evaluate the traditional trigonometric functions below.

1. $sin(45°)$

2. $cos(45°)$

3. $cos(60°)$ 4. $sin(30°)$

5. $sin(17°)$ 6. $cos(110°)$

(Computing Generalized Sine and Cosine Functions)

Evaluate the generalized trigonometric functions below.

1. $sin_{75°}(45°)$ 2. $cos_{45°}(45°)$

3. $cos_{120°}(60°)$ 4. $sin_{130°}(90°)$

5. $sin_{58°}(27°)$ 6. $cos_{50°}(110°)$

Solving for Segment Lengths Using Generalized Trigonometric Functions)

Using the definition of the generalized trigonometric functions, compute the specified length of the side of the triangle when the following information is known.

1. Compute length of side opposite angle α, when be $\alpha = 25°$, $\theta = 110°$ and length of side opposite angle θ is 10.
2. Compute length of side opposite angle θ, when be $\alpha = 45°$, $\theta = 75°$ and length of side opposite angle α is 8.

CHAPTER 2

Matrix Algebra

In this chapter, we introduce the system of matrix algebra as a physical example of an algebraic group and an algebraic ring. This section defines the binary matrix operations of addition, multiplication and scalar multiplication. In these definitions, the operations are defined in terms of the elements of the matrix. The first section of this chapter is devoted to basic definitions of matrices and concepts that will prove useful in the study of matrices and matrix algebra.

2.1 Basic Definitions

Before we venture into the exciting world of matrix algebra, we need some basic definitions to jump-start our knowledge base. These definitions will serve to define a standard point of reference for the reader.

Definition 2.1.1 An $m \times n$ matrix is a rectangular array of numbers arranged into 'm' rows and 'n' columns. This matrix has a total of $m \cdot n$ (interpreted as the product of m times n) numbers. The elements of the matrix are represented by doubly subscripted values such as $[v_{ij}]$ where $i = 1, \ldots, m$ and $j = 1, \ldots, n$.

An example of a three by four (3×4) matrix is $\boldsymbol{A} = \begin{bmatrix} 1 & 2 & 3 & 11 \\ 0 & 4 & 9 & 8 \\ -1 & 5 & 7 & 6 \end{bmatrix}$. Using the notation that $\boldsymbol{A} = \lfloor a_{ij} \rfloor$, we have that $a_{1,3} = 3$.

Some special cases of the matrix are the row matrix, the column matrix and the square matrix. The row matrix is a $1 \times n$ matrix. The column matrix is a $m \times 1$ matrix. The square matrix is the special case where the number of columns and number rows of a matrix are equal. The formal definitions of these matrices are expressed in Definitions 2.2 through 2.4 below.

Definition 2.1.2 The row matrix is a matrix consisting of a single row of numbers. In terms of an $m \times n$ matrix the dimension of the row matrix has the dimensions $1 \times n$. The elements of the row matrix are denoted by he double subscripted values $[v_{1j}]$, where $j = 1, \ldots, n$.

Definition 2.1.3 The column matrix is a matrix consisting of a single column of numbers. In terms of an $m \times n$ matrix the dimension of the row matrix has the dimensions $m \times 1$. The elements of the row matrix are denoted by the double subscripted values $[v_{i1}]$, where $i = 1, \ldots, m$.

Definition 2.1.4 The square matrix is an $m \times n$ matrix such that $m = n$. In other words, a square matrix has an equivalent number of rows and columns.

Note: Throughout the majority of the remaining chapters, when we reference a matrix in all likelihood we will be referencing a square matrix. If the matrix is not a square matrix, then its dimensions will be explicitly referenced.

Definition 2.1.5 Two matrices $\boldsymbol{A} = \left[a_{ij}\right]$ and $\boldsymbol{B} = \left[b_{ij}\right]$ are said to be equal if for each **i** and **j** combination we have $a_{ij} = b_{ij}$.

Some additional definitions for some more specific matrices will prove useful in some complex computations. These definitions are for zero matrix, diagonal matrix, elementary matrices, row echelon form of a matrix and a triangular matrix.

Definition 2.1.6 The zero matrix is the matrix for which all of its elements have the value of zero. Throughout the remainder of this book we will refer to the $m \times n$ zero matrix as $\varnothing_{\boldsymbol{mn}}$. Since the value of each element of $\varnothing_{\boldsymbol{mn}}$ is zero we have

$$\varnothing_{\boldsymbol{mn}} = [0] \ .$$

Definition 2.1.7 The diagonal matrix is a square matrix such that for $\boldsymbol{D}_{mn} = [d_{ij}]$, we have $d_{ij} = 0$ for $i \neq j$.

Definition 2.1.8 An upper triangular matrix is a matrix $U = [u_{ij}]$ where $u_{ij} = 0$ for $i > j$. A lower triangular matrix is a matrix $L = [b_{ij}]$ where $b_{ij} = 0$ for $i < j$.

Definition 2.1.9 The elementary matrix is a square matrix such that, when it is multiplied to another matrix, it performs elementary row (or column) transformations of the matrix. The elementary operations are (Type I) exchanging position of two rows (or two columns), (Type II) multiplying elements of a single row (or column) by a scalar value and (Type III) adding corresponding elements of one row (or column) to the corresponding elements of another.

The elementary matrices will be discussed in more detail in section 2.5. These matrices form the cornerstone of matrix algebra and will be used to introduce more advanced concepts such as matrix inverses.

Definition 2.1.10 A matrix B is said to be row equivalent to the matrix A if B can be derived from A through a finite sequence of elementary row operations.

Definition 2.1.11 A matrix B is said to be column equivalent to the matrix A if B can be derived from A through a finite sequence of elementary column operations.

More generally, if we have a matrix that is both row equivalent and column equivalent to a second matrix then we state that the original matrix is equivalent to the second matrix. The formal definition for matrices being equivalent is given in the following definition.

Definition 2.1.12 A matrix B is said to be equivalent to the matrix A if there are two matrices $P \text{ and } Q$, both are derived as finite products of elementary matrices (Matrix multiplication will be defined in section 2.3.) such that $B = P \times A \times Q$.

Definition 2.1.13 The row echelon form of a matrix is a matrix such that (1) the top **k** rows of the matrix are non-zero row matrices, (2) for **r** < **k** the element in column **r** of the **r**th row is one and the preceding elements in the row are equal to zero and (3) if there are remaining rows (beyond the initial **k**) they are zero row matrices.

Definition 2.1.14 The column echelon form of a matrix is a matrix such that (1) the lifetimes **k** columns of the matrix are non-zero column matrices, (2) for **r** < **k** the element in row **r** of the **r**th column is one and the preceding elements in the column are equal to zero and (3) if there are remaining columns (beyond the initial **k**) they are zero column matrices.

Another term used when discussing echelon forms of matrices is that of "canonical form of a matrix". A matrix B is said to be the row canonical form of the matrix A if B is row equivalent to the matrix A and B is in row echelon form. Similarly a matrix B is said to be the column canonical form of the matrix A if B is column equivalent to the matrix A and B is in column echelon form.

Definition 2.1.15 The transpose of a matrix is the newly formed matrix derived from the existing matrix by interchanging row values with corresponding column values. In matrix notation, the matrix $B = [b_{ij}]$ is the transpose of matrix $A = [a_{ij}]$ if $b_{ij} = [a_{ji}]$. Nomenclature for a matrix transpose is A^T.

Definition 2.1.16 A symmetric matrix is a square matrix A, such that

$$A^T = A \text{ (i.e. } a_{ij} = a_{ji}\text{)}.$$

A concept that is closely akin to the concept of symmetry is that of skewed symmetry. The definition of skewed symmetry is shown in the following definition.

Definition 2.1.17 A skewed symmetric square matrix is a matrix A, such that $A^T = -A$ (i.e. $a_{ij} = -a_{ji}$). Note that from this definition, the elements along the diagonal (i = j) of a skew symmetric matrix must be zero.

Definition 2.1.18 The product of a scalar (real number) multiplied times a matrix is defined a the matrix with elements equivalent the scalar multiplied times each element of the original matrix. In matrix notation, we write $c \cdot A = [c \cdot a_{ij}]$.

The definitions presented in this section are referenced many times throughout this book and will aid the reader in understanding more complex topics covered later in the book.

Exercises

Perform the following matrix operations specified in the exercises below. Refer to the definitions given in this section.

1. $[6, 12, -1, 5]^T =$

2. $6 \cdot [10, -6, -4] =$

3.

$$3 \cdot \begin{bmatrix} 1 & 0 & 0 \\ 2 & 4 & 0 \\ 4 & 16 & 7 \\ 8 & 32 & 9 \end{bmatrix} =$$

4.

$$\begin{bmatrix} 3 & 7 & 13 \\ 10 & 2 & 20 \\ 4 & 7 & 1 \end{bmatrix}^T =$$

5.

$$5 \cdot \begin{bmatrix} 3 & 3 & 3 \\ 7 & 4 & 2 \\ 6 & 1 & 8 \end{bmatrix} =$$

6.

$$2 \cdot \begin{bmatrix} -12 & -3 \\ 4 & -9 \\ 5 & 8 \end{bmatrix}^T =$$

7.

$$12 \cdot \begin{bmatrix} -3 & 1 & -4 \\ 0 & -5 & 7 \end{bmatrix} =$$

8.

$$-2 \cdot \begin{bmatrix} 13 & 5 & 3 & -3 \\ 0 & 9 & 1 & 8 \\ 0 & 0 & -7 & 1 \\ 0 & 0 & 0 & -4 \end{bmatrix}^T =$$

9.

$$\begin{bmatrix} 6 & -11 \\ -11 & 9 \end{bmatrix}^T =$$

10.

$$\begin{bmatrix} 16 & -8 \\ 8 & 16 \end{bmatrix}^T =$$

2.2 *Matrix Addition*

The concept of addition for matrices is the binary operation that adds the corresponding elements of the matrices. The formal definition of matrix addition is the Definition 2.2.1.

Definition 2.2.1 For any two $m \times n$ matrices $\boldsymbol{A} = [a_{ij}]$ and $\boldsymbol{B} = [b_{ij}]$, matrix addition is denoted by the sum of the matrices is denoted by $\boldsymbol{A} + \boldsymbol{B} = [a_{ij} + b_{ij}]$.

A requirement for adding two matrices is that both matrices be of the same order; $m \times n$ in Definition 2.2.1.

Having defined the binary operation of addition for the set of $m \times n$ matrices, we are ready to investigate the set of $m \times n$ matrices and the binary matrix addition as the algebraic system "the group." The group is a non-empty set on which a binary operation is defined and together they satisfy certain properties. The properties are stated as axioms and are shown in the for a group below.

Definition 2.2.2 A group **G** is a non-empty set of elements for which a binary operation $\oplus$ is defined and the following axioms are true.

Axiom1. The closure law for the binary operation is satisfied for all elements in G. For all pairs A, B contained in G then $A \oplus B$ is also contained in G.

Axiom2. The associative law for the binary operation is satisfied for all elements in G. For all triplets A, B, C contained in G then the equality $(A \oplus B) \oplus C = A \oplus (B \oplus C)$ is true.

Axiom3. There exists an identity element contained in G for the operation $\oplus$. The existence of an identity element in G implies that there is an E contained in G such that for all elements in A we have the following relationship involving the element E: $E \oplus A = A \oplus E = A$.

Axiom4. For each element in G, there exist an inverse element also in G under the operation $\oplus$. The existence of the inverse element under this operation implies that for each element A in G, there exist an element B also in G such that the equality $A \oplus B = B \oplus A = E$ is true.

In chapter one, we used the property of groups to prove certain theorems relating to real numbers. We now investigate the set of matrices that forms a group under matrix addition. With the definition of a group defined, we are now equipped with the tools needed to determine if the binary operation of matrix addition and some specific set of matrices form a group. Two sets of matrices come to mind, when we think in terms of matrix addition. Since by the Definition 2.2.1 the matrices must be identical in size, we consider the set of $m \times n$ matrices. Another set would be the global set of matrices; the set of all real matrices (the set of all matrices having only real elements). It turns out that both of these sets of matrices form a group under the binary operation of matrix addition. This is shown for $m \times n$ matrices in Theorem 2.2.3 below. The proof for the set of all real matrices follows a very similar method of proof.

Theorem 2.2.3 Under the binary operation of matrix addition operator, the set **G** of matrices forms a group.

Proof: To prove that G, the set of $m \times n$ matrices, forms a group under matrix addition we must show that G satisfies axioms 1 through 4 of Definition 2.2.2. Since for any $m \times n$ matrix, we can construct an example that is not null, we have that the set **G** is not empty.

Axiom 1 - To prove axiom 1, let two $m \times n$ matrices $\boldsymbol{A} = [a_{ij}]$ and $\boldsymbol{B} = [b_{ij}]$. Then by Definition 2.2.1, the sum $\boldsymbol{A} + \boldsymbol{B} = [a_{ij} + b_{ij}]$. The resultant matrix is an $m \times n$ matrix and therefore it is contained in G. Hence the law of closure is satisfied for G under matrix addition.

Axiom 2 - To prove axiom 2, let **A**, **B** and **C** be arbitrary elements of G ($m \times n$ matrices). By definition of matrix addition we have $(\boldsymbol{A} + \boldsymbol{B}) + \boldsymbol{C} = [(a_{ij} + b_{ij}) + c_{ij}]$. Since the set of real numbers form a group under addition, the associative law for addition is true; therefore we get $(a_{ij} + b_{ij}) + c_{ij} = a_{ij} + (b_{ij} + c_{ij})$. Hence by definition of matrix addition we have the equality $(\boldsymbol{A} + \boldsymbol{B}) + \boldsymbol{C} = \boldsymbol{A} + (\boldsymbol{B} + \boldsymbol{C})$. Hence the associative law of addition holds for $m \times n$ matrices.

Axiom 3 - To prove axiom 3, we must verify the existence of an identity element. We begin by assuming that there is a matrix **E** contained in G that satisfies the relation $\boldsymbol{A} + \boldsymbol{E} = \boldsymbol{E} + \boldsymbol{A} = \boldsymbol{A}$ for any arbitrary element **A** in G. For this matrix equality to hold true, we must have the following equality among the real elements of both matrices:

$a_{ij} + e_{ij} = e_{ij} + a_{ij} = a_{ij}$. For this equality to hold for real numbers, we must have $e_{ij} = 0$ for all $i = 1, \ldots, m$ and $j = 1, \ldots, n$. This states that such a matrix exists and it is the zero matrix (All elements of matrix are zero.). Therefore the additive identity element of group G is $\varnothing_{\boldsymbol{mn}}$.

Axiom 4 - To prove axiom 4, we must verify the existence of an additive inverse of every element contained in G. If we can show that any arbitrary element of G has an additive inverse then that would be sufficient proof. To begin the proof, let's assume that for any arbitrary matrix **A** contained in G we can find a matrix **B** also contained in G that satisfies the following relationship $\boldsymbol{A} + \boldsymbol{B} = \boldsymbol{B} + \boldsymbol{A} = \varnothing_{\boldsymbol{mn}}$. In terms of the real components of the matrix, we have the following relationship:

$a_{ij} + b_{ij} = b_{ij} + a_{ij} = 0$. This implies that each b_{ij} is the corresponding additive inverse of a_{ij}. For real numbers the additive inverse of a given number is minus one times the value of the given number. Hence b_{ij} = $-a_{ij}$. In matrix notation we rephrase this relationship as $\boldsymbol{B} = [-a_{ij}]$.

Using the definition of scalar multiplying a matrix we have $\boldsymbol{B} = -\boldsymbol{A}$. Therefore the additive inverse for any matrix in G is minus one (-1) times that matrix.

Since G is non-empty and all four axioms of Definition 2.2.2 are satisfied, we have proven that G forms a group under matrix addition.

Example

Using the definition of matrix addition compute $\begin{bmatrix} -2 & 0 \\ 5 & 4 \\ 2 & 3 \end{bmatrix} + \begin{bmatrix} 3 & 2 \\ -1 & 4 \\ 0 & 4 \end{bmatrix} = \begin{bmatrix} 1 & 2 \\ 4 & 8 \\ 2 & 7 \end{bmatrix}$

Exercises

Using the definition of the binary operation of matrix addition, evaluate the expressions in exercises 1 through 10 below.

1. $\begin{bmatrix} -2 & 0 \\ 5 & 4 \\ 2 & 3 \end{bmatrix} + \begin{bmatrix} 3 & 2 \\ -1 & 4 \\ 0 & 4 \end{bmatrix}$

2. $\begin{bmatrix} 3 & -5 \\ 4 & 8 \end{bmatrix} + \begin{bmatrix} 7 & 8 \\ -4 & -9 \end{bmatrix}$

3. $\begin{bmatrix} 1 & -4 & -2 \\ 0 & 6 & 2 \\ -3 & -8 & -3 \end{bmatrix} + \begin{bmatrix} 0 & 4 & 2 \\ 0 & -5 & -2 \\ 3 & & 4 \end{bmatrix}$

4. $\begin{bmatrix} 10 \\ 9 \\ -7 \end{bmatrix} + \begin{bmatrix} 2 \\ 6 \\ 20 \end{bmatrix}$

5. $\begin{bmatrix} 11 & 6 \\ 35 & 11 \\ 10 & -50 \\ 20 & -9 \end{bmatrix} + \begin{bmatrix} 4 & 4 \\ -25 & -9 \\ -8 & 30 \\ 5 & 8 \end{bmatrix}$

6. $\begin{bmatrix} -28 & 54 & 33 \\ -30 & -7 & 47 \end{bmatrix} + \begin{bmatrix} 8 & 16 & 8 \\ 15 & -13 & 3 \end{bmatrix}$

7. $\begin{bmatrix} -1 & -24 \\ 3 & 8 \end{bmatrix} + 3 \cdot \begin{bmatrix} 7 & 12 \\ 9 & 0 \end{bmatrix}$

8. $2 \cdot \begin{bmatrix} -40 \\ 14 \\ 0 \end{bmatrix} + 4 \cdot \begin{bmatrix} 20 \\ -8 \\ -22 \end{bmatrix}$

9. $15 \cdot \begin{bmatrix} -2 & 0 \\ 5 & 1 \\ 2 & 0 \end{bmatrix} + 21 \cdot \begin{bmatrix} 0 & 2 \\ -1 & -2 \\ 0 & 1 \end{bmatrix}$

10. $3 \cdot \begin{bmatrix} 3 & 0 & 0 \\ 2 & 3 & 0 \\ 1 & 2 & 3 \end{bmatrix} - \begin{bmatrix} 1 & -1 & -3 \\ 1 & 2 & 4 \\ 2 & 0 & -1 \end{bmatrix}$

11. Show that any square matrix can be composed (expressed as a sum) of a symmetric matrix and a skew symmetric matrix.

12. Show that the symmetric and skew symmetric composition of the matrix **A** is unique.

2.3 *Matrix Multiplication*

As with real numbers, the next concept under matrix algebra is the concept of matrix multiplication. Matrix by scalar multiplication was defined in section 2.1. This section introduces matrix by matrix multiplication and gives a working definition of matrix multiplication.

Definition 2.3.1 For an $m \times k$ matrix denoted as $\boldsymbol{A} = [a_{ij}]$ where $i = 1, \ldots, m$, $j = 1, \ldots, k$ and for a $k \times n$ matrix denoted as $\boldsymbol{B} = [b_{ij}]$ where $i = 1, \ldots, k$ and $j = 1, \ldots, m$, the product of the two matrices is denoted by $\boldsymbol{C} = \boldsymbol{A} \times \boldsymbol{B} = [c_{ij}]$. The resultant matrix product, **C**, has the elements defined as

$$c_{ij} = \sum_{l=1}^{k} a_{il} \cdot b_{lj} .$$

This definition imposes the requirement that the number of columns of the left matrix (A in Definition 2.3.1) of the product must be equivalent to the number of rows of the right matrix (B in Definition 2.3.1). To illustrate matrix multiplication consider the 3×2 matrix $\begin{bmatrix} 1 & 0 \\ 2 & 3 \\ 0 & 1 \end{bmatrix}$ and the 2×1 matrix $\begin{bmatrix} -1 \\ 2 \end{bmatrix}$.

Using Definition 2.3.1, we compute the product

$$\boldsymbol{A} \times \boldsymbol{B} = \begin{bmatrix} (1)(-1) + (0)(2) \\ (2)(-1) + (3)(2) \\ (0)(-1) + (1)(2) \end{bmatrix} = \begin{bmatrix} -1 + 0 \\ -2 + 6 \\ 0 + 2 \end{bmatrix} = \begin{bmatrix} -1 \\ 4 \\ 2 \end{bmatrix} .$$

Having shown that the set of matrices form a group under matrix addition and having defined a second binary operation 'matrix multiplication, the next logical extension is to determine if under matrix addition and matrix multiplication we can establish a set of matrices that form a ring. In chapter 1 of this book, we stated without proof that the real numbers form a ring relative to the binary operations of addition and multiplication. As promised in chapter one, we will now give a definition for a ring (using a general approach).

Definition 2.3.2 A ring R is a non-empty set of elements for which two binary operations are defined addition $\oplus$ and multiplication $\otimes$ and the axioms 1 through 8 below are all true.

Axiom1. The closure law for the binary addition operation is satisfied for all elements in R. For all pairs A, B contained in R then $A \oplus B$ is also contained in R.

Axiom2. The associative law for the binary operation is satisfied for all elements in R. For all triplets A, B, C contained in R then the equality $(A \oplus B) \oplus C = A \oplus (B \oplus C)$ is true.

Axiom3. There exists an identity element contained in R for the operation $\oplus$. The existence of an identity element in R implies that there is an E contained in R such that for any element A in R, we have the following relationship involving the element E: $E \oplus A = A \oplus E = A$.

Axiom4. For each element in R, there exist an additive inverse element also in R under the operation $\oplus$. The existence of the additive inverse element under this operation implies that for each element A in R, there exist an element B also in R such that the equality

$A \oplus B = B \oplus A = E$ is true.

Axiom5. For the binary additive operation $\oplus$ in R, the commutative law holds for all elements contained in R. That is to say that for each **A** and **B** contained R, the equality $A \oplus B = B \oplus A$ is true.

Axiom6. For the binary multiplication operation $\otimes$, the closure law is satisfied for all elements in R. The restatement of this axiom states that for all **A** and **B** contained in R, the resultant product $A \otimes B$ is also contained in R.

Axiom7. For the binary multiplication operation $\otimes$, the associative law is satisfied for all elements in R. The restatement of this axiom states that for all **A**, **B** and **C** contained in R, the equality $(A \otimes B) \otimes C = A \otimes (B \otimes C)$ is true.

Axiom8. For the binary addition operation $\oplus$ and the binary multiplication operation $\otimes$, the distributive law is satisfied for all elements contained in R. The restatement of this axiom states that for all **A**, **B** and **C** contained in R, both of the following equalities are true:

$$A \otimes (B \oplus C) = (A \otimes B) \oplus (A \otimes C) \text{ and}$$

$$(A \oplus B) \otimes C = (A \otimes C) \oplus (B \otimes C) .$$

Upon closer examination of Definition 2.3.2, axioms 1 through 4 indicate that the non-empty set R is a group under the binary addition operation. Hence to show that a set forms a ring, we must first show that it is a group under the addition operation and secondly we must show that axioms 5 through 8 are true. Having defined both matrix addition and matrix multiplication, we are now ready to embark upon the mission of determining whether a set of matrices forms a ring.

If we follow the lead shown in section 2.2, then we will consider the set of $m \times n$ matrices. However as we attempt to apply the definition for matrix multiplication to this set, we discover that multiplication is only defined over this set if 'm' equals 'n'. Hence we must restrict our attention to the set of $n \times n$ matrices (square matrices).

Theorem 2.3.3 Using Definition 2.2.1 and Definition 2.3.1 respectively for matrix addition and matrix multiplication, the set of $n \times n$ matrices forms a ring under the binary operations of matrix addition and matrix multiplication.

Proof: To prove this theorem, it is necessary to show that axioms 1 through 8 of Definition 2.3.2 are satisfied for the set of $n \times n$ matrices. Since the square matrix is a special case of the $m \times n$ matrix, we can state that axioms 1 through 4 hold by Theorem 2.2.3.

Axiom 1 - Closure Law is true by Theorem 2.2.3 for $m \times n$ matrices.

Axiom 2 - Associative Law is true by Theorem 2.2.3 for $m \times n$ matrices.

Axiom 3 - Existence of Zero Identity Element by Theorem 2.2.3 for $m \times n$ matrices.

Axiom 4 - Additive Inverse Exist by Theorem 2.2.3 for $m \times n$ matrices.

Axiom 5 - To prove the Commutative Law of Addition, select two $n \times n$ matrices $\boldsymbol{A} = [a_{ij}]$ and $\boldsymbol{B} = [b_{ij}]$. The sum of these two matrices is defined by Definition 2.2.1 to be $A + B = [a_{ij} + b_{ij}]$. Since addition of real numbers is a commutative operation we have

$(a_{ij} + b_{ij}) = (b_{ij} + a_{ij})$. By definition of matrix addition we have

$A + B = [a_{ij} + b_{ij}] = [b_{ij} + a_{ij}] = B + A$. Hence matrix addition is a commutative property for the set R.

Axiom 6 - To prove the Closure Law of Multiplication for the set R, select two $n \times n$ matrices $A = [a_{ij}]$ and $B = [b_{ij}]$. Since the number of rows for one matrix is equivalent to the number of columns of the remaining matrix, by using Definition 2.3.1 we have the existence of the products $A \times B$ and $B \times A$. Both of these products are $n \times n$ matrices and are therefore contained in R. Hence the Closure Law holds for multiplication in the set R.

Axiom 7 - To prove the Associative Law of Multiplication for the set R, select three $n \times n$ matrices $A = [a_{ij}]$, $B = [b_{ij}]$ and $C = [c_{ij}]$. Applying Definition 2.3.1 twice we have the following equality for the products of the three matrices (first **A** times **B** then (**A** times **B**) times **C**):

$$(A \times B) \times C = \left[\sum_{l=1}^{n} \left(\sum_{k=1}^{n} a_{ik} \cdot b_{kl} \right) \cdot c_{lj} \right] .$$

Applying the distributive property of real numbers to the right hand side of the above equation we get the following equality:

$$\left[\sum_{l=1}^{n} \left(\sum_{k=1}^{n} a_{ik} \cdot b_{kl} \right) \cdot c_{lj} \right] = \left[\sum_{l=1}^{n} \sum_{k=1}^{n} (a_{ik} \cdot b_{kl}) \cdot c_{lj} \right]$$

.

Now applying the associative property of real numbers to the right side of the above equation we get

$$\left[\sum_{l=1}^{n}\left(\sum_{k=1}^{n} a_{ik}\cdot b_{kl}\right)\cdot c_{lj}\right] = \left[\sum_{l=1}^{n}\sum_{k=1}^{n} a_{ik}\cdot\left(b_{kl}\cdot c_{lj}\right)\right]$$

.

Rearranging the summations on the right side of the above equations and applying the distributive property of real numbers to the right side of the above equation we get

$$\left[\sum_{l=1}^{n}\left(\sum_{k=1}^{n} a_{ik}\cdot b_{kl}\right)\cdot c_{lj}\right] = \left[\sum_{k=1}^{n} a_{ik}\cdot \sum_{l=1}^{n}\left(b_{kl}\cdot c_{lj}\right)\right]$$

. By Definition 2.3.1, the right side of the equation is the product $\boldsymbol{A} \times (\boldsymbol{B} \times \boldsymbol{C})$. Therefore we have the equality $(\boldsymbol{A} \times \boldsymbol{B}) \times \boldsymbol{C} = \boldsymbol{A} \times (\boldsymbol{B} \times \boldsymbol{C})$. Hence the associative property is true for R, the set of $n \times n$ matrices.

Axiom 8 - To prove the Distributive Law for the set R, select three $n \times n$ matrices $\boldsymbol{A} = \left[a_{ij}\right]$, $\boldsymbol{B} = \left[b_{ij}\right]$ and $\boldsymbol{C} = \left[c_{ij}\right]$. Using matrix addition we produce the sum of **B** plus **C** and then multiply **A** times the resulting sum. This operation gives us the following equality:

$$\boldsymbol{A} \times (\boldsymbol{B} + \boldsymbol{C}) = \left[\sum_{k=1}^{n} a_{ik}\cdot\left(b_{kj} + c_{kj}\right)\right] .$$

Applying the distributive law for real numbers to the right side of this equation we get the following equation:

$$\boldsymbol{A} \times (\boldsymbol{B} + \boldsymbol{C}) = \left[\sum_{k=1}^{n} \left(a_{ik}\cdot b_{kj} + a_{ik}\cdot c_{kj}\right)\right] .$$

Now we apply the associative property of real numbers for the terms of the summation on the right side of this equation. The repetitive use of the associative property of addition of real numbers gives us the following equation:

$$A \times (B + C) = \left[\sum_{k=1}^{n} (a_{ik} \cdot b_{kj}) + \sum_{k=1}^{n} (a_{ik} \cdot c_{kj}) \right] .$$

By Definition 2.3.1, for multiplication of matrices, we can rewrite the right side of this equation as follows: $A \times (B + C) = A \times B + A \times C$.

The proof that $(A + B) \times C = A \times C + B \times C$ follows a similar reasoning as shown above.

From these two equalities, the distributive law is true for the set R using the binary operations of matrix addition and matrix multiplication.

Hence the set **R** of $n \times n$ matrices forms a ring with the binary operations of matrix addition and matrix multiplication.

Example

Defining the matrix $B = \begin{bmatrix} 1 & 0 & -1 \\ 0 & 1 & 2 \end{bmatrix} \times \begin{bmatrix} 2 & 4 \\ 4 & 1 \\ 2 & 0 \end{bmatrix}$, use the definition of binary matrix multiplication to determine the value of the matrix entries.

1. Letting $B = [b_{ij}]$ for {i = 1, 2 and j = 1,2}, we can use the definition of binary matrix multiplication to compute each element of the matrix.
2. $b_{11} = 1 \cdot 2 + 0 \cdot 4 - 1 \cdot 2 = 0$

3. $b_{12} = 1 \cdot 4 + 0 \cdot 1 - 1 \cdot 0 = 4$

4. $b_{21} = 0 \cdot 2 + 1 \cdot 4 + 2 \cdot 2 = 8$

5. $b_{12} = 0 \cdot 4 + 1 \cdot 1 + 2 \cdot 0 = 1$

6. Hence we have $B = \begin{bmatrix} 0 & 4 \\ 8 & 1 \end{bmatrix}$.

Exercises

For the following matrix pairs, use the definition of matrix multiplication to perform the indicated operation.

1. $\begin{bmatrix} 1 & 0 & 0 \\ 0 & 0 & 1 \\ 0 & 1 & 0 \end{bmatrix} \times \begin{bmatrix} 1 & 1 & 1 \\ 2 & 2 & 2 \\ 3 & 3 & 3 \end{bmatrix}$

2. $\begin{bmatrix} -2 & -1 & 0 & 1 & 2 \\ 0 & 1 & 2 & 3 & 4 \end{bmatrix} \times \begin{bmatrix} 3 & 6 \\ 7 & -1 \\ -4 & 9 \\ -6 & -5 \end{bmatrix}$

3. $\begin{bmatrix} 1 & 2 & 3 \end{bmatrix} \times \begin{bmatrix} 1 \\ 2 \\ 3 \end{bmatrix}$

4. $\begin{bmatrix} 4 \\ 7 \\ -1 \end{bmatrix} \times \begin{bmatrix} 2 & 3 & 2 \end{bmatrix}$

5. $\begin{bmatrix} 0 & 1 & 4 & 6 \\ 1 & 5 & 3 & 7 \\ 12 & 4 & -5 & -8 \end{bmatrix} \times \begin{bmatrix} 1 \\ -2 \\ -4 \\ 3 \end{bmatrix}$

6. $\begin{bmatrix} 1 & 2 & 3 \\ 1 & 2 & 3 \\ 1 & 2 & 3 \end{bmatrix} \times \begin{bmatrix} 1 & 0 & 0 \\ 0 & 0 & 1 \\ 0 & 1 & 0 \end{bmatrix}$

7. $\begin{bmatrix} 1 & 10 \\ 10 & 1 \end{bmatrix} \times \begin{bmatrix} 5 & 8 & 11 & 20 \\ 6 & -2 & -9 & 0 \end{bmatrix}$

8. $\begin{bmatrix} 1 & 10 \\ 10 & 1 \end{bmatrix} \times \begin{bmatrix} 5 & 8 & 11 & 20 \\ 6 & -2 & -9 & 0 \end{bmatrix}$

9. $\begin{bmatrix} 0.14 & -0.12 & 1.75 \\ 0.02 & -0.02 & 0.22 \\ 0.07 & -0.06 & 0.88 \end{bmatrix} \times \begin{bmatrix} 0.14 & -0.12 & 1.75 \\ 0.02 & -0.02 & 0.22 \\ 0.07 & -0.06 & 0.88 \end{bmatrix}$

10. $\begin{bmatrix} -2 & 0 & -4 \\ -0.72 & -1.37 & 0.79 \\ -0.21 & 0.18 & -3.63 \end{bmatrix} \times \begin{bmatrix} 0.14 & -0.12 & 1.75 \\ 0.02 & -0.02 & 0.22 \\ 0.07 & -0.06 & 0.88 \end{bmatrix}$

11. $\begin{bmatrix} 0.59 & 0.36 & -1.26 \end{bmatrix} \times \begin{bmatrix} -2 & 0 & -4 \\ -0.72 & -1.37 & 0.79 \\ -0.21 & 0.18 & -3.63 \end{bmatrix}$

2.4 Methodology For Solving Linear Equations

Probably the first introduction to solving a system of linear equations for students comes in basic algebra courses. Let's consider the following system of linear equations to solve for the single point of intersection of two lines.

(1a) $2x - 4y = 1$

(2a) $3x - 2y = -4.$

To solve this equation, we can divide the first equation by 2 and then multiply it by -3 and add it to the second equation to replace the second equation by the resultant equation. These steps will appear as follows:

(1b) $x - 2y = 1/2$ (The first step is to divide the first equation by 3.)

(2b) $3x - 2y = -4$

(1c) $x - 2y = 1/2$

(2c) $4y = -11/2$ (Next add -3 times the first equation to the second equation.).

To complete solving this system of equations, we divide equation (2c) by 4 and then add two times the resultant equation to equation (1c). These steps appear as follows:

(1d) $x - 2y = 1/2$

(2d) $y = -11/8$ (Divide this equation by 4.)

(1e) $x \quad = -9/4$ (Add 2 times equation 2d to equation 1d.)

(2e) $y = -11/8.$

Hence the point of intersection for the two lines given by equations 1a and 1b is (-9/4, -11/8).

Each of the operations used in the steps above are called elementary operations. There are three elementary operations that are possible. Two elementary operations (multiply a row by a scalar and adding scalar times one row to another) are illustrated in solving the above equations. In some instances one row may need to be switched with another row to enable a methodical approach to solving the equations. The third elementary operation is that of switching rows.

The three elementary matrix operations can be carried using square matrices. These matrices are called elementary matrices as described in Definition 2.4.1 below.

Definition 2.4.1 A square $m \times m$ matrix $\boldsymbol{E}$ is an elementary matrix if it accomplishes an elementary row operation on an $m \times n$ matrix $\boldsymbol{A_{mn}}$ through the following multiplication $\boldsymbol{E} \times \boldsymbol{A_{mn}}$. A square $n \times n$ matrix $\boldsymbol{E}$ is an elementary matrix if it accomplishes an elementary column operation on an $m \times n$ matrix $\boldsymbol{A_{mn}}$ through the following multiplication $\boldsymbol{A_{mn}} \times \boldsymbol{E}$. Such matrices are classified according to the operation performed on the row (or column) of the matrix. A Type I elementary matrix exchanges rows (or exchanges columns) of the targeted matrix. A Type II elementary matrix multiplies a single row (or column) of the targeted matrix by a number other than 1. A Type III elementary matrix adds a multiple of one row (or column) to another row (or column) of the targeted matrix.

Examples of a Type I elementary matrix of dimension 4×4 are the following:

$$\begin{bmatrix} 0 & 1 & 0 & 0 \\ 1 & 0 & 0 & 0 \\ 0 & 0 & 1 & 0 \\ 0 & 0 & 0 & 1 \end{bmatrix}$$

This matrix exchanges row 1 with row 2 (or column 1 with column 2) of the targeted matrix.

This matrix exchanges row 2 with row 4 (or column 2 with column 4) of the targeted matrix.

$$\begin{bmatrix} 1 & 0 & 0 & 0 \\ 0 & 0 & 0 & 1 \\ 0 & 0 & 1 & 0 \\ 0 & 1 & 0 & 0 \end{bmatrix}$$

Examples of a Type II elementary matrix of dimension 4×4 are the following:

$$\begin{bmatrix} \frac{1}{3} & 0 & 0 & 0 \\ 0 & 1 & 0 & 0 \\ 0 & 0 & 1 & 0 \\ 0 & 0 & 0 & 1 \end{bmatrix}$$

This matrix multiplies row 1(or column 1) of the targeted matrix by 1/3.

$$\begin{bmatrix} 1 & 0 & 0 & 0 \\ 0 & 1 & 0 & 0 \\ 0 & 0 & -2 & 0 \\ 0 & 0 & 0 & 1 \end{bmatrix}$$

This matrix multiplies row 3 (or column 3) of the targeted matrix by -2.

An example of a Type III elementary matrix of dimension 4×4 is the following:

$$\begin{bmatrix} 1 & 0 & 0 & 0 \\ -2 & 1 & 0 & 0 \\ 0 & 0 & 1 & 0 \\ 0 & 0 & 0 & 1 \end{bmatrix}$$

This matrix adds -2 times row 1(or column 1) to row 2 (or column 2) of the targeted matrix.

Now that we have formalized the elementary operations in terms of elementary operations, let's revisit the earlier system of linear equations. By expressing equations (1a) and (1b) as a matrix operation we have the following matrix equation:

1. $$\begin{bmatrix} 2 & -4 \\ 3 & -2 \end{bmatrix} \begin{bmatrix} x \\ y \end{bmatrix} = \begin{bmatrix} 1 \\ -4 \end{bmatrix}.$$

By multiplying both sides of equation with the appropriate elementary matrix and applying the associative law on the left side of the equation, we reduce equations 2a and 2b to the following matrix equation:

2. $$\begin{bmatrix} 1 & -2 \\ 3 & -2 \end{bmatrix} \begin{bmatrix} x \\ y \end{bmatrix} = \begin{bmatrix} 1/2 \\ -4 \end{bmatrix}, \; \boldsymbol{E} = \begin{bmatrix} 1/2 & 0 \\ 0 & 1 \end{bmatrix}.$$

The next step in the equations (3a) and (3b) is to remove the 'x' expression from equation (3b). In matrix notation this is accomplished by the following multiplication:

3. $\begin{bmatrix} 1 & -2 \\ 0 & 4 \end{bmatrix} \begin{bmatrix} x \\ y \end{bmatrix} = \begin{bmatrix} 1/2 \\ -11/2 \end{bmatrix}, \quad \boldsymbol{E} = \begin{bmatrix} 1 & 0 \\ -3 & 1 \end{bmatrix}$.

The next step in solving the equation is to make the coefficient of "y" a value of 1. This is shown in equation (4a) and (4b) above. This is accomplished in matrix notation with the following elementary matrix multiplication:

4. $\begin{bmatrix} 1 & -2 \\ 0 & 1 \end{bmatrix} \begin{bmatrix} x \\ y \end{bmatrix} = \begin{bmatrix} 1/2 \\ -11/8 \end{bmatrix}, \quad \boldsymbol{E} = \begin{bmatrix} 1 & 0 \\ 0 & 1/4 \end{bmatrix}$.

The final step in this process is to remove the "y" term from the first equation. This is accomplished in equations (5a) and (5b). The corresponding matrix operation is a Type III elementary matrix operation as follows:

5. $\begin{bmatrix} 1 & 0 \\ 0 & 1 \end{bmatrix} \begin{bmatrix} x \\ y \end{bmatrix} = \begin{bmatrix} -9/4 \\ -11/8 \end{bmatrix}, \quad \boldsymbol{E} = \begin{bmatrix} 1 & 2 \\ 0 & 1 \end{bmatrix}$.

Using Definition 2.2.1, for multiplication of two matrices, the product $\begin{bmatrix} 1 & 0 \\ 0 & 1 \end{bmatrix} \begin{bmatrix} x \\ y \end{bmatrix}$ yields the results $\begin{bmatrix} x \\ y \end{bmatrix}$. Hence the solution to the above equation is $\begin{bmatrix} x \\ y \end{bmatrix} = \begin{bmatrix} -9/4 \\ -11/8 \end{bmatrix}$.

The diagonal matrix with all 1's along the diagonal is a special matrix. It is the multiplication identity matrix and is formally defined in Definition 2.4.2 below.

Definition 2.4.2 The identity matrix I_n of order "n" is an $n \times n$ diagonal matrix such that all diagonal elements have a value of 1. Notation for the identity matrix is the following $I_n = \lfloor \delta_{ij} \rfloor$ where $\delta_{ij} = 0;(i \neq j)$ and $\delta_{ii} = 1$.

As shown in section 2.1 the definitions that are derived from the elementary row operations of the matrices are row equivalent matrices, column equivalent matrices, row echelon form of matrix and reduced row echelon form of matrix (row canonical form). Other definitions that follow from the definition of elementary row (or column) operations are non-singular matrix and singular matrix.

Definition 2.4.3 A square matrix is said to be non-singular if it is equivalent to the Identity matrix.

Definition 2.4.4 A square matrix is said to be singular if it is row equivalent to a matrix that is in row echelon form having at least one row containing all zeros.

Example

Given the following linear system of equations with three unknown variables and three equations, use the concepts learned in this section to solve for the three unknown variables.

$$2x + 3y - 4z = 12$$

$$-x + 6y - 2z = 5$$

$$12x - y + 5z = 8$$

We begin solving for the unknown variables by transcribing these equations into matrix notation and then preceding as discussed in this chapter to solve a system of linear equations.

1. The matrix notation for this system of equations is the following:

$$\begin{bmatrix} 2 & 3 & -4 \\ -1 & 6 & -2 \\ 12 & -1 & 5 \end{bmatrix} \times \begin{bmatrix} x \\ y \\ z \end{bmatrix} = \begin{bmatrix} 12 \\ 5 \\ 8 \end{bmatrix}.$$

2. The first elementary row operation is a type II elementary matrix operation. We multiply row one by 0.5 (1/2). This operation gives the results $\begin{bmatrix} 1 & 3/2 & -2 \\ -1 & 6 & -2 \\ 12 & -1 & 5 \end{bmatrix} \times \begin{bmatrix} x \\ y \\ z \end{bmatrix} = \begin{bmatrix} 6 \\ 5 \\ 8 \end{bmatrix}$.

3. Next operations are a type III, type II and type III elementary matrix operations. We add row one to row two, multiply row one by six (6) and then add row one to row three. The results of these operations yields

$$\begin{bmatrix} 1 & \frac{3}{2} & -2 \\ 0 & \frac{15}{2} & -4 \\ 0 & -19 & 29 \end{bmatrix} \times \begin{bmatrix} x \\ y \\ z \end{bmatrix} = \begin{bmatrix} 6 \\ 11 \\ -64 \end{bmatrix}.$$

4. Next we multiply row two by (2/15) to get

$$\begin{bmatrix} 1 & \frac{3}{2} & -2 \\ 0 & 1 & -\frac{8}{15} \\ 0 & -19 & 29 \end{bmatrix} \times \begin{bmatrix} x \\ y \\ z \end{bmatrix} = \begin{bmatrix} 6 \\ \frac{22}{15} \\ -64 \end{bmatrix}.$$

5. Like step 3, we multiply row two by the appropriate values and add it to rows one and three to result in zero values in the second column of rows one and three. Performing the appropriate elementary matrix operations we get

$$\begin{bmatrix} 1 & 0 & -\frac{6}{5} \\ 0 & 1 & -\frac{8}{15} \\ 0 & 0 & \frac{283}{15} \end{bmatrix} \times \begin{bmatrix} x \\ y \\ z \end{bmatrix} = \begin{bmatrix} \frac{29}{5} \\ \frac{22}{15} \\ \frac{542}{15} \end{bmatrix} .$$

6. Now we multiply row three by (15/283) to get

$$\begin{bmatrix} 1 & 0 & -\frac{6}{5} \\ 0 & 1 & -\frac{8}{15} \\ 0 & 0 & 1 \end{bmatrix} \times \begin{bmatrix} x \\ y \\ z \end{bmatrix} = \begin{bmatrix} \frac{29}{5} \\ \frac{22}{15} \\ \frac{542}{283} \end{bmatrix} .$$

7. Finally, with the appropriate type II and type III elementary matrix operations, we can replace the nonzero values in the third column for rows one and two with zero values. These operations yield $\begin{bmatrix} 1 & 0 & 0 \\ 0 & 1 & 0 \\ 0 & 0 & 1 \end{bmatrix} \times \begin{bmatrix} x \\ y \\ z \end{bmatrix} = \begin{bmatrix} \frac{24467}{1415} \\ \frac{10562}{4245} \\ \frac{542}{283} \end{bmatrix}$ and hence we have $\begin{bmatrix} x \\ y \\ z \end{bmatrix} = \begin{bmatrix} \frac{24467}{1415} \\ \frac{10562}{4245} \\ \frac{542}{283} \end{bmatrix}$.

Exercises

For the system of equations given in the exercises below, use the row reduction method discussed in this section to find the unknown quantities in the matrix equation.

1. $\begin{bmatrix} 2 & 4 \\ 0 & -1 \end{bmatrix} \times \begin{bmatrix} x \\ y \end{bmatrix} = \begin{bmatrix} 1 \\ -3 \end{bmatrix}$

2. $\begin{bmatrix} 1 & 25 \\ 5 & 9 \end{bmatrix} \times \begin{bmatrix} x \\ y \end{bmatrix} = \begin{bmatrix} 0 \\ 5 \end{bmatrix}$

3. $\begin{bmatrix} 10 & 2 \\ -1 & 5 \end{bmatrix} \times \begin{bmatrix} x \\ y \end{bmatrix} = \begin{bmatrix} 1 \\ 1 \end{bmatrix}$

4. $\begin{bmatrix} 1 & -2 \\ -1 & 1 \end{bmatrix} \times \begin{bmatrix} x \\ y \end{bmatrix} = \begin{bmatrix} 9 \\ 49 \end{bmatrix}$

5. $\begin{bmatrix} 3 & 1 & -2 \\ -3 & 0 & -4 \\ 0 & 2 & 3 \end{bmatrix} \times \begin{bmatrix} x \\ y \\ z \end{bmatrix} = \begin{bmatrix} 1 \\ 3 \\ 9 \end{bmatrix}$

6. $\begin{bmatrix} 2 & 4 & 6 \\ -5 & -9 & 4 \\ -2 & -3 & 0 \end{bmatrix} \times \begin{bmatrix} x \\ y \\ z \end{bmatrix} = \begin{bmatrix} 9 \\ 3 \\ 1 \end{bmatrix}$

7. $\begin{bmatrix} 0 & 1 & 0 \\ 1 & -3 & 9 \\ 4 & -10 & -20 \end{bmatrix} \times \begin{bmatrix} x \\ y \\ z \end{bmatrix} = \begin{bmatrix} -1 \\ 3 \\ -5 \end{bmatrix}$

8. $\begin{bmatrix} 0 & 1 & 3 \\ -2 & -3 & -2 \\ 1 & 0 & 2 \end{bmatrix} \times \begin{bmatrix} x \\ y \\ z \end{bmatrix} = \begin{bmatrix} -1 \\ 4 \\ 1 \end{bmatrix}$

9. $\begin{bmatrix} 1 & 2 & 4 \\ 2 & 4 & 1 \\ 4 & 1 & 2 \end{bmatrix} \times A^{-1} = \begin{bmatrix} 1 & 0 & 0 \\ 0 & 1 & 0 \\ 0 & 0 & 1 \end{bmatrix}$

10. $A^{-1} \times \begin{bmatrix} 0 & 1 & 1 \\ 3 & 6 & 9 \\ 2 & 0 & 2 \end{bmatrix} = \begin{bmatrix} 1 & 0 & 0 \\ 0 & 1 & 0 \\ 0 & 0 & 1 \end{bmatrix}$

2.5 *Matrix Inverse*

In the previous section, 2.4, we introduced a square matrix; $\boldsymbol{I}_n$. We illustrated that a solution of 'n' linear equations having 'n' unknowns has a unique solution if the matrix representing the system of linear equations is row equivalent to the identity matrix. To formalize the matrix approach to solving such equations we introduce the concept of an inverse of a non-singular matrix.

Definition 2.5.1 The inverse matrix of a non-singular matrix $\boldsymbol{A}_{nn}$, is a matrix $\boldsymbol{B}_{nn}$ such that the following equation is true: $\boldsymbol{A}_{nn} \times \boldsymbol{B}_{nn} = \boldsymbol{B}_{nn} \times \boldsymbol{A}_{nn} = \boldsymbol{I}_n$.

We have shown in the Gauss-Jordan reduction method (or Gaussian Elimination method) that there is a matrix $\boldsymbol{P}_{nn}$ defined as a product of elementary matrices that satisfy the equation $\boldsymbol{P}_{nn} \times \boldsymbol{A}_{nn} = \boldsymbol{I}_n$, for the non-singular matrix $\boldsymbol{A}_{nn}$ representing a system of 'n' unknown variables in 'n' equations. This approach gives us a matrix that multiplies $\boldsymbol{A}_{nn}$ from the left and yields the identity matrix, so what if we multiplied the same matrix to $\boldsymbol{A}_{nn}$ from the right. Do we still obtain the identity matrix? The theorems below illustrate that this is the case, and hence we have a systematic approach to computing the inverse of a matrix.

Before we tackle the problem of showing that a product is the inverse of the matrix, we shall investigate the properties of the elementary matrices.

Theorem 2.5.2 Each elementary matrix is non-singular and the inverse of an elementary matrix is a matrix of the same type. In matrix terminology, if the elementary matrix is denoted by $\boldsymbol{E}$, then we denote the inverse as $\boldsymbol{E}^{-1}$ and we have

$$\boldsymbol{E}^{-1} \times \boldsymbol{E} = \boldsymbol{E} \times \boldsymbol{E}^{-1} = \boldsymbol{I}_n \quad .$$

Proof: To prove this theorem, we will address each type of elementary matrix separately.

(Type I) Let the matrix $\boldsymbol{E}_{r \leftrightarrow s}$, represent the interchange of rows 'r' and 's' in the targeted matrix. Then if we apply the same matrix to the resultant exchange then we effectively have the original matrix back. In matrix terminology this is represented as $\boldsymbol{E}_{r \leftrightarrow s} \times (\boldsymbol{E}_{r \leftrightarrow s} \times \boldsymbol{A}) = \boldsymbol{A}$. Since this property is true for all matrices **A**, we can apply it to the identity matrix which would yield $\boldsymbol{E}_{r \leftrightarrow s} \times (\boldsymbol{E}_{r \leftrightarrow s} \times \boldsymbol{I}_n) = \boldsymbol{I}_n$. After evaluating the expression within the parenthesis (using definition of identity matrix), have that $\boldsymbol{E}_{r \leftrightarrow s} \times \boldsymbol{E}_{r \leftrightarrow s} = \boldsymbol{I}_n$. Hence we can state that the matrix. $\boldsymbol{E}_{r \leftrightarrow s}$ is it's own inverse; $\boldsymbol{E}_{r \leftrightarrow s}^{-1} = \boldsymbol{E}_{r \leftrightarrow s}$.

(Type II) Let the matrix $\boldsymbol{E}_{c \cdot r}$, represent the-multiplication of row 'r' by the scalar **c** in the targeted matrix. Then if we multiply row 'r' of the resultant product by the reciprocal scalar '1/c' then we effectively have the original matrix back. In matrix terminology this is represented as $\boldsymbol{E}_{(1/c) \cdot r} \times (\boldsymbol{E}_{c \cdot r} \times \boldsymbol{A}) = \boldsymbol{A}$. Since this property is true for all matrices **A**, we can apply it to the identity matrix which would yield $\boldsymbol{E}_{(1/c) \cdot r} \times (\boldsymbol{E}_{c \cdot r} \times \boldsymbol{I}_n) = \boldsymbol{I}_n$. After evaluating the expression within the parenthesis (using definition of identity matrix), have that $E_{(1/c) \cdot r} \times E_{c \cdot r} = I_n$. Since **c** is an arbitrary scalar, we can substitute **1/c** for the value of **c** and then we also have the equation $E_{c \cdot r} \times E_{(1/c) \cdot r} = I_n$. By definition of inverse of a matrix, we have $E^{-1}_{c \cdot r} = E_{(1/c) \cdot r}$.

(Type III) Let the matrix $E_{c \cdot r + s}$, represent the addition of **c** times row **r** to row **s** in the targeted matrix. Then if we multiply **-c** times row **r** and add it to row s of the resultant matrix then we effectively have the original matrix back. In matrix terminology this is represented as $E_{(-c) \cdot r + s} \times (E_{c \cdot r + s} \times A) = A$. Since this property is true for all matrices **A**, we can apply it to the identity matrix which would yield $E_{(-c) \cdot r + s} \times (E_{c \cdot r + s} \times I_n) = I_n$. After evaluating the expression within the parenthesis (using definition of identity matrix), have that $E_{(-c) \cdot r + s} \times E_{c \cdot r + s} = I_n$. Since 'c' is an arbitrary scalar, we can substitute '-c' for the value of 'c' and then we also have the equation

$E_{c \cdot r + s} \times E_{(-c) \cdot r + s} = I_n$. By definition of inverse of a matrix, we have $E^{-1}_{c \cdot r + s} = E_{(-c) \cdot r + s}$.

Hence the theorem is proven for each of the three types of elementary matrices. Now let us consider the non-singular matrix A_{nn}. It is clear from the definition of non-singular matrices that we can find a product $P = E_1 \times E_2 \times \ldots \times E_k$ of elementary matrices such that $P \times A_{nn} = I_n$. This suggests the possibility that P is an inverse for the matrix A_{nn}.

Theorem 2.5.3 If A_{nn} is a non-singular matrix, then there exists a matrix $P = E_1 \times E_2 \times \ldots \times E_k$ such that $P \times A_{nn} = A_{nn} \times P = I_n$ (that is to say that P is the inverse of the matrix). Furthermore, the inverse is unique such that if B is an inverse of A_{nn}, then $B = P$.

Proof: Since A_{nn} is a non-singular matrix, using Definition 2.4.3, their exist a product of elementary matrices $P = E_1 \times E_2 \times \ldots \times E_k$, such that $E_1 \times E_2 \times \ldots \times E_k \times A_{nn} = I_n$. Now let us multiply both equations from the left by E_1^{-1} and from the right by E_1 to get the following:
$E_1^{-1} \times (E_1 \times E_2 \times \ldots \times E_k \times A_{nn}) \times E_1 = E_1^{-1} \times I_n \times E_1$
which yields $(E_2 \times \ldots \times E_k) \times A_{nn} \times (E_1) = I_n$ using the associative properties of matrix multiplication within the ring of $n \times n$ matrices. By repeating this step k-1 additional times for elementary matrices E_2 through E_k , we have the following equality: $A_{nn} \times (E_1 \times E_2 \times \ldots \times E_k) = I_n$. Hence

we have the relationship $P \times A_{nn} = A_{nn} \times P = I_n$. Therefore P is an inverse of the matrix A_{nn}.

For the uniqueness part of the theorem, let's consider a matrix B that is also an inverse of the matrix A_{nn}. Then by definition of an inverse, we have $B \times A_{nn} = I_n$. But recall that P is also an inverse of the matrix A_{nn}. By multiplying this equation by P from the right we get the following equation: $(B \times A_{nn}) \times P = I_n \times P$. Applying the associative property to the left hand side of the equation and the definition of the identity matrix for the right hand side of the equation we get $B = P$. Hence there is only one inverse for a non-singular matrix.

Other properties of non-singular matrices that will be helpful as the discussion of linear algebra progresses are the given in the theorems that follow in this section. These theorems will be given without proof. They will be left as exercises for the reader.

Theorem 2.5.4 Any non-singular matrix can be expressed as a finite product of elementary matrices.

Theorem 2.5.5 The inverse of an inverse of a matrix is equal to the original matrix. In matrix notation we have the equation $(A^{-1})^{-1} = A$.

Theorem 2.5.6 The inverse of the product $A \times B$ is the product $B^{-1} \times A^{-1}$. In matrix notation we have the equation $(A \times B)^{-1} = B^{-1} \times A^{-1}$.

Theorem 2.5.7 The inverse of the transpose of a matrix is the transpose of the inverse of the matrix. In matrix notation we have the equation $(A^T)^{-1} = \left(A^{-1}\right)^T$.

Exercises

Using the Gauss-Jordan reduction method (or method illustrated in exercise 9 and 10 of section 2.4) compute the inverse for the matrices in exercises 1 through 6.

1. $\begin{bmatrix} 1 & -2 & -7 \\ 0 & 2 & 4 \\ 0 & 0 & 3 \end{bmatrix}$

2. $\begin{bmatrix} 1 & 2 & 4 \\ 2 & 1 & 2 \\ 4 & 2 & 1 \end{bmatrix}$

3. $\begin{bmatrix} 1 & -1 & 2 \\ 3 & -4 & 0 \\ 5 & -5 & -5 \end{bmatrix}$

4. $\begin{bmatrix} 0 & 5 & 20 \\ -2 & 0 & 30 \\ 0 & 0 & 1 \end{bmatrix}$

5. $\begin{bmatrix} 1 & 2 & 4 & 8 \\ 2 & 1 & 2 & 4 \\ 4 & 2 & 1 & 2 \\ 8 & 4 & 2 & 1 \end{bmatrix}$

6. $\begin{bmatrix} 2 & -4 & -5 & 17 \\ 0 & 0 & -5 & 5 \\ 0 & 3 & 3 & 6 \\ 1 & -2 & 2 & 1 \end{bmatrix}$

7. For a non-singular matrix **A**, prove theorem 2.5.4 (i.e. show that it can be expressed as a finite product of elementary matrices).
8. For a non-singular matrix A, prove theorem 2.5.5 (i.e. show that the inverse of the inverse of a matrix is equal to the original matrix).
9. Prove theorem 2.5.6.
10. Prove theorem 2.5.7.

This Page Intentionally Left Blank

CHAPTER 3

Real Vector Spaces and Geometry

For the previous chapter, we discussed the algebraic system of groups and rings as they apply to square matrices. But as we look at the set of all matrices, we note that matrices that are not square cannot satisfy the multiplication axioms for rings. However, in this chapter we introduce the concept of vector spaces and we show through examples that $m \times n$ matrices may form a vector space. Vectors are widely used in physics to represent forces, velocities and other physical phenomena having direction and magnitude. In this chapter we introduce the general concept of vector spaces and develop some theorems about vector spaces and illustrate two and three dimensional coordinate systems.

Traditionally in physics, vectors are represented as a single array in row form enclosed by parenthesis; $(x_1, \ldots, x_n)$. For this book, a column matrix will be used to represent a vector; $\begin{bmatrix} x_1 \\ \ldots \\ x_n \end{bmatrix}$.

3.1 Real Vector Spaces

In a first course of university physics, force and velocity are physical manifestations having both magnitude and direction. To represent such phenomena, directed line segments are used. Mathematically, the directed line segment, may be described using vectors. A similar phenomena of a directed line segment is a point in a coordinate system. Mathematically, we show that a directed line segments and points in a coordinate system have similar behaviors. Their mathematical equivalence will be illustrated in this chapter.

In a plane (or two dimensional space), we represent a point $\boldsymbol{P} = (x,y)$ as one that is a distance of **x** units along x-axis from a fixed point **O** and **y** units along the y-axis from the fixed point **O**. The fixed point **O** is called the origin of the coordinate system. The axes labeled x-axis and y-axis are formed by the intersection of two perpendicular lines. The graphical representation of a point **P** as described here is shown in Figure 3 below.

Figure 3 Representation of 2-Dimensional Coordinates

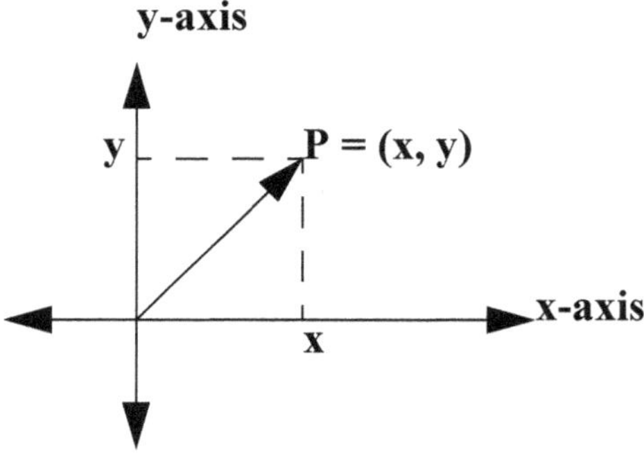

To formalize the discussion of vectors, we define the general concept of vector spaces and show that physical vectors and points in n-dimensional space are examples of a vector space.

Definition 3.1.1 A real vector space is a non-empty set **V** having the binary operation of addition and the binary operation of scalar multiplication and it satisfies the following axioms:

Axiom1. The closure law for the binary addition is satisfied for all elements in **V**. For all pairs A, B contained in **V** then $A \oplus B$ is also contained in **V**.

Axiom2. The commutative law for binary addition is satisfied for all elements in **V**. For all pairs A, B contained in **V**, the equality $A \oplus B = B \oplus A$ is true.

Axiom3. The associative law for binary addition is satisfied for all elements in **V**. For all triplets A, B, C contained in **V** then the equality $(A \oplus B) \oplus C = A \oplus (B \oplus C)$ is true.

Axiom4. There exists an identity element contained in **V** for the addition operation $\oplus$. The existence of an identity element in **V** implies that there is a $\varnothing$ contained in **V** such that for all elements in A we have the following relationship involving the element $\varnothing$:

$\varnothing \oplus A = A \oplus \varnothing = A$.

Axiom5. There exists a unique additive inverse in **V** for all elements contained in **V**. The existence of an additive inverse implies the existence of B for each A contained in **V** such that the equality $B \oplus A = A \oplus B = \varnothing$ is true.

In addition to the above axioms for binary addition, the following axioms for scalar multiplication are satisfied:

Axiom6. The closure law under real scalar multiplication is satisfied for all elements contained in **V**. This implies that if **A** is an element of **V**, then $c \cdot \boldsymbol{A}$ is an element of **V**; where c is a real scalar.

Axiom7. For the binary addition operation $\oplus$ and scalar multiplication operation, the distributive law is satisfied for all elements contained in **V**. The restatement of this axiom states that, for all **A** and **B** contained in **V**, the following equality is true: $c \cdot (\boldsymbol{A} \oplus \boldsymbol{B}) = (c \cdot \boldsymbol{A}) \oplus (c \cdot \boldsymbol{B})$.

Axiom8. For real number addition, binary addition operation $\oplus$ and scalar multiplication, the distributive law for real numbers is satisfied when multiplying by an element of **V**. The restatement of this axiom states that, for all **A** contained in **V** and for any two real numbers c and d, the following equation is true:

$$(c + d) \cdot \boldsymbol{A} = (c \cdot \boldsymbol{A}) \oplus (d \cdot \boldsymbol{A}) \ .$$

Axiom9. For scalar multiplication of an element **V** by a product of real numbers, the associative law is satisfied. The restatement of this axiom states that, for all A contained in V and any two real numbers c and d, the following equation is true: $c \cdot (d \cdot \boldsymbol{A}) = (c \cdot d) \cdot \boldsymbol{A}$.

Axiom10. The real number 1 (one) when multiplied to any element of **V**, yields the original element. A restatement of this axiom states that, for any element **A** contained in **V**, the following equality is true: $1 \cdot \boldsymbol{A} = \boldsymbol{A}$.

Definition 3.1.2 Each element of a vector space is called a vector.

Example 1: $\boldsymbol{R^n}$

Consider the n-tuples represented by the $n \times 1$ column matrix and satisfying the definitions of matrix addition and scalar-matrix multiplication. Then we will show that the set $\boldsymbol{R}^n$ of all n-tuples represented as $n \times 1$ column matrices form a vector space.

Under the binary operation of matrix addition, the set $\boldsymbol{R}^n$ forms an algebraic group (See theorem 1 of chapter 2). This implies that axioms 1, 3, 4 and 5 of Definition 3.1.1 is satisfied for the set $\boldsymbol{R}^n$. To show that it is a commutative group, consider two arbitrary elements $\boldsymbol{A} = \begin{bmatrix} a_1 \\ \dots \\ a_n \end{bmatrix}$ and $\boldsymbol{B} = \begin{bmatrix} b_1 \\ \dots \\ b_n \end{bmatrix}$ contained in $\boldsymbol{R}^n$.

The sum of the two matrices is $\boldsymbol{A} + \boldsymbol{B} = \begin{bmatrix} a_1 \\ \dots \\ a_n \end{bmatrix} + \begin{bmatrix} b_1 \\ \dots \\ b_n \end{bmatrix} = \begin{bmatrix} a_1 + b_1 \\ \dots \\ a_n + b_n \end{bmatrix}$

by definition 2.2.1 of chapter 2. Since the commutative property holds for real numbers we have

$\begin{bmatrix} a_1 + b_1 \\ \dots \\ a_n + b_n \end{bmatrix} = \begin{bmatrix} b_1 + a_1 \\ \dots \\ b_n + a_n \end{bmatrix} = \begin{bmatrix} b_1 \\ \dots \\ b_n \end{bmatrix} + \begin{bmatrix} a_1 \\ \dots \\ a_n \end{bmatrix} = \boldsymbol{B} + \boldsymbol{A}$. Therefore we have $\boldsymbol{A} + \boldsymbol{B} = \boldsymbol{B} + \boldsymbol{A}$ and hence axiom 2 (commutative property) holds for the set $\boldsymbol{R}^n$.

The proofs for axioms 5 through 10 are shown below for the set $\boldsymbol{R}^n$.

(Axiom 6) Let $\boldsymbol{A} = \begin{bmatrix} a_1 \\ \dots \\ a_n \end{bmatrix}$ be an arbitrary element of $\boldsymbol{R}^n$. Then for any arbitrary real number 'c' we have

$c \cdot \boldsymbol{A} = c \cdot \begin{bmatrix} a_1 \\ \dots \\ a_n \end{bmatrix} = \begin{bmatrix} c \cdot a_1 \\ \dots \\ c \cdot a_n \end{bmatrix}$. Since the resultant matrix is an $n \times 1$ column matrix, it is an element of $\boldsymbol{R}^n$. Hence axiom 5 the closure law under scalar multiplication is satisfied.

(Axiom 7) Let $\boldsymbol{A} = \begin{bmatrix} a_1 \\ \dots \\ a_n \end{bmatrix}$ and $\boldsymbol{B} = \begin{bmatrix} b_1 \\ \dots \\ b_n \end{bmatrix}$ be arbitrary elements of $\boldsymbol{R}^n$. For any arbitrary real number 'c' and using the definition of $\boldsymbol{A} + \boldsymbol{B}$ we have the following computation:

$c \cdot (\boldsymbol{A} + \boldsymbol{B}) = c \cdot \begin{bmatrix} a_1 + b_1 \\ \dots \\ a_n + b_n \end{bmatrix}$. By definition 2.1.9 of chapter 2, this equation becomes $c \cdot (\boldsymbol{A} + \boldsymbol{B}) = \begin{bmatrix} c \cdot (a_1 + b_1) \\ \cdots \\ c \cdot (a_n + b_n) \end{bmatrix}$. Using the distributive property of real numbers and the definition of $\boldsymbol{A} + \boldsymbol{B}$, the equation now becomes

$$c \cdot (\boldsymbol{A} + \boldsymbol{B}) = \begin{bmatrix} c \cdot a_1 + c \cdot b_1 \\ \cdots \\ c \cdot a_n + c \cdot b_n \end{bmatrix} = \begin{bmatrix} c \cdot a_1 \\ \cdots \\ c \cdot a_n \end{bmatrix} + \begin{bmatrix} c \cdot b_1 \\ \cdots \\ c \cdot b_n \end{bmatrix} = c \cdot \boldsymbol{A} + c \cdot \boldsymbol{B} \quad .$$

Hence axiom 7 is true for the set $\boldsymbol{R}^n$.

(Axiom 8) Let $\boldsymbol{A} = \begin{bmatrix} a_1 \\ \cdots \\ a_n \end{bmatrix}$ be an arbitrary element of $\boldsymbol{R}^n$ and let 'c' and 'd' be any given real numbers. Using the distributive property of real numbers we get the following equation:

$$(c + d) \cdot \boldsymbol{A} = \begin{bmatrix} (c + d) \cdot a_1 \\ \cdots \\ (c + d) \cdot a_n \end{bmatrix} = \begin{bmatrix} c \cdot a_1 \\ \cdots \\ c \cdot a_n \end{bmatrix} + \begin{bmatrix} d \cdot a_1 \\ \cdots \\ d \cdot a_n \end{bmatrix} = c \cdot \boldsymbol{A} + d \cdot \boldsymbol{A} \quad .$$

Hence axiom 8 for $\boldsymbol{R}^n$ is satisfied.

(Axiom 9) Let $A = \begin{bmatrix} a_1 \\ \dots \\ a_n \end{bmatrix}$ be an arbitrary element of $\boldsymbol{R}^n$ and let 'c' and 'd' be any given real numbers. Using the associative property of real numbers we get the following equation:

$$c \cdot (d \cdot A) = \begin{bmatrix} c \cdot (d \cdot a_1) \\ \dots \\ c \cdot (d \cdot a_n) \end{bmatrix} = \begin{bmatrix} (c \cdot d) \cdot a_1 \\ \dots \\ (c \cdot d) \cdot a_n \end{bmatrix} = (c \cdot d) \cdot A$$

. Hence axiom 9 for $\boldsymbol{R}^n$ is satisfied.

(Axiom 10) Let $A = \begin{bmatrix} a_1 \\ \dots \\ a_n \end{bmatrix}$ be an arbitrary element of $\boldsymbol{R}^n$.

Using the multiplicative identity property of real numbers we get the following equation: $1 \cdot A = \begin{bmatrix} 1 \cdot a_1 \\ \dots \\ 1 \cdot a_n \end{bmatrix} = \begin{bmatrix} a_1 \\ \dots \\ a_n \end{bmatrix} = A$. Hence axiom 10 for $\boldsymbol{R}^n$ is satisfied.

Since all 10 axioms for $\boldsymbol{R}^n$ are satisfied, we can conclude that it is a vector space under matrix addition and real number (scalar) multiplication.

Example 2: $\boldsymbol{R}_n$

The set of ordered n-tuples represented by a row matrix is denoted as $A = \begin{bmatrix} a_1 & \dots & a_n \end{bmatrix}$. More detailed information concerning ordered n-tuples will be given later in this section. Knowing that row matrices are used to represent the elements in $\boldsymbol{R}_n$, the proof that it forms a vector space under matrix addition and scalar multiplication follows a similar proof as shown in example 1.

Example 3: ${}_m\boldsymbol{R}_n$

The set ${}_m\boldsymbol{R}_n$ represents the set of $m \times n$ matrices. Under matrix addition and scalar multiplication, this set is shown to be a vector space using the same reasoning as shown in example 1. The verification of the 10 axioms that verifies that ${}_m\boldsymbol{R}_n$ is a vector space is left as an exercise for the reader.

Example 4: $p_n(t) = a_n t^n + a_{n-1} t^{n-1} + \dots + a_0$

Polynomials of degree "n" of variable "t" form a vector space. It is left as an exercise for the reader to prove.

Some properties of vector spaces that are useful to know are shown in Theorem 3.1.3 below.

Theorem 3.1.3 The following properties are true for any vector space **V**:

1. For any vector **A** in **V** we have $0 \cdot \boldsymbol{A} = \varnothing$.

2. For any scalar 'c' the equation $c \cdot \varnothing = \varnothing$ is true.

3. For any scalar 'c' and vector **A**, if the equation $c \cdot \boldsymbol{A} = \varnothing$ is true, then either $c = 0$ or $\boldsymbol{A} = \varnothing$.

4. For any vector **A** in **V** we have $(-1) \cdot \boldsymbol{A} = -\boldsymbol{A}$.

Proof: For property one (1) of this theorem, we begin by citing axiom 10 of definition 3.1.1; $1 \cdot \boldsymbol{A} = \boldsymbol{A}$ for any **A** in **V**. Using the additive identity property of zero, for real numbers, we can write

$(1+0) \cdot \boldsymbol{A} = \boldsymbol{A}$ or $(0+1) \cdot \boldsymbol{A} = \boldsymbol{A}$ .From axiom 7, using distributive property of real numbers, we have the following equalities $0 \cdot \boldsymbol{A} + \boldsymbol{A} = \boldsymbol{A} + 0 \cdot \boldsymbol{A} = \boldsymbol{A}$. Since this is true for all vectors **A** in **V**, we must have $0 \cdot \boldsymbol{A} = \varnothing$.

For property two (2) of this theorem, use axiom 4 of definition 3.1.1 to get the equality

$c \cdot (\boldsymbol{A} + \varnothing) = c \cdot (\varnothing + \boldsymbol{A}) = c \cdot \boldsymbol{A}$ .Property 6, we get

$c \cdot \boldsymbol{A} + c \cdot \varnothing = c \cdot \varnothing + c \cdot \boldsymbol{A} = c \cdot \boldsymbol{A}$. Since the vector **A** is arbitrary (and hence $c \cdot \boldsymbol{A}$ is arbitrary), axiom 4 of definition 3.1.1 implies that $c \cdot \varnothing = \varnothing$.

For property three (3) of this theorem, first assume that $c \neq 0$. Applying axiom 9 of definition 3.1.1, we get the equality

$(1/c) \cdot (c \cdot A) = \left(\frac{c}{c} \cdot A\right) = A = \varnothing$. Now let's assume that $A \neq \varnothing$, then if $c \neq 0$, we must have $A = \varnothing$ (as shown in previous sentence). This provides a contradiction and hence when $A \neq \varnothing$, we must have $c = 0$.

Since $1 + (-1) = 0$ and the application of property 1 of this theorem, we get the equality $(1 + (-1)) \cdot A = \varnothing$. Applying axiom 8, we get $A + (-1) \cdot A = \varnothing$. By the uniqueness of the additive inverse of a vector, we have that $(-1) \cdot A$ is the additive of **A** or $(-1) \cdot A = -A$.

Having observed several examples of vector spaces, lets consider the more specialized case of R^n, for each entry $A = \begin{bmatrix} 0 \\ a_2 \\ \dots \\ a_n \end{bmatrix}$ of the specialized set, the first element is zero. It can be shown that this set is closed under matrix addition. This set has the special distinction of being a subset of R^n and being a vector space under the binary operation of matrix addition and scalar multiplication. This leads to the following definition of subspaces.

Definition 3.1.4 Let **V** be a vector space and let **W** be a subset of **V**. If **W** is a vector space relative to the same binary operations of addition and scalar multiplication, then **W** is called a subspace of **V**.

Note: Every vector space has at least two subspaces; the set itself and the set comprised of only the zero vector.

Exercises

1. Show that the set ${}_mR_n$ (the set of **m x n** matrices of real numbers) is a vector space.

2. Show that the set of polynomials $p_n(t) = a_n t^n + a_{n-1} t^{n-1} + \dots + a_0$ is a vector space.

3. Let $S(\underset{\sim}{A}')$ be a subset of R^n. We define it to be the set of **n x 1** matrices (column matrices) $\underset{\sim}{X}$ such that $\underset{\sim}{A}' \times \underset{\sim}{X} = 0$. Show that this set forms a vector space.

4. In exercise three above, let $n = 3$ and let the determining matrix of the set be $\underset{\sim}{A}' = \begin{bmatrix} 4 & -1 & -2 \end{bmatrix}$. Compute the value of $\underset{\sim}{A}' \times \begin{bmatrix} 2 \\ 2 \\ 3 \end{bmatrix}$. Is the column matrix $\begin{bmatrix} 2 \\ 2 \\ 3 \end{bmatrix}$ an element of the space $S(\underset{\sim}{A}')$?

5. Given the column matrix contained in the family of **3 x 1** matrices of form $\begin{bmatrix} 2x \\ 4y \\ 3 \end{bmatrix}$; under what conditions will it be an element of the vector space $S(\begin{bmatrix} 1 & -1 & 2 \end{bmatrix})$?

6. Given the column matrix contained in the family of **3 x 1** matrices of form $\begin{bmatrix} 3x \\ 2y \\ 4x \end{bmatrix}$; under what conditions will it be an element of the vector space $S([1\ 1\ -1])$?

7. Let **A** be a **k x m** real matrix. Show that the set of **k x n** matrices that are expressed as the product $M = A \times X$ (where X is an arbitrary **m x n** matrix) form a vector space.

8. Now suppose that we have **k 'm x n'** real matrices $(A_1, A_2, \ldots, A_k)$. If we define the set $S = \left\{ \sum_{i=1}^{k} a_i \cdot A_i \right\}$ all possible linear combinations of these matrices; show that the set S forms a vector space.

3.2 Linear Independence, Bases and Coordinate Systems

As we completed section 3.1, we ended with exercise number eight, in which we defined the elements of a space as the sum of a scalar products of a finite number of elements. It turns out that each of the elements is also an element of the space. In vector spaces we call such a sum a linear combination. See definition 3.2.1 for a formal definition of linear combination.

Definition 3.2.1 A vector α is a linear combination of the vectors in a finite set $S = \{\alpha_1, \alpha_2, \ldots, \alpha_n\}$ of vectors in the vector space **V**, if

$$\alpha = \sum_{i=1}^{n} a_i \cdot \alpha_i$$, were the a_i's are real numbers.

If in exercise 8 of the previous section, one of the matrices A_1 could be expressed as a linear combination of the other matrices, then we could have expressed the all elements of the space with at least one fewer matrix. The general concept of all vectors in a vector space being defined as a linear combination of a finite number of vectors in the vector spaces, introduces another concept that of a finite number of vectors in the space spanning the vector space. The concept of one or more of the original vectors being a linear combination of the others is a concept of dependence or independence as formalized in the following definitions.

Definition 3.2.2 A vector space **V** is spanned by a set of vectors **S**, if each vector contained in **V** is a linear combination of the set of vectors in **S**. In some instances we may say that **S** spans the vector space **V**.

Although the vectors contained in a set S spans the vector space **V**, it raises the possibility that one or more of the vectors in S could be expressed as a linear combination of the remaining vectors in the set S. This implies that a subset of the finite set S could also span the vector space **V**. We formalize this concept in the following definition. The definition also introduces the concept, when none of the vectors in the finite set S can be expressed as a linear combination of the remaining vectors in the finite set.

Definition 3.2.3 A set $S = \{\alpha_1, \alpha_2, \ldots, \alpha_n\}$ of distinct vectors of the vector space V is called linearly dependent if there exist a set of real numbers $a_1, \ldots, a_n$ such that $0 = \sum_{i=1}^{n} a_i \cdot \alpha_i$ and all $a_i's$ is not equal to zero. The set is called linearly independent if $0 = \sum_{i=1}^{n} a_i \cdot \alpha_i$ implies that all $a_i's$ is equal to zero.

With the definitions of independence and dependence, we show the relationship of finite subsets and supersets when one is either independent or dependent.

Theorem 3.2.4 If S_1 and S_2 are finite subsets of a vector space such that S_1 is a subset of S_2, then the following two properties hold:

(1) If S_1 is linearly dependent then S_2 is linearly dependent.

(2) If S_2 is linearly independent then S_1 is linearly independent.

Proof: The proof for this theorem is left as an exercise for the reader.

Theorem 3.2.4 is useful for identifying a vector sub-space of the vector space spanned by a finite set of linearly independent vectors. Recall exercise 8 of previous section that a space defined as all possible linear combinations of the finite set of vectors forms a vector space (Provided that certain operations such as scalar multiplication and binary addition are well defined.). Now we introduce the concept of a vector space spanned by a finite set of linearly independent vectors.

Definition 3.2.5 A set $S = \{\alpha_1, \alpha_2, \ldots, \alpha_n\}$ of distinct vectors of the vector space **V** forms a basis for **V** if S spans **V** and S is linearly independent.

Now that we have identified the concept of a basis of a vector space, a question that arises is the uniqueness of the representation of a vector in the vector space as a linear combination of the finite set of vectors forming the basis.

Theorem 3.2.6 If the set $S = \{\alpha_1, \alpha_2, \ldots, \alpha_n\}$ of distinct vectors of the vector space **V** forms a basis for **V** then each vector of **V** is expressed uniquely as a linear combination of vectors in S.

Proof: To prove this theorem we assume that a given vector α is representable as $\alpha = \sum_{i=1}^{n} a_i \cdot \alpha_i$ and $\alpha = \sum_{i=1}^{n} b_i \cdot \alpha_i$ using the set $S = \{\alpha_1, \alpha_2, \ldots, \alpha_n\}$ which forms a basis for the vector space **V**. By setting the representations equal to each other and collecting all terms to the left side of the equation, we get $\sum_{i=1}^{n} (b_i - a_i) \cdot \alpha_i = 0$. From statement of theorem, the set $S = \{\alpha_1, \alpha_2, \ldots, \alpha_n\}$ is a linearly independent set and hence all coefficients must be equal to zero by definition 3.2.3. Hence we have shown that for each vector in the vector space, the representation with a given basis is unique.

A key element in the above proof is that the vectors in the finite set are linearly independent vectors spanning a space. If the vectors spanning the vector space are not linearly independent, then the uniqueness property is not there and we may have several representations of a given vector in the space. The proof for the theorem below considers this possibility.

Theorem 3.2.7 If the set $S = \{\alpha_1, \alpha_2, \ldots, \alpha_n\}$ of nonzero vectors, contained in the vector space **V,** spans **V** then there is a subset of **S** that is a basis for **V**.

Proof: To begin this proof, we introduce the notation of two mutually exclusive sets of finite vectors S_m and T_{n-m} such that $S = S_m \cup T_{n-m}$. In each case the set S_m contains '**m**' vectors and forms increasing subsets such that $S_1 \subset S_2 \subset \ldots \subset S_n = S$. Similarly, with the T_{n-m} sets we have $T_1 \subset T_2 \subset \ldots \subset T_n = S$.

We define the matrix T_1 to be the set containing one vector of the set S such that the vector can be expressed as a linear combination of the remaining vectors in S. If we can find no such vector in S, then the set must be a linearly independent set of vectors that span **V** and hence it is a basis for **V**. If we can find such a vector, then the set S_{n-1} is defined to be all vectors in S that are not in T_1. Since the set S spans **V** and the vectors in T_1 are linear combinations of the elements of S_{n-1}, we must have that S_{n-1} spans the vector space **V**. We continue this logic until we find the subset S_m such that it is a linearly independent set of vectors that span **V**.

Hence we have shown that there is a subset of S that is a basis of **V**.

Since this book is a course in linear algebra and matrix theory, it is only natural that we use the concepts of matrices that have been introduced to this point to assist in proving additional properties of vector spaces. Let's examine the relationship of a finite subset to another finite set contained in the vector space **V** in terms of the linear equations. As specified in definition 3.2.1, a vector may be expressed as a linear combination of a finite set of vectors in **V**. Such a linear combination for a single vector α is expressed using the summation notation $\alpha = \sum_{i=1}^{n} a_i \cdot \alpha_i$ .Placing the finite set of vectors into a column matrix, we replace the summation notation with a matrix notation:

$\alpha = \begin{bmatrix} a_1 & \dots & a_n \end{bmatrix} \cdot \begin{bmatrix} \alpha_1 \\ \dots \\ \alpha_n \end{bmatrix}$. To shorten this notation further, we might use the symbol $\underset{\sim}{K'}$ to represent the row matrix of real numbers. Hence we rewrite the expression as follows: $\alpha = \underset{\sim}{K'} \cdot \begin{bmatrix} \alpha_1 \\ \dots \\ \alpha_n \end{bmatrix}$. Now we generalize this expression by considering a finite set of vectors $\boldsymbol{T} = \{\beta_1, \beta_2, \dots, \beta_r\}$ that is spanned by another finite set of vectors $S = \{\alpha_1, \alpha_2, \dots, \alpha_n\}$. By definition of spanning and linear combinations we have the following set of equations:

$\begin{bmatrix} \beta_1 = \sum_{i=1}^{n} a_{1i} \cdot \alpha_i \\ \dots \\ \beta_r = \sum_{i=1}^{n} a_{ri} \cdot \alpha_i \end{bmatrix}$. These equations may be represented using the matrix equation $\begin{bmatrix}\beta_1\\ \dots\\ \beta_r\end{bmatrix} = \begin{bmatrix}a_{11} & \dots & a_{1n}\\ \dots & \dots & \dots\\ a_{r1} & \dots & a_{rn}\end{bmatrix} \times \begin{bmatrix}\alpha_1\\ \dots\\ \alpha_n\end{bmatrix}$. Using the symbol **A** to represent the **r-by-n** matrix in this equation we now define the spanning of a finite set **T** by another finite set **S** if there is an **r-by-n** matrix **A** such that

$\begin{bmatrix}\beta_1\\ \dots\\ \beta_r\end{bmatrix} = A \times \begin{bmatrix}\alpha_1\\ \dots\\ \alpha_n\end{bmatrix}$. To simplify the notation further we will let the equation

$\boldsymbol{T} = A \times S$ represent the above matrix equation. By restating the definition of linear combination using this more general approach, we formally state the definition of spanning below.

Definition 3.2.8 Given two finite sets of vectors $S = \{\alpha_1, \alpha_2, \dots, \alpha_n\}$ and $\boldsymbol{T} = \{\beta_1, \beta_2, \dots, \beta_r\}$. Then **S** spans **T** if there is an **r-by-n** matrix **A** such that $\begin{bmatrix}\beta_1\\ \dots\\ \beta_r\end{bmatrix} = A \times \begin{bmatrix}\alpha_1\\ \dots\\ \alpha_n\end{bmatrix}$. This equation is symbolically represented as $\boldsymbol{T} = A \times S$.

Now that we are defining linear combinations in terms of matrices, it is only natural that we are interested in expressing dependence or independence relationships of the two distinct finite sets in terms of the matrix relating the two sets.

Corollary 3.2.9 Given the linearly independent finite set $S = \{\alpha_1, \alpha_2, \ldots, \alpha_n\}$ that spans the finite set $\boldsymbol{T} = \{\beta_1, \beta_2, \ldots, \beta_r\}$ of the vector space **V.** If the matrix **A**, such that $\boldsymbol{T} = A \times S$, is equivalent to a row echelon form having at least one zero row matrix, then the set **T** is linearly dependent.

Proof: If matrix **A** is equivalent to a matrix in row echelon form having a least one zero row matrix, then there is a non-singular matrix $P = \prod E_j$ such that $P \times A = B$ and the matrix **B** is in row echelon form with at least one zero row. After the multiplication we have the equation $P \times \boldsymbol{T} = B \times S$. Since the matrix P is non-singular none of its rows is a zero row and hence by definition of $P \times \boldsymbol{T}$, we have at least one linear combination of the vectors in T that is equal to zero. By definition of linearly dependence we conclude that T is linearly dependent.

Contrary to this corollary, we ask the question "what if the matrix A is non singular"? The answer to this question is given using the following corollary.

Corollary 3.2.10 Given the linearly independent finite set $S = \{\alpha_1, \alpha_2, \ldots, \alpha_n\}$ that spans the finite set $\boldsymbol{T} = \{\beta_1, \beta_2, \ldots, \beta_r\}$ of the vector space **V.** If the matrix **A**, such that $\boldsymbol{T} = A \times S$, is **not** equivalent to a row echelon form with at least one zero row, then the set **T** is linearly **independent**.

Proof: The proof of this corollary is left as an exercise for the reader.

Theorem 3.2.11 Given the two sets $S = \{\alpha_1, \alpha_2, \ldots, \alpha_n\}$ and $T = \{\beta_1, \beta_2, \ldots, \beta_r\}$ of the vector space **V** such that **S** forms a basis for **V** and **T** is a set of linearly independent vectors. Then the following inequality is true: $r \le n$.

Proof: Since S forms a basis for **V,** all elements of the set **T** are linear combinations of the elements of S. Therefore we have

$T = A \times S$, (A is a **r-by-n** matrix). If $r > n$, that is more rows than columns, then we can reduce A to row echelon form with **r-n** zero rows matrices. From corollary 3.2.9 this implies that T is a set of linearly dependent vectors. This contradicts the statement of the theorem that T is a linearly independent set. Hence we must have $r \le n$.

Corollary 3.2.12 Given the two sets $S = \{\alpha_1, \alpha_2, \ldots, \alpha_n\}$ and $T = \{\beta_1, \beta_2, \ldots, \beta_r\}$ of the vector space **V** such that both **S** and **T** form a basis for **V**. Then the following equality is true: r = n.

Proof: Since S is a basis of V and T is a linearly independent set (definition of basis), we must have $r \le n$ from theorem 3.2.11. However T is a basis of V and S is a linearly independent set (definition of basis), we must have $n \le r$ from theorem 3.2.11. Both equalities can only be true if $r = n$.

The above corollary illustrates that the number of elements in the basis of a vector space is a unique value. This leads to the definition stated below.

Definition 3.2.13 The dimension of a nonzero vector space **V** is the number of vectors contained in a basis for the vector space **V**. Mathematically we express the dimension of V as dim(**V**).

Theorem 3.2.14 If **m** is less than **n** and $S = \{\alpha_1, \alpha_2, \ldots, \alpha_m\}$ is a linearly independent set of vectors in the n-dimensional vector space **V**, then there exist vectors $\{\beta_1, \beta_2, \ldots, \beta_{n-m}\}$ such that $\boldsymbol{T} = \{\alpha_1, \alpha_2, \ldots, \alpha_m, \beta_1, \beta_2, \ldots, \beta_{n-m}\}$ is a basis for **V**.

Proof: By definition of vector space, all linear combinations of the vectors in S are contained in V. Hence if we define a new vector space V_0 as the set of all vectors that are linear combinations of the set of vectors contained in S, it is a subspace of the vector space V. Since $V_0 \subseteq V$ and $dim(V_0) < dim(V)$, we have that there are elements in V that are not V_0. Therefore we can find a vector β_1 contained in V but not in V_0. Hence the set of vectors $\boldsymbol{T}_1 = \{\alpha_1, \alpha_2, \ldots, \alpha_m, \beta_1\}$ is a linearly independent set of V.

By continuing this process for each independent and finite set $\boldsymbol{T}_k = \{\alpha_1, \alpha_2, \ldots, \alpha_m, \beta_1, \beta_2, \ldots, \beta_k\}$, we can define a vector space V_k such that $V_k \subseteq V$ and $dim(V_k) = \boldsymbol{m} + \boldsymbol{k}$. When we have $\boldsymbol{k} = \boldsymbol{n} - \boldsymbol{m}$, we get $dim(V_k) = n$ and therefore $\boldsymbol{T} = \{\alpha_1, \alpha_2, \ldots, \alpha_m, \beta_1, \beta_2, \ldots, \beta_{n-m}\}$ is a basis for V. Hence we have shown that there exist vectors $\{\beta_1, \beta_2, \ldots, \beta_{n-m}\}$ such that $\boldsymbol{T} = \{\alpha_1, \alpha_2, \ldots, \alpha_m, \beta_1, \beta_2, \ldots, \beta_{n-m}\}$ is a basis for **V**.

Theorem 3.2.15 Let the set $S = \{\alpha_1, \alpha_2, \ldots, \alpha_n\}$ be a subset of the n-dimensional vector space **V**. Then the following properties are true for the subset **S**:

1. If S is a linearly independent set of vectors, then S is a basis for **V**.
2. If S spans **V**, then S is a basis for **V**.

Proof: **Property 1**

To prove property 1 use the fact that dim(V) = **n** implies that there is a finite set T of **n** vectors contained in V such that **T** is a basis for V. From the definition of basis we have a matrix A such that $S = A \times \boldsymbol{T}$. Since the set S is linearly independent and A is an **n-by-n** matrix, the matrix must be non-singular (not equivalent to a row echelon form and is **n-by-n**) according to corollary 3.2.10. This implies that A^{-1} exists such that $\boldsymbol{T} = A^{-1} \times S$. For any arbitrary vector α in V, we have a none zero row vector $\underset{\sim}{K}'$ such that $\alpha = \underset{\sim}{K}' \cdot \boldsymbol{T}$. Since the set T is spanned by the set S, we make a substitution to rewrite the following the representation of α as

$\alpha = \underset{\sim}{K}' \cdot (A^{-1} \times S)$ or $\alpha = (\underset{\sim}{K}' \cdot A^{-1}) \cdot S$. Hence the vector space V is spanned by S and by definition it is a basis for V.

Property 2

If S spans the n-dimensional vector space V and the basis is the finite set T of **n** vectors contained in V, then we have a matrix A such that $\boldsymbol{T} = A \times S$. If A is singular, then there is a non-singular matrix P which is the product of elementary matrices such that $P \times \boldsymbol{T} = B \times S$ and B is in row echelon form with at least one zero row matrix. Since P is singular, this implies that there is a non zero linear combination of the basis vectors equal to zero. This presents a contradiction since T forms a basis for V. Hence the matrix A must be non-singular and $S = A^{-1} \times \boldsymbol{T}$. From corollary 3.2.10, S is independent and hence S forms a basis of V.

As seen earlier in this section not all sets that span a vector space are linearly independent. In some instances, we may have a linearly dependent set that spans a vector space. The following definition gives the definition of a special subset of such a finite set of vectors that span the vector space.

Definition 3.2.16 A maximal independent subset of the set S of vectors in the vector space **V** is the linearly independent subset T in S such that no other linearly independent subset of S has more vectors than T.

Theorem 3.2.17 If S is a finite subset of the vector space **V** and S spans **V**, then a maximal independent subset of S is a basis for **V**.

Proof: Let T be a maximal independent subset of the set S. Also we define the set U to be the vectors contained in S but not in T. By definition 3.2.17 this is the largest linearly independent subset of S. This implies that any of the elements of U when added to the set T would create a linearly dependent set. Hence each element contained in U is a linear combination of the vectors contained in T. Using the matrix notation for a linear combination in matrix notation we write $U = A \cdot \boldsymbol{T}$.

Since the finite set S spans the vector space **V,** we can express any arbitrary vector in **V** as a linear combination of the elements in U and T. Hence we have the following expression;

$\alpha = \underset{\sim}{K'} \cdot T + \underset{\sim}{J'} \cdot U$. Since U is spanned by the set T, we make the above substitution for U to give the following expression: $\alpha = \underset{\sim}{K'} \cdot T + (\underset{\sim}{J'} \cdot A) \cdot T$. Using distributed property of matrices we have $\alpha = (\underset{\sim}{K'} + (\underset{\sim}{J'} \cdot A)) \cdot T$ and since α is an arbitrary vector contained in V, we have that T spans V. Hence the maximal independent subset of a set that spans the vector space is a basis for the vector space.

Exercises

1. Prove theorem 3.2.4. That is for the finite subsets S_1 and S_2 of a vector space such that S_1 is a subset of S_2, then the following two properties hold:

    (1) If S_1 is linearly dependent then S_2 is linearly dependent.

    (2) If S_2 is linearly independent then S_1 is linearly independent.

2. Prove corollary 3.2.10. Given the linearly independent finite set $S = \{\alpha_1, \alpha_2, \ldots, \alpha_n\}$ that spans the finite set $\boldsymbol{T} = \{\beta_1, \beta_2, \ldots, \beta_r\}$ of the vector space **V.** If the matrix **A**, such that $\boldsymbol{T} = A \times S$, is not equivalent to a row echelon form with at least one zero row, then the set **T** is linearly independent.

3. Let S and $\boldsymbol{T}$ be finite subsets of a vector space such that $\boldsymbol{T}$ consists of three vectors and S has four linearly independent vectors. If the vectors in $\boldsymbol{T}$ are linear combinations of the elements in S according to the transformation $T = \begin{bmatrix} 3 & 1 & -4 & 5 \\ -3 & -2 & 10 & -7 \\ 6 & 4 & -20 & 14 \end{bmatrix} \times S$, determine if $\boldsymbol{T}$ is linearly dependent or independent and explain why.

4. Let S and T be finite subsets of a vector space such that T consists of four vectors and S has 4 linearly independent vectors. If the vectors in T are linear combinations of the elements in S according to the transformation $T = \begin{bmatrix} 1 & 2 & 4 & 0 \\ 2 & 0 & 1 & 3 \\ -4 & -2 & -1 & -3 \\ -12 & -2 & 4 & -12 \end{bmatrix} \times S$, determine if T is linearly dependent or independent and explain why.

5. Verify that the three polynomials shown in the following column matrix $\begin{bmatrix} t^2 + 3 \cdot t - 5 \\ -4 \cdot t^2 \\ t + 1 \end{bmatrix}$ span the vector space of polynomials having form

$$P_2(t) = a_2 \cdot t^2 + a_1 \cdot t + a_0 .$$

6. Given the finite subsets $\begin{bmatrix} t^2 + 3 \cdot t - 5 \\ -4 \cdot t^2 \\ t + 1 \end{bmatrix}$ and $\begin{bmatrix} t^2 + 3 \cdot t - 5 \\ -4 \cdot t^2 \\ t + 1 \end{bmatrix}$ in the vector spaces of polynomials $P_2(t) = a_2 \cdot t^2 + a_1 \cdot t + a_0$.

3.3 Ordered Basis And Isomorphisms

The discussion of a basis for an n-dimensional vector space, in the preceding sections, did not consider the order of vectors in the basis. Consider the finite sets $S = \{\alpha_1, \alpha_2, \alpha_3\}$ and $\boldsymbol{T} = \{\beta_1, \beta_2, \beta_3\}$ in the 3-dimensional vector space **V**, where S is a basis of **V**. We also have the matrix $A = \begin{bmatrix} 1 & -1 & 4 \\ 2 & 0 & -2 \\ 5 & -6 & -1 \end{bmatrix}$ such that $\begin{bmatrix} \beta_1 \\ \beta_2 \\ \beta_3 \end{bmatrix} = \begin{bmatrix} 1 & -1 & 4 \\ 2 & 0 & -2 \\ 5 & -6 & -1 \end{bmatrix} \times \begin{bmatrix} \alpha_1 \\ \alpha_2 \\ \alpha_3 \end{bmatrix}$. If we switch the positions of vectors one and two in the subset S, then we have $\begin{bmatrix} \beta_1 \\ \beta_2 \\ \beta_3 \end{bmatrix} = \begin{bmatrix} -1 & 1 & 4 \\ 0 & 2 & -2 \\ -6 & 5 & -1 \end{bmatrix} \times \begin{bmatrix} \alpha_2 \\ \alpha_1 \\ \alpha_3 \end{bmatrix}$. Notice that the first and second columns are interchanged in the second equation. Hence depending on the order of the vectors in the basis, the matrix may change. Hence the matrices that define the linear combinations from a linearly independent finite set to another finite set, depending on order, are equivalent matrices through the elementary matrix operation of interchanging columns. By associating a particular arrangement of the linearly independent vectors of a finite set of vectors with all linear combinations, then the matrix denoting linear combinations is always unique. This prompts us to state the following definition.

Definition 3.3.1 Let a finite subset of an n-dimensional vector space be a basis for the vector space, such that the order of the vectors that form the basis set are always in a fixed order, we refer to such a basis as an ordered basis.

Now that we have a definition for ordered n-dimensional vector spaces, we can now categorically state that the linear representation of a vector in the n-dimensional vector space is unique. From this uniqueness property, we define the coefficients associated with each basis vector in a linear representation of an arbitrary vector in the vector space to be the coordinates of the represented vector. Using vector notation, we state the formal definition of coordinates and coordinate systems.

Definition 3.3.2 Consider an n-dimensional ordered vector space V with the finite ordered subset $S = \{\alpha_1, \alpha_2, \ldots, \alpha_n\}$ as a basis for it. If a vector α contained in V is represented as $\alpha = \sum_{i=1}^{n} a_i \cdot \alpha_i$, then the coefficients $a_1, a_2, \ldots, a_n$ are called vector coordinates for the vector α. For a given ordered basis we simplify the representation of a vector by introducing the notation of the coordinates in a row matrix such as $\alpha = [a_1, a_2, \ldots, a_n]_S$. This notation is equivalent to $\alpha = [a_1, a_2, \ldots, a_n] \cdot \begin{bmatrix} \alpha_1 \\ \alpha_2 \\ \ldots \\ \alpha_n \end{bmatrix}$. We refer to this system of notations as a coordinate system for the vector space.

In the previous section we discussed the transformation from one basis to another basis within the same vector space. For intra-vector space transformation the binary operations of vector addition and scalar multiplication were well defined within the vector space. When we consider transformation from one vector space to another vector space, the binary operations may not be the same. Hence for inter-vector space transformations, we must consider how vector addition and scalar multiplication is defined in each. An example of inter-vector space transformation is shown in the following example.

Example

Remembering the trigonometric functions from the previous chapter, we define a 2-dimensional vector space using distance from origin and angle from a fixed segment. The space as defined here is a vector space and the proof is left to the reader. Consider the vector space W where vectors in W have the form (r,θ) ; r is a real number and θ is an angle having a value between 0 and 2π radians inclusive. For this vector space addition of two vectors in the space is defined to as follows: $(r_1,\theta_1) \oplus (r_2,\theta_2) = (r_S,\theta_S)$ where

$$r_S = \sqrt{r_1^{\,2} + r_2^{\,2} + 2 \cdot r_1 \cdot r_2 \cdot cos(\theta_2 - \theta_1)}$$ and

$$\theta_S = asin\left(\frac{r_1 \cdot sin(\theta_1) + r_2 \cdot sin(\theta_2)}{r_S}\right)$$

. Scalar multiplication for the vector space W is defined for the scalar value c:

$$c \bullet (r,\theta) = (|c| \cdot r,\theta_c) \quad \text{where} \quad \theta_c = \begin{cases} \theta & if \;\; c > 0 \\ 0 & if \; (c = 0) \\ \theta + \pi & if \;\; c > 0 \end{cases}$$

. Note that for this definition of scalar multiplication; zero times a vector yields $0 \bullet (r,\theta) = (0,0)$.

To construct a basis for the vector space W we consider the subset, containing two linearly independent vectors namely $S = \left\{(1,0), (1,\frac{\pi}{2})\right\}$. These two vectors are shown to span the vector space W since

$(r,\theta) = a \bullet (1,0) \oplus b \bullet (1,\frac{\pi}{2})$ when $a = r \cdot cos(\theta)$ and

$b = r \cdot sin(\theta)$. Another basis for this vector space is the two element set

$T = \left\{(1,\pi), (1,\frac{\pi}{4})\right\}$. Using the concepts shown in the previous section we

can relate these two linearly independent sets using the following matrix equa-

tion: $\begin{bmatrix}(1,\pi)\\(1,\frac{\pi}{4})\end{bmatrix} = \begin{bmatrix}-1 & 0\\ \frac{\sqrt{2}}{2} & \frac{\sqrt{2}}{2}\end{bmatrix} \bullet \begin{bmatrix}(1,0)\\(1,\frac{\pi}{2})\end{bmatrix}$. Using the inverse of the matrix, the solu-

tion of the S vectors in terms of the T vectors is

$$\begin{bmatrix}(1,0)\\(1,\frac{\pi}{2})\end{bmatrix} = \begin{bmatrix}-1 & 0\\ 1 & \frac{2}{\sqrt{2}}\end{bmatrix} \bullet \begin{bmatrix}(1,\pi)\\(1,\frac{\pi}{4})\end{bmatrix}.$$

Having established properties of the 2-dimensional vector space W (2-dimensional polar coordinate system), we are now ready to illustrate inter-vector space transformation from the R^2 to W. Let (x,y) be an arbitrary vector in the vector space R^2. We define the mapping M() such that

$M(x, y) = (\sqrt{x^2+y^2}, asin\left(\frac{y}{\sqrt{x^2+y^2}}\right))$. As we examine the properties of

this mapping, we have $M(c \cdot (x, y)) = (c \cdot \sqrt{x^2+y^2}, asin\left(\frac{y}{\sqrt{x^2+y^2}}\right))$.

Using the property of scalar multiplication in the vector space W, we have

$M(c \cdot (x, y)) = c \bullet M(x, y)$. The next question of interest is how does the sum of two vectors in vector space R^2 behave when transformed to the vector space W by the mapping M(). To observe this behavior we select two arbitrary vectors in R^2; namely (x_1,y_1) and (x_2,y_2). Using the definition of this mapping M() we have $M((x_1 + x_2, y_1 + y_2)) = (r_S,\theta_S)$ where $r_S = \sqrt{(x_1 + x_2)^2 + (y_1 + y_2)^2}$ and $\theta_S = asin\left(\frac{y_1 + y_2}{r_S}\right)$. Now we introduce the following notation:

$$r_1 = \sqrt{(x_1)^2 + (y_1)^2} \ , \ r_2 = \sqrt{(x_2)^2 + (y_2)^2} \ ,$$

$$x_1 = r_1 \cdot cos(\theta_1) \ , \ x_2 = r_2 \cdot cos(\theta_2) \ ,$$

$$y_1 = r_1 \cdot sin(\theta_1) \text{ and } y_2 = r_2 \cdot sin(\theta_2) .$$

Using this notation and rearranging the terms for the values of r_S and θ_S we get the following: $r_S = \sqrt{(r_1)^2 + (r_2)^2 + 2 \cdot r_1 \cdot r_2 \cdot cos(\theta_2 - \theta_1)}$ and $\theta_S = asin\left(\frac{r_1 \cdot sin(\theta_1) + r_2 \cdot sin(\theta_2)}{r_S}\right)$. Hence we have by definition of $\oplus$ the equation $M((x_1 + x_2, y_1 + y_2)) = (r_S,\theta_S) = (r_1,\theta_1) \oplus (r_2,\theta_2)$. By definition of M() we get $M((x_1 + x_2, y_1 + y_2)) = M((x_1,y_1)) \oplus M((x_2,y_2))$.

These properties of M() illustrate that for any linear combination of the basis in R^2 it has a unique linear combination in W. This leads to the following definition for vector spaces.

Definition 3.3.3 Let V be a real vector space with operations $+$ and $\cdot$. Let W be a real vector space with operations $\oplus$ and $\bullet$. Let the one-to-one function M() map V onto W which satisfy the following properties:

1. $M(\alpha + \beta) = M(\alpha) \oplus M(\beta)$ for α, β in V.

2. $M(c \cdot \alpha) = c \bullet M(\alpha)$ for α in V and c a real number.

Such a function is called an isomorphism of **V** onto **W** and **V** is said to be isomorphic to **W**. We also say that **V** and **W** are isomorphic.

Some properties of **n**-dimensional real vector spaces are illustrated in the theorems and corollaries stated below.

Theorem 3.3.4 If V is an **n**-dimensional real vector space, then V is isomorphic to $\boldsymbol{R}^n$.

Proof: Let V be an **n**-dimensional real vector space with operations of vector addition denoted as $\oplus$ and scalar multiplication denoted as $\bullet$. Furthermore a basis for V is the subset $S = \{\alpha_1, \alpha_2, \ldots, \alpha_1\}$. Now define the function M() mapping

V onto $\mathbf{R^n}$ (with basis $\{\underset{\sim}{e}_1, \underset{\sim}{e}_2, \ldots, \underset{\sim}{e}_n\}$) such that

$$M(a_1 \bullet \alpha_1 \oplus \ldots \oplus a_n \bullet \alpha_n) = a_1 \cdot \underset{\sim}{e}_1 + \ldots + a_n \cdot \underset{\sim}{e}_n \cdot$$

To show that this function satisfies condition one of definition 3.3.3, we introduce two vectors in the vector space V; namely

$\alpha = a_1 \bullet \alpha_1 \oplus \ldots \oplus a_n \bullet \alpha_n$ and

$\beta = b_1 \bullet \alpha_1 \oplus \ldots \oplus b_n \bullet \alpha_n$. By definition of vector addition, we have the following expression

$\alpha + \beta = (a_1 + b_1) \bullet \alpha_1 \oplus \ldots \oplus (a_n + b_n) \bullet \alpha_n$. From the definition of $M()$ we have

$M(\alpha + \beta) = (a_1 + b_1) \cdot \underset{\sim}{e}_1 + \ldots + (a_n + b_n) \cdot \underset{\sim}{e}_n$ and using the property that $\mathbf{R^n}$ is a vector space we have

$$M(\alpha + \beta) = M(\alpha) \oplus M(\beta) \cdot$$

For condition two of definition 3.3.3 we multiply the arbitrary vector

$\alpha = a_1 \bullet \alpha_1 \oplus \ldots \oplus a_n \bullet \alpha_n$ by the scalar c to get

$c \bullet \alpha = (c \cdot a_1) \bullet \alpha_1 \oplus \ldots \oplus (c \cdot a_n) \bullet \alpha_n$. This leads to the following expression: $M(c \bullet \alpha) = (c \cdot a_1) \cdot \underset{\sim}{e}_1 + \ldots + (c \cdot a_n) \cdot \underset{\sim}{e}_n$

and hence $M(c \bullet \alpha) = c \cdot M(\alpha)$.

Now that both conditions in definition 3.3.3 are proven true, we can conclude that any **n**-dimensional vector space is isomorphic to $\mathbf{R^n}$.

Theorem 3.3.5 If U,V and W are vector spaces, then the following properties are true:

3. Every vector space V is isomorphic to itself.
4. If V and W are vector spaces such that V is isomorphic to W, then W is isomorphic to V.
5. If U,V and W are vector spaces such that U is isomorphic to V and V is isomorphic to W, then U is isomorphic to W.

Proof: The proof of this theorem is left as an exercise for the reader.

Corollary 3.3.6 If V and W are **n**-dimensional vector spaces, then they are isomorphic.

Proof: The proof of this corollary is left as an exercise for the reader.

Theorem 3.3.7 Two finite dimensional vector spaces are isomorphic if and only if their dimensions are equal.

Proof: From corollary 3.3.6, if the dimensions of the vector spaces V and W are equal then V and W are isomorphic. Hence the "if" portion of this theorem is true.

To begin the proof for the "only if" part of this theorem, we start with two isomorphic vector spaces V and W of dimensions **k** and **n** respectively. Since V is **n**-dimensional, there is a finite set $S = \{\alpha_1, \alpha_2, \ldots, \alpha_k\}$ which is a basis for V. By definition of V and W isomorphic we have a one-to-one mapping M() such that for every vector $\beta \in W$, we have a corresponding vector $\alpha \in V$ such that $\beta = M(\alpha)$. Since all vectors in V are a linear combination of the basis vectors we have $\alpha = a_1 \cdot \alpha_1 + \ldots + a_k \cdot \alpha_k$ and $\beta = a_1 \cdot M(\alpha_1) + \ldots + a_k \cdot M(\alpha_k)$. Since all vectors in W can be expressed as a linear combination of these vectors, we can conclude that the set $T = \{M(\alpha_1), M(\alpha_2), \ldots, M(\alpha_k)\}$ spans the vector space W. This implies that $n \leq k$. Applying this same methodology to by mapping W into V, we can conclude that $k \leq n$. Both of these conditions can only be true if $k = n$. Hence two vector spaces are isomorphic only if their dimensions are equal and the theorem is proven true.

The following corollary is a special case of the theorem above.

Corollary 3.3.8 If **V** is a finite dimensional space which is isomorphic to $\boldsymbol{R}^n$, then dim **V** = **n**.

Proof: Apply theorem 3.3.7 to prove this corollary.

Exercises

1. In the example of this section using the two-dimensional polar vector space, we defined the space W (and assumed it was a vector space) where elements in W have the form (r,θ) ; r is a non-negative real number and θ is an angle having a value between 0 and 2π radians inclusive. For this space addition of two elements in the space is defined as follows:

   $(r_1,\theta_1) \oplus (r_2,\theta_2) = (r_S,\theta_S)$ where

$r_S = \sqrt{r_1^{\ 2} + r_2^{\ 2} + 2 \cdot r_1 \cdot r_2 \cdot cos(\theta_2 - \theta_1)}$ and

$\theta_S = asin\left(\dfrac{r_1 \cdot sin(\theta_1) + r_2 \cdot sin(\theta_2)}{r_S}\right)$. Scalar multiplication for the

space W is defined for the scalar value c: $c \bullet (r,\theta) = (|c| \cdot r, \theta_c)$ where

$\theta_c = \begin{cases} \theta & if \ \ c > 0 \\ 0 & if \ (c = 0) \\ \theta + \pi & if \ \ c < 0 \end{cases}$. Note that for this definition of scalar multi-

plication; zero times a vector yields $0 \bullet (r,\theta) = (0,0)$. Show that this space is a vector space.

2. Prove statements 1, 2 and 3 of theorem 3.3.5.

3. Prove corollary 3.3.6, that is, if V and W are **n**-dimensional vector spaces, then they are isomorphic.

4. Defining the space W, where elements in W have the form (R,θ) ; R is a non-negative real number and θ is an angle having a value between 0 and 2π radians inclusive. For this space addition of two elements in the space is

   defined as follows: $(R_1,\theta_1) \oplus (R_2,\theta_2) = (R_S,\theta_S)$ where

   $R_S = \sqrt{R_1^{\ 2} + R_2^{\ 2} + 2 \cdot f(\theta_1, \theta_2)}$ and

$\theta_S = atan(g(\theta_1, \theta_2, R_1, R_2))$. The functions f() and g() are defined as follows:

$$f(\theta_1, \theta_2) = (1-\beta^2) \cdot \frac{(1-\beta^2) \cdot cos(\theta_1) \cdot cos(\theta_2) + sin(\theta_1) \cdot sin(\theta_2)}{(1-\beta \cdot cos(\theta_1)) \cdot (1-\beta \cdot cos(\theta_2))}$$

and $g(\theta_1, \theta_2, R_1, R_2) = $ ...

$$\frac{\dfrac{sin(\theta_1)}{(1-\beta \cdot cos(\theta_1))} + \dfrac{sin(\theta_2)}{(1-\beta \cdot cos(\theta_2))}}{\dfrac{cos(\theta_1)}{(1-\beta \cdot cos(\theta_1))} + \dfrac{cos(\theta_2)}{(1-\beta \cdot cos(\theta_2))} + \dfrac{\beta}{(1-\beta^2)} \cdot (R_1 + R_2)}.$$

The scalar multiplication for the space W is defined for the scalar value c to be defined as follows: $c \bullet (R,\theta) = (|c| \cdot R, \theta_c)$ where the angle θ_c is defined to be $\theta_c = \begin{cases} \theta & if \quad c > 0 \\ 0 & if \ (c = 0) \\ \theta + \pi & if \quad c < 0 \end{cases}$. Note that for this definition of scalar multiplication; zero times a vector yields $0 \bullet (R,\theta) = (0,0)$. The parameter β is computed as follows: $\beta = \sqrt{1-\alpha^2}$ and α represents the ratio of minor axis to major axis of an ellipse. The major axis is denoted as R and the minor axis as $\alpha \cdot R$. Show that this space is a vector space.

5. Show that the vector space, defined in exercise above, satisfy the conditions of being isomorphic to the R^2 vector space with vectors of the form (x, y) with the one-to-one mapping $M(x, y) = \left(\sqrt{x^2 + \frac{y^2}{\alpha^2}}, atan\left(\frac{y}{x + \beta \cdot R}\right)\right)$.

3.4 Basic Geometry And Trigonometry Of $\mathbf{R}^2$

Now that the general subject of vector spaces has been discussed, let's look at the specific example such as a space in a plane.

This section steps back from the general topic of vector spaces and examines some properties of the 2-dimensional vector space R^2 .

3.4.1 Cartesian Coordinates

In elementary geometry we learned that two intersecting lines determines a plane. Now we consider the two dimensional plane that is spanned by two perpendicular line segments $e_1 = (1,0)$ and $e_2 = (0,1)$. For any point $p = (x,y)$ contained in the two dimensional space, we have the linear combination $p = x \cdot e_1 + y \cdot e_2$. This representation of a coordinate or point in a plane is illustrated in Figure 4 below.

Figure 4 Illustration of Coordinates in Plane

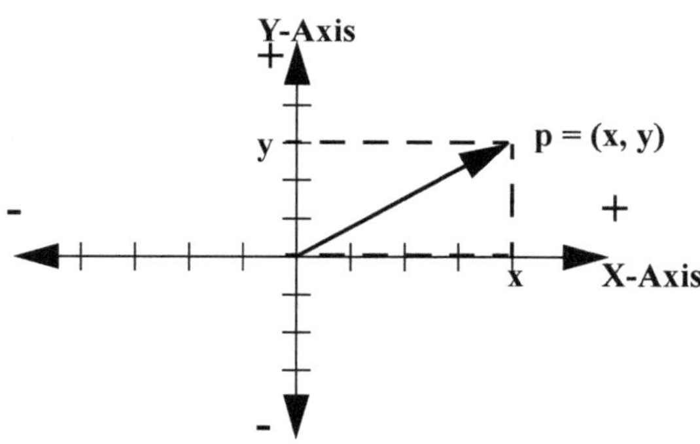

As shown in Figure 4, the vertical axis is called the **Y-Axis** and the horizontal axis is called the **X-Axis**. The unit vector along the **X-Axis** is denoted as $e_1 = (1,0)$, while the unit vector along the **Y-Axis** is denoted as $e_2 = (0,1)$.

Definition 3.4.2 A cartesian coordinate system is a coordinate system that has as its basis independent and intersecting lines at a single point. The single point of intersection is considered the origin of the coordinate system.

The selection of e_1 and e_2 as the basis of a plane is not by accident. This basis was chosen because the vectors divide the plane into four equal quadrants. Since a line contains an angle of $180°$, each quadrant formed by the adjacent rays of the quadrant and the point of intersection of the two lines form an angle of $90°$. Such basis are called orthogonal basis. This forms a special case of a cartesian coordinate system; it is called a rectangular coordinate system. For a rectangular coordinate system, the right triangle plays an important part in defining properties of a vector. Before continuing further, some properties of triangles and rectangles need to be discussed.

(Computation Of Areas)

Postulate 3 The area of a rectangle having a height of ‘h’ and a base length of ‘b’ is defined as $h \times b$.

Figure 5 Area Of Rectangle

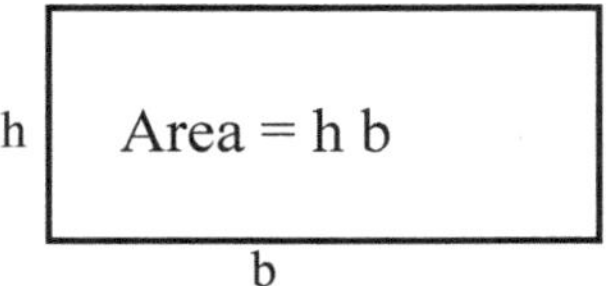

Corollary 3.4.3 The area of a right triangle having legs adjacent to right angle of length h and b is equal to

$$\left(\frac{1}{2}\right) \times (h \times b) \ .$$

Proof: To prove this corollary we use the figure above and draw a line along the diagonal of the rectangle to form two triangles with heights equal to h and bases equal to b.

Figure 6 Forming Triangles from Rectangle

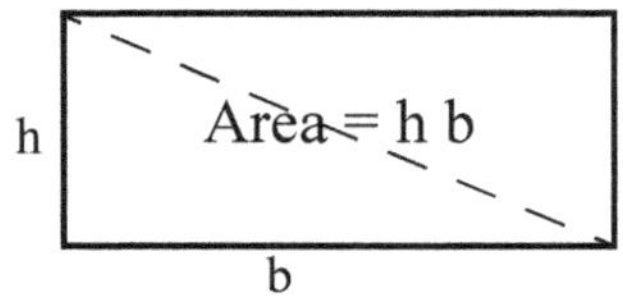

h Area = h b

b

Since the two triangles have one side in common (the diagonal of rectangle), all corresponding sides are equal and hence by the Side-Side-Side postulate of geometry, the two triangles are equal and have equal areas. Hence the area of the right triangle is equal to one-half the area of the rectangle; that is

$$\left(\frac{1}{2}\right) \times (h \times b) \ .$$

Theorem 3.4.4 The area of a parallelogram with height of 'h' and the length of base 'b' is equal to $h \times b$.

Proof: We prove this theorem by adding equal right triangles to each side of the parallelogram to get the following figure.

Figure 7 Forming Rectangle from Parallelogram

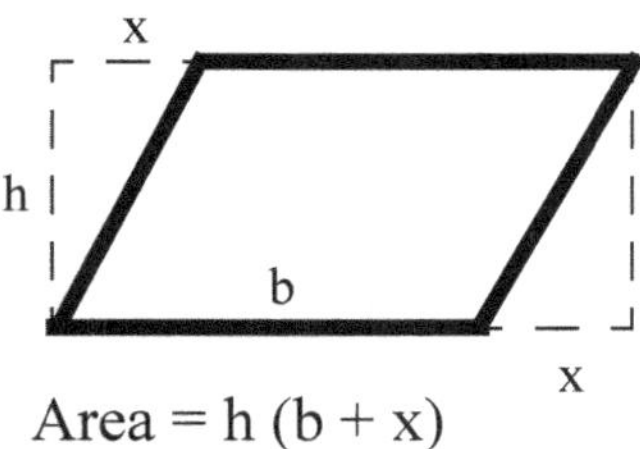

For the newly created rectangle, the area is computed to be $h \times (b + x)$. Since the areas of each right triangle is computed to be $\left(\frac{1}{2}\right) \times (h \times x)$ according to corollary 3.4.3, we subtract area of both triangles $(h \times x)$ from the area of the newly formed rectangle to get the area of the parallelogram. This area is computed to be $h \times b$.

Theorem 3.4.5 The area of a trapezoid with height 'h', upper base with length of 'b' and lower base with length of 'B' is equal to $h \times (b + B)/2$.

Proof: A trapezoid is a four sided polygon with parallel bases as shown in the figure below. By constructing the appropriate triangles within the trapezoid, we partition the trapezoid into two right triangles and one rectangle. From the figure below the area of the trapezoid is computed to be the sum of the area of the two triangles and the area of the rectangle. This sum is computed to be

$$h \times \frac{(b-x)}{2} + h \times \frac{(B-x)}{2} + h \times x \ .$$

Figure 8 Partitioning a Trapezoid

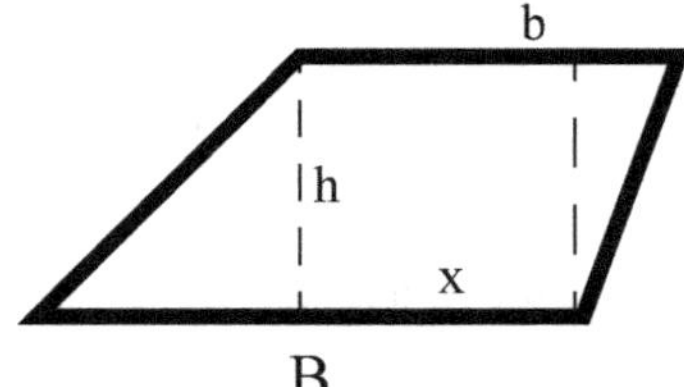

Now we combine like terms in the above expression to obtain a simple expression for the area of a trapezoid. Namely we get $h \times \dfrac{(b+B)}{2}$.

Theorem 3.4.6 The area of a triangle with height ‘h’ and a base with length of ‘b’ is equal to $\left(\frac{1}{2}\right) \times (h \times b)$.

Proof: From a triangle with base of length b and height of length h, we construct a trapezoid with upper base of length B, lower base of length b and height of length h. This figure is shown below. The area for the newly formed trapezoid is computed to be $h \times \dfrac{(b+B)}{2}$. The area of the right triangle added to the original triangle has area given by $h \times \dfrac{B}{2}$. Hence the area of the original triangle is equal to the area of the trapezoid minus the area of the right triangle. The subtraction of these areas gives the results $h \times \frac{(b+B)}{2} - h \times \frac{B}{2}$. Therefore the area of the original triangle is $\left(\frac{1}{2}\right) \times (h \times b)$.

Figure 9 Building a Trapezoid from Triangles

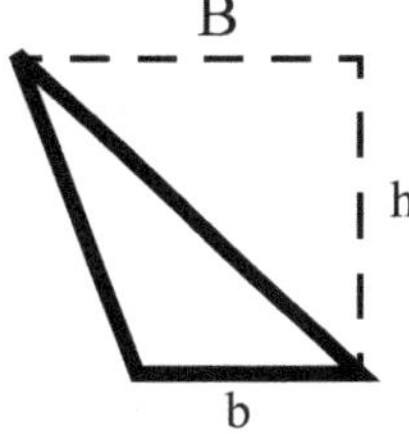

(Similar Triangles)

Definition 3.4.7 Two distinct triangles are similar if for each angle contained in one triangle, there is a corresponding equivalent angle in the other triangle.

Theorem 3.4.8 Given two distinct right triangles that are similar, if the corresponding legs of the triangles are of length H and B for the first triangle and h and b for the second triangle, then we have the following equation: $\frac{H}{B} = \frac{h}{b}$.

Proof: By assuming that the triangle with legs of length h and b is the smaller of the two triangles, we can inscribe the smaller triangle within the larger triangle as shown in Figure 10 below.

Figure 10 Similar Right Triangles

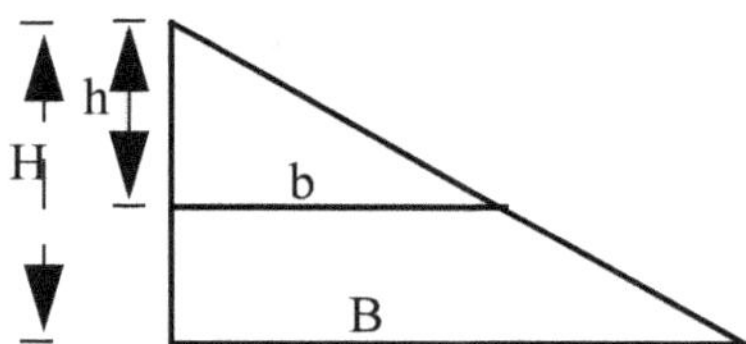

To prove this theorem, let's take a look at the area of the larger triangle. Applying corollary 4, the area of the right triangle is $\left(\frac{1}{2}\right) \times (H \times B)$. In Figure 10, note that the larger triangle is constructed from the smaller right triangle and a trapezoid. Applying corollary 4 again, the area of the smaller right triangle is $\left(\frac{1}{2}\right) \times (h \times b)$. Applying Theorem 3.4.5, the area of the inscribed trapezoid is $\left(\frac{1}{2}\right) \times (H-h) \times (B+b)$ .Since the smaller triangle and the trapezoid are completely contained within the larger triangle and their areas do not intersect, then the sum of their areas is equivalent to the area of the larger triangle. This gives us the following equation:

$$\left(\frac{1}{2}\right) \times (H \times B) = \left(\frac{1}{2}\right) \times (h \times b) + \left(\frac{1}{2}\right) \times (H - h) \times (B + b) \ .$$

Multiplying both sides of equation by 2 we have the following equation:

$$(H \times B) = (h \times b) + (H - h) \times (B + b) \ .$$

Using the distributive property of numbers, we expand the right side of the equation as follows:

$$(H \times B) = (h \times b) + (H \times B) + (H \times b) - (h \times B) - (h \times b) \ .$$

Using the associative property of numbers, we eliminate like terms in the expression to get the following equation: $(h \times B) = (H \times b)$ or $\frac{H}{B} = \frac{h}{b}$.

Having proved Theorem 3.4.8, we are now ready to prove the more general relationship for arbitrary similar triangles.

Theorem 3.4.9 Given two distinct similar triangles as shown in Figure 11, the following relationship for the length of segments opposite corresponding angles is true: $\frac{A}{a} = \frac{B}{b} = \frac{C}{c}$.

Proof:

Figure 11 Similar Triangles with Arbitrary Angles

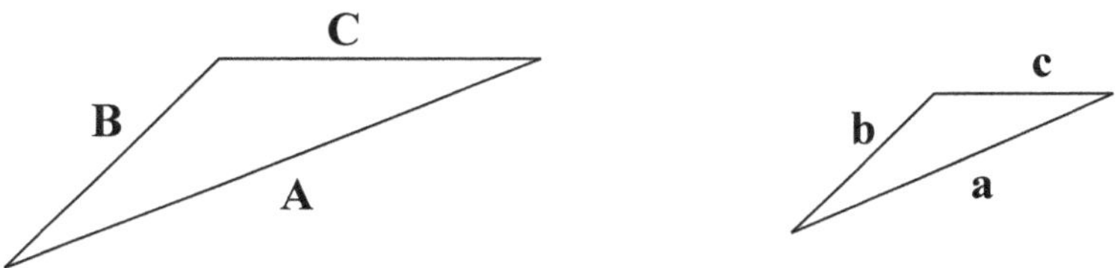

To prove the statement of Theorem 3.4.9, we inscribe the smaller of the similar triangles within the larger triangles to obtain the figure shown in Figure 6.

Figure 12 Inscribing Smaller Triangle Within Larger Triangle

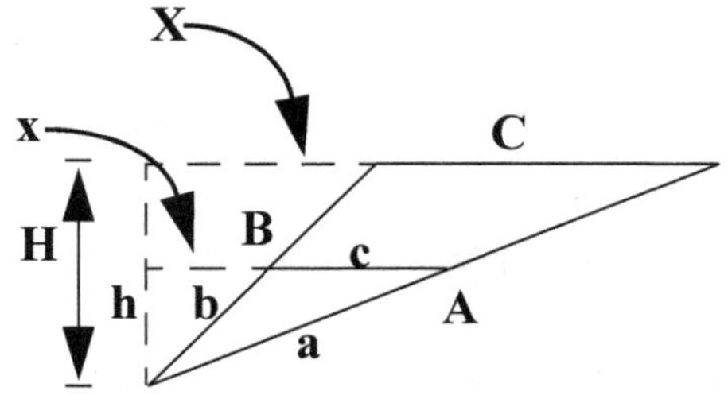

Constructing the height of both triangles, H and h, we have the relationships

(1) $\frac{H}{(C+X)} = \frac{h}{(c+x)}$ and (2) $\frac{H}{X} = \frac{h}{x}$ from Theorem 3.4.8 for similar right triangles. From equation (2) above we can express X in terms of x as

(2a) $X = \frac{H}{h} \times x$. Rearranging the terms in equation (1) above, we get the

equation (1a) $H \times (c+x) = h \times (C+X)$. Substituting for X, using equation (2a), we get the following equation from (1a):

$H \times (c+x) = h \times \left(C + \frac{H}{h} \times x\right)$. This equation gives us the following

relationship $H \times c = h \times C$ or $\frac{H}{C} = \frac{h}{c}$. By multiplying the left side of this equation by $\left(\frac{1}{2}\right) \times \frac{C}{C}$ and the right side by $\left(\frac{1}{2}\right) \times \frac{c}{c}$, we get the following equation: $\frac{[AREA]}{C^2} = \frac{[area]}{c^2}$ where [AREA] equals area of big triangle and [area] equals area of little triangle.

If we had drawn a perpendicular line to line with length 'A' Figure 12 would have the appearance as shown in Figure 13.

Figure 13 Perpendicular Line To Segment **A**

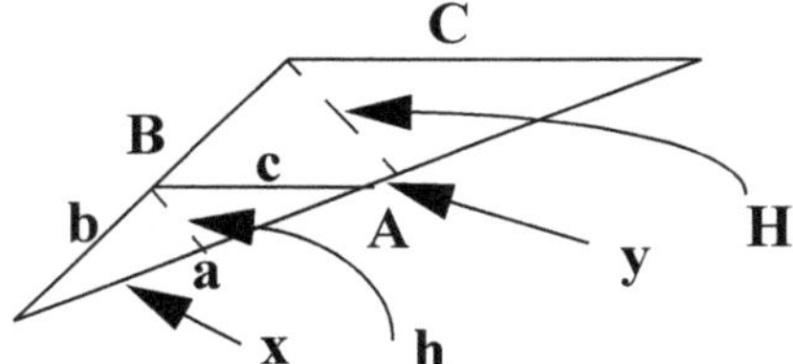

Applying Theorem 3.4.8 for similar right triangles in Figure 13 we get the following equations:

(1) $\frac{H}{(a+y)} = \frac{h}{x}$ and (2) $\frac{H}{(A-(a+y))} = \frac{h}{(a-x)}$.

From equation (1), we can express $(a+y)$ in terms of x so that we have

(1a) $(a+y) = \left(\frac{H}{h}\right) \times x$. Rearranging the terms in equation (2) and making the substitution for $(a+y)$ as given in (1a), we get the equation

(2a) $H \times (a-x) = h \times \left(A - \left(\frac{H}{h}\right) \times x\right)$. Using the distributive property of real numbers and combining like terms we get the equation

(2b) $\boldsymbol{H} \times a = \boldsymbol{h} \times A$ or $\frac{\boldsymbol{H}}{A} = \frac{\boldsymbol{h}}{a}$. By multiplying the left side of this equation by $\left(\frac{1}{2}\right) \times \frac{A}{A}$ and the right side by $\left(\frac{1}{2}\right) \times \frac{a}{a}$, we get the following equation: $\frac{[AREA]}{A^2} = \frac{[area]}{a^2}$ where [AREA] equals area of big triangle and [area] equals area of little triangle.

Combining the previous equation and this equation, we get the following equalities: $\frac{[AREA]}{[area]} = \frac{A^2}{a^2} = \frac{C^2}{c^2}$. Hence we can conclude that

$\frac{A}{a} = \frac{C}{c}$. A similar argument as shown for lengths A and C can also be shown for B. Hence we have proven that $\frac{A}{a} = \frac{B}{b} = \frac{C}{c}$.

Before leaving properties of triangles, we give the proof for the most famous of all theorems in geometry; the pythagorean theorem.

(Pythagorean Theorem)

Theorem 3.4.10 Given a right triangle with legs of length A and B and with hypotenuse of length C, then the sum of the squares of the length of the legs is equal to the square of the length of the hypotenuse.

Proof:

Figure 14 Proving Pythagorean Theorem

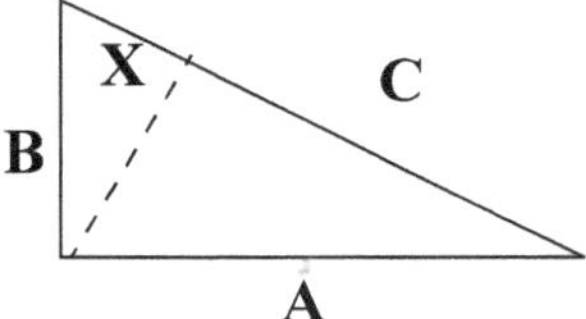

Using the similar right triangles constructed in Figure 14 and Theorem 3.4.9 for similar triangles, the following equations are derived:

(1) $\frac{B}{X} = \frac{C}{B}$ and (2) $\frac{A}{(C-X)} = \frac{C}{A}$.

Through multiplication, equation (1) becomes (1a) $B^2 = C \times X$.

Using the distributive law of real numbers on equation (2), the equation

becomes (2a) $A^2 = C^2 - C \times X$.

Combining equations (1a) and (2a) gives the statement of the theorem:

$$A^2 + B^2 = C^2 .$$

Now let's revisit the coordinate system for 2-dimensional space (plane). As was stated earlier, the choice of $e_1 = (\mathbf{1,0})$ and $e_2 = (\mathbf{0,1})$ as the basis for this space since they are orthogonal. Since a vector defined as $p = x \cdot e_1 + y \cdot e_2$ in a rectangular coordinate system forms the hypotenuse of a triangle, as shown in Figure 4, we can use the pythagorean theorem to define the length of the vector.

Definition 3.4.11 The length of a vector (from the origin) with coordinates $p = (x, y)$, or $p = x \cdot e_1 + y \cdot e_2$ is defined to be $\sqrt{x^2 + y^2}$.

3.4.2 Sine and Cosine Functions

Consider two distinct similar right triangles having corresponding sides of A, B, C and a, b, c. Let **C** and **c** be the respective hypotenuse of each right triangle. From Theorem 3.4.9 the following equalities are found to be true: $\frac{A}{a} = \frac{B}{b} = \frac{C}{c}$. If we examine the ratios $\frac{A}{C}$ and $\frac{B}{C}$, we have that $\frac{A}{C} = \frac{a}{c}$ and $\frac{B}{C} = \frac{b}{c}$. Hence the ratio of the length of each leg to the length of the hypotenuse is a constant value. If the angle opposite the segment of length **A** is α , then we define the sine of the angle α to be the ratio of the length of the leg opposite the angle to the length of the hypotenuse. The functional expression for the sine function is $sin(\alpha)$. In terms of the right triangle, we have $sin(\alpha) = \frac{A}{C}$. The ratio of the length of the segment adjacent to the angle to the length of the hypotenuse is called the cosine of the angle. The functional expression for the sine function is $cos(\alpha)$. In terms of the right triangle, we have $cos(\alpha) = \frac{B}{C}$.

If we apply the pythagorean theorem to the sine and cosine functions for a single angle α , we get the relationship $sin(\alpha)^2 + cos(\alpha)^2 = \left(\frac{A}{C}\right)^2 + \left(\frac{B}{C}\right)^2 = 1$. Since the square of a number is non-negative, this relationship implies that $sin(\alpha)^2 \leq 1$ and $cos(\alpha)^2 \leq 1$.

If we restrict ourselves to these definitions of sine and cosine, then it is clear that the angle α must be an acute angle and that the function is a non-negative number. To remove these restrictions, the following definition, based on the rectangular coordinate system, gives a general definition of the sine and cosine function.

Definition 3.4.3 Given a vector represented as the following linear combination $x \cdot e_1 + y \cdot e_2$ of the basis of the rectangular coordinate system, the sine of the angle α between the vector and the positive side of the X-Axis line segment as shown in Figure 9 is defined to be $sin(\alpha) = \frac{y}{r}$ where $r = \sqrt{x^2 + y^2}$ is the length of the vector as defined in Definition 3.4.11.

Figure 15 Angles of Vectors in Plane

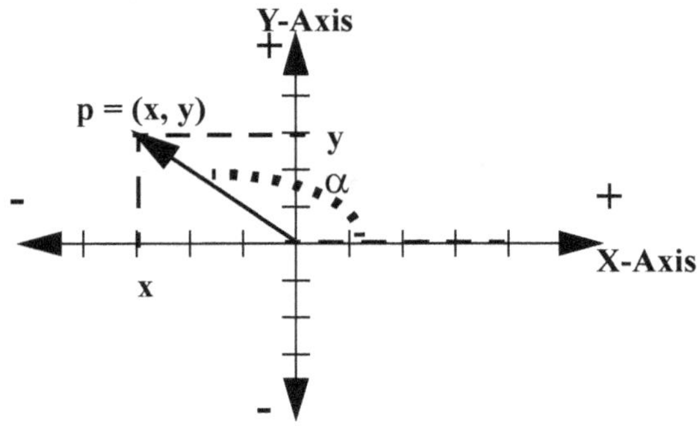

Definition 3.4.4 Given a vector represented as the following linear combination $x \cdot \vec{e_1} + y \cdot \vec{e_2}$ of the basis of the rectangular coordinate system, the cosine of the angle α between the vector and the positive side of the X-Axis line segment as shown in Figure 9 is defined to be $cos(\alpha) = \frac{x}{r}$ where

$r = \sqrt{x^2 + y^2}$ is the length of the vector as defined in Definition 3.4.11.

The above definitions allow for angles that are greater than 90 degrees as well as angles that are both positive and negative. Positive angles move in a counter-clockwise direction from the positive side of the X-Axis, while negative angles are taken to move in a clockwise direction.

3.4.3 Trigonometric Functions of Sums & Differences

Before leaving this section, the relationships of the sine and cosine of a sum of angles to the sine and cosine of the individual angles is derived. The importance of these relationships to linear algebra concepts will become apparent as we discuss the properties of vectors in later chapters.

Theorem 3.4.4 If α and β are two acute angles such that $\alpha + \beta < 90°$, then the following relationships are true:

(1) $sin(\alpha + \beta) = sin(\alpha) \times cos(\beta) + sin(\beta) \times cos(\alpha)$ and

(2) $cos(\alpha + \beta) = cos(\alpha) \times cos(\beta) - sin(\alpha) \times sin(\beta)$.

Proof: We begin this proof by constructing a right triangle with one acute angle of $\alpha + \beta$. This right triangle is shown in Figure 16.

Figure 16 Acute Angles in Right Triangle

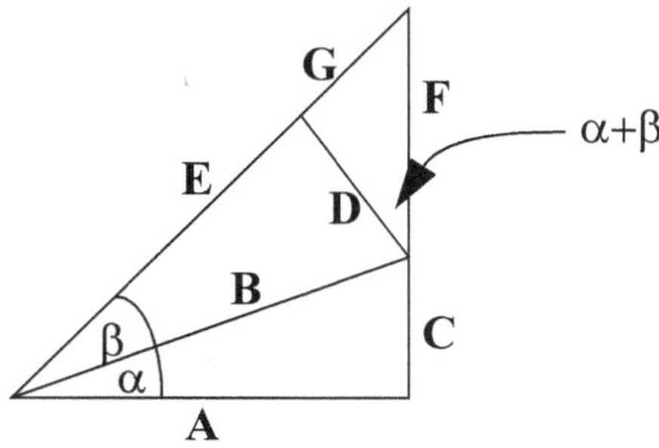

Using the length designation for each segment, we can proceed to express each of the segment lengths shown in Figure 16 in terms of segment of length **A**. The relationships, using the sine and cosine functions and the concept of similar triangles are as follows:

$$B = \frac{A}{cos(\alpha)}$$

$$C = \frac{A \times sin(\alpha)}{cos(\alpha)}$$

$$D = \frac{A \times sin(\beta)}{cos(\alpha)}$$

$$E = \frac{A \times cos(\beta)}{cos(\alpha)}$$

$$F = \frac{A \times sin(\beta)}{cos(\alpha) \times cos(\alpha + \beta)}$$

$$G = \frac{A \times sin(\beta) \times sin(\alpha + \beta)}{cos(\alpha) \times cos(\alpha + \beta)}$$

Having expressed each of the line segment lengths in terms of the length A, we are now able to establish equations that will allow us to solve for the sine and cosine of the sum in terms of the sine and cosine of the individual angles. We begin by putting the $cos(\alpha + \beta)$ in terms of the lengths of the line segments of the large right triangle. Using this approach, we get

$cos(\alpha + \beta) = \frac{A}{E + G}$. Substituting the expressions for the respective lengths and eliminating like terms we get the following results:

$$cos(\alpha + \beta) \times \left(cos(\beta) + \frac{sin(\beta) \times sin(\alpha + \beta)}{cos(\alpha + \beta)} \right) = cos(\alpha)$$ .Applying the distributive law for real numbers, we get the following equation:

(1) $$cos(\alpha) = cos(\beta) \times cos(\alpha + \beta) + sin(\beta) \times sin(\alpha + \beta)$$.

Now we put the $\boldsymbol{sin}(\alpha+\beta)$ in terms of the lengths of the line segments of the large right triangle. Using this approach, we get $\boldsymbol{sin}(\alpha+\beta) = \dfrac{\boldsymbol{C}+\boldsymbol{F}}{\boldsymbol{E}+\boldsymbol{G}}$.

From the derivation of equation (1) above, we have $\boldsymbol{cos}(\alpha+\beta) = \dfrac{A}{E+G}$.

Therefore this equation becomes $\boldsymbol{sin}(\alpha+\beta) = \dfrac{\boldsymbol{C}+\boldsymbol{F}}{\boldsymbol{A}} \times \boldsymbol{cos}(\alpha+\beta)$.

Substituting the expressions for the respective lengths and eliminating like terms we get the following results:

$\boldsymbol{sin}(\alpha+\beta) = \dfrac{\boldsymbol{sin}(\alpha)\times\boldsymbol{cos}(\alpha+\beta)}{\boldsymbol{cos}(\alpha)} + \dfrac{\boldsymbol{sin}(\beta)}{\boldsymbol{cos}(\alpha)}$ .Applying the distributive law for real numbers and rearranging terms, we get the following equation:

(2) $\boldsymbol{sin}(\beta) = \boldsymbol{cos}(\alpha)\times\boldsymbol{sin}(\alpha+\beta) - \boldsymbol{sin}(\alpha)\times\boldsymbol{cos}(\alpha+\beta)$.

From equations (1) and (2) we solve for $\boldsymbol{sin}(\alpha+\beta)$ and $\boldsymbol{cos}(\alpha+\beta)$ to obtain the following equations:

(1a) $\boldsymbol{sin}(\alpha+\beta) = \dfrac{\boldsymbol{sin}(\alpha)\times\boldsymbol{cos}(\alpha)+\boldsymbol{sin}(\beta)\times\boldsymbol{cos}(\beta)}{\boldsymbol{sin}(\alpha)\times\boldsymbol{sin}(\beta)+\boldsymbol{cos}(\alpha)\times\boldsymbol{cos}(\beta)}$ and

(2a) $\boldsymbol{cos}(\alpha+\beta) = \dfrac{\boldsymbol{cos}(\alpha)^2-\boldsymbol{sin}(\beta)^2}{\boldsymbol{sin}(\alpha)\times\boldsymbol{sin}(\beta)+\boldsymbol{cos}(\alpha)\times\boldsymbol{cos}(\beta)}$. It is left as an exercise to show that both equations reduce to the following equations (1b) and (2b). (Hint: Use the fact that $\boldsymbol{sin}(\alpha)^2+\boldsymbol{cos}(\alpha)^2 = 1$ and $\boldsymbol{sin}(\beta)^2+\boldsymbol{cos}(\beta)^2 = 1$.)

(Results)

(1b) $sin(\alpha + \beta) = sin(\alpha) \times cos(\beta) + sin(\beta) \times cos(\alpha)$

(2b) $cos(\alpha + \beta) = cos(\alpha) \times cos(\beta) - sin(\alpha) \times sin(\beta)$

Hence for acute angles whose sum is less than 90 degrees, we have a working formula for the sine and cosine in terms of the individual angle sine and cosine functions.

Now we expound on whether or not the formulas derived in Theorem 3.4.4 will hold if any of the restrictions are removed. The answer to this question is yes. However, we will take this one step at a time, first we remove the restriction that the sum of the angles must be less than 90 degrees. Hence we have another theorem.

Theorem 3.4.5 If α and β are two acute angles, then the following relationships are true:

(1) $sin(\alpha + \beta) = sin(\alpha) \times cos(\beta) + sin(\beta) \times cos(\alpha)$ and

(2) $cos(\alpha + \beta) = cos(\alpha) \times cos(\beta) - sin(\alpha) \times sin(\beta)$.

Proof: We begin this proof by constructing at triangle with one obtuse angle of $\alpha + \beta$. This triangle is shown in Figure 17.

Figure 17 Obtuse Angle in Triangle

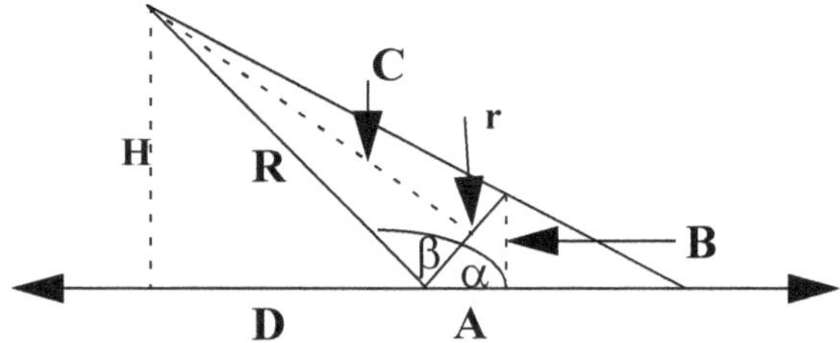

The dotted lines with lengths of **R** and **r** in Figure 17 represent the height of the smaller triangles contained in the large triangle. The dotted line, with length **H**, is the height of the large triangle. The lengths of **R** and **r** are unknown quantities that depend on the initial selection of the vector with length **R**. We can derive the other lengths using the sine and cosine functions as shown in the formulas below:

$\boldsymbol{A} = \boldsymbol{r} \times \boldsymbol{cos}(\alpha)$, $\boldsymbol{B} = \boldsymbol{r} \times \boldsymbol{sin}(\alpha)$,

$\boldsymbol{C} = \boldsymbol{R} \times \boldsymbol{sin}(\beta)$, $D = -\boldsymbol{R} \times \boldsymbol{cos}(\alpha + \beta)$ and

$H = \boldsymbol{R} \times \boldsymbol{sin}(\alpha + \beta)$. We place a negative sign in the expression for D, since by definition, the cosine of this function will be negative and the length must be positive.

We now derive a relationship for the sine/cosine of the sum of the angles and the sine/cosine of the individual angles by computing the area of the trapezoid formed by the two parallel line segments of lengths **H** and **B**. We compute the area using Theorem 3.4.5 to get the following area $\frac{1}{2} \times (H + B) \times (A + D)$. An alternate approach to computing the area is to compute the sum of the areas of the triangles that make up the trapezoid. This approach yields the following area

$$\frac{1}{2} \times \boldsymbol{H} \times \boldsymbol{D} + \frac{1}{2} \times \boldsymbol{C} \times \boldsymbol{r} + \frac{1}{2} \times \boldsymbol{A} \times \boldsymbol{B} .$$

Since the area of the trapezoid is constant regardless of how we compute the area, we have the following equality:

$$\boldsymbol{H} \times \boldsymbol{D} + \boldsymbol{C} \times \boldsymbol{r} + \boldsymbol{A} \times \boldsymbol{B} = (H + B) \times (A + D) \ .$$

Expanding the right side of the equation and then eliminating similar terms we have the resulting equation: $\boldsymbol{C} \times \boldsymbol{r} = B \times D + H \times A$.

Making the appropriate substitutions and dividing both sides of the equation by **r** and **R** we get the following equation:

$$\textbf{(1a)} \quad \boldsymbol{sin}(\beta) = -\boldsymbol{sin}(\alpha) \times \boldsymbol{cos}(\alpha + \beta) + \boldsymbol{cos}(\alpha) \times \boldsymbol{sin}(\alpha + \beta) \ .$$

To arrive at equation (1a), we started with angle α and then added angle β .

If we reverse this order by starting with α then adding β , we get the following equation using the same logic as before:

$$\textbf{(2a)} \quad \boldsymbol{sin}(\alpha) = -\boldsymbol{sin}(\beta) \times \boldsymbol{cos}(\alpha + \beta) + \boldsymbol{cos}(\beta) \times \boldsymbol{sin}(\alpha + \beta) \ .$$

Solving this linear equation for the sine and cosine of the sums, we get the following equations:

$$\textbf{(1a)} \ \boldsymbol{sin}(\alpha + \beta) = \frac{\boldsymbol{sin}(\beta)^2 - \boldsymbol{sin}(\alpha)^2}{\boldsymbol{sin}(\beta) \times \boldsymbol{cos}(\alpha) - \boldsymbol{sin}(\alpha) \times \boldsymbol{cos}(\beta)} \quad \text{and}$$

$$\textbf{(2a)} \quad \boldsymbol{cos}(\alpha + \beta) = \frac{\boldsymbol{sin}(\alpha) \times \boldsymbol{cos}(\alpha) - \boldsymbol{sin}(\beta) \times \boldsymbol{cos}(\beta)}{\boldsymbol{sin}(\alpha) \times \boldsymbol{cos}(\beta) - \boldsymbol{sin}(\beta) \times \boldsymbol{cos}(\alpha)}$$

. It is left as an exercise to show that both equations reduce to the following equations (1b) and (2b). (Hint: Use the fact that $sin(\alpha)^2 + cos(\alpha)^2 = 1$ and

$sin(\beta)^2 + cos(\beta)^2 = 1$.)

(Results)

(1b) $sin(\alpha + \beta) = sin(\alpha) \times cos(\beta) + sin(\beta) \times cos(\alpha)$

(2b) $cos(\alpha + \beta) = cos(\alpha) \times cos(\beta) - sin(\alpha) \times sin(\beta)$

Hence for acute angles whose sum is greater than or equal to 90 degrees, we have a working formula for the sine and cosine in terms of the individual angle sine and cosine functions.

Next we consider the case with one angle being acute and the other being obtuse, but their sum is less than 180 degrees. Similar to the previous conditions we have the theorem below.

Theorem 3.4.6 If α is an obtuse angle and β is an acute angles such that the sum is less than 180 degrees, then the following relationships are true:

(1) $sin(\alpha + \beta) = sin(\alpha) \times cos(\beta) + sin(\beta) \times cos(\alpha)$ and

(2) $cos(\alpha + \beta) = cos(\alpha) \times cos(\beta) - sin(\alpha) \times sin(\beta)$.

Proof: The proof of this theorem is left as an exercise.

Finally we are ready to prove the general theorem for all angles without restrictions or conditions.

Theorem 3.4.7 For any given angles α and β the following relationships are true:

(1) $\boldsymbol{sin}(\alpha+\beta) = \boldsymbol{sin}(\alpha) \times \boldsymbol{cos}(\beta) + \boldsymbol{sin}(\beta) \times \boldsymbol{cos}(\alpha)$ and

(2) $\boldsymbol{cos}(\alpha+\beta) = \boldsymbol{cos}(\alpha) \times \boldsymbol{cos}(\beta) - \boldsymbol{sin}(\alpha) \times \boldsymbol{sin}(\beta)$.

Proof: Again we leave this as an exercise for the reader.

For the difference of two angles, the application of Theorem 22 for the sum of the angles α and $\beta-\alpha$ gives the sine and cosine of β as a linear combination of sine and cosine of $\beta-\alpha$. By solving the linear equations we derive.

Theorem 3.4.8 For any given angles α and β the following relationships are true:

(1) $\boldsymbol{sin}(\beta-\alpha) = \boldsymbol{sin}(\beta) \times \boldsymbol{cos}(\alpha) - \boldsymbol{sin}(\alpha) \times \boldsymbol{cos}(\beta)$ and

(2) $\boldsymbol{cos}(\beta-\alpha) = \boldsymbol{cos}(\alpha) \times \boldsymbol{cos}(\beta) + \boldsymbol{sin}(\alpha) \times \boldsymbol{sin}(\beta)$.

Exercises

Using the descriptions in exercises one through five, compute the area of each polygon described.

1. Rectangle with base = 4 feet and height = 80 feet.

2. Rectangle with base = 20 feet and height

3. Parallelogram with base = 12 meters and height = 7 meters

4. Triangle with base = 13 centimeters and height = 26 centimeters

5. Trapezoid with lower base = 10 meters, upper base = 12 meters and height = 11 meters

---

For exercises 6 through 10, multi-sides polygons are shown. Use areas of rectangles, parallelograms, triangles and trapezoids to compute the composite areas of the displayed polygons.

6. Given the 6 sided figure and the dimensions specified, compute the area of the polygon. Note that each of the inscribed triangles are equilateral triangles and hence each angle has a

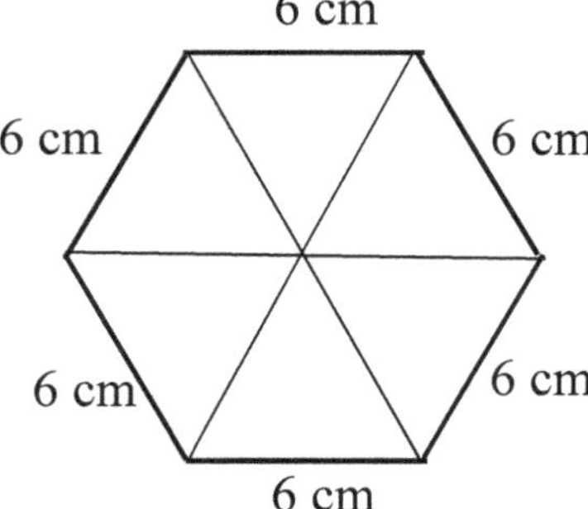

measure of 60 degrees.

7. Compute the area of the following figure. Note that each triangle is an isosceles triangle with angle opposite the base having a measure of 36 degrees.

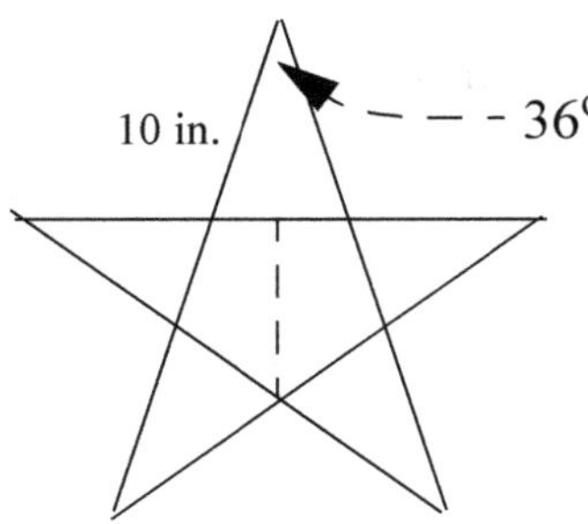

8. Compute the area of the following figure:

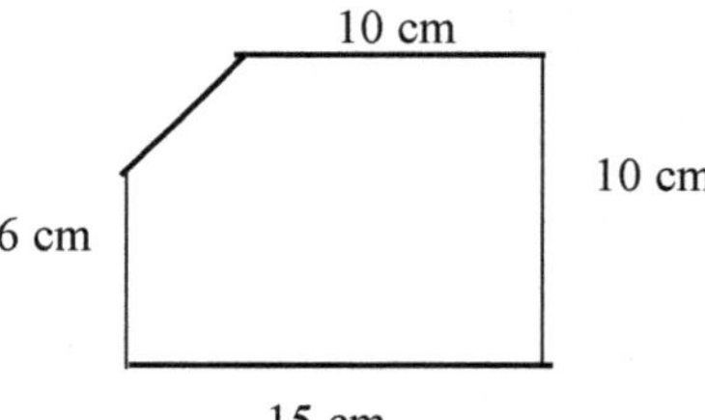

9. Compute the area of the following figure:

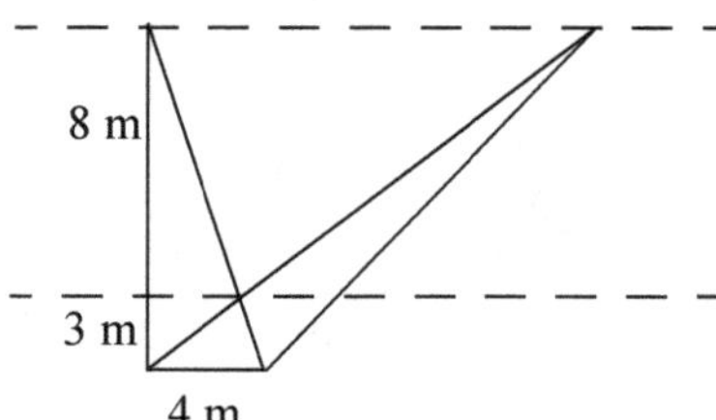

10. Compute the area of the following figure:

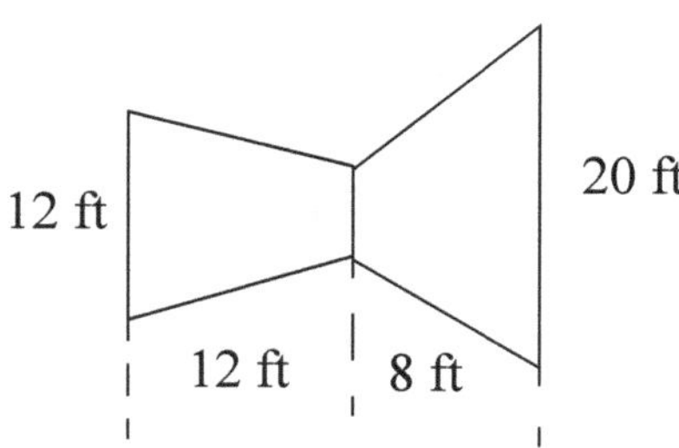

11. Show that the expression $\dfrac{\sin(\alpha) \times \cos(\alpha) + \sin(\beta) \times \cos(\beta)}{\sin(\alpha) \times \sin(\beta) + \cos(\alpha) \times \cos(\beta)}$ reduces to the expression $\sin(\alpha) \times \cos(\beta) + \sin(\beta) \times \cos(\alpha)$.

12. Show that the expression $\dfrac{\cos(\alpha)^2 - \sin(\beta)^2}{\sin(\alpha) \times \sin(\beta) + \cos(\alpha) \times \cos(\beta)}$ reduces to the expression $\cos(\alpha) \times \cos(\beta) - \sin(\alpha) \times \sin(\beta)$.

13. Using the expression for $\sin(\alpha + \beta)$, derive an expression for $\sin(2 \cdot \alpha)$ in terms of $\sin(\alpha)$ and $\cos(\alpha)$.

14. Using the expression for $\cos(\alpha + \beta)$, derive an expression for $\cos(2 \cdot \alpha)$ in terms of $\sin(\alpha)$ and $\cos(\alpha)$.

15. Give expressions for $\cos(\alpha/2)$ and $\sin(\alpha/2)$ in terms of $\cos(\alpha)$.

3.5 *Basic Geometry/Trigonometry For* $\mathbf{R}^3$

The previous section discussed a two dimensional vector space using a cartesian coordinate system. In this section, we discuss a three dimensional vector space using cartesian coordinate system. More specifically we will discuss a three dimensional rectangular coordinate system that is isomorphic to $\boldsymbol{R}^3$. The basis for this vector space under the rectangular coordinate system are the following:

$\vec{e_1} = \begin{bmatrix} 1 \\ 0 \\ 0 \end{bmatrix}$, $\vec{e_2} = \begin{bmatrix} 0 \\ 1 \\ 0 \end{bmatrix}$ and $\vec{e_3} = \begin{bmatrix} 0 \\ 0 \\ 1 \end{bmatrix}$. Since any arbitrary vector in $\boldsymbol{R}^3$ can be expressed as a linear combination of these three vectors,

$\begin{bmatrix} x \\ y \\ z \end{bmatrix} = x \cdot \vec{e_1} + y \cdot \vec{e_2} + z \cdot \vec{e_3}$, they form a basis for $\boldsymbol{R}^3$.

3.5.1 Alternative Basis for $\mathbf{R}^3$

Using the matrix notation for expressing linear combinations (shown in section 3.2), we can represent any vector as the matrix product $\begin{bmatrix} x \\ y \\ z \end{bmatrix} = \begin{bmatrix} x & y & z \end{bmatrix} \times \begin{bmatrix} \vec{e_1} \\ \vec{e_2} \\ \vec{e_3} \end{bmatrix}$.

We are interested in representing all vectors in $\boldsymbol{R}^3$ using another basis. Let's represent the vectors of this alternative basis as the following: $\vec{f_1} = \begin{bmatrix} x_1 \\ y_1 \\ z_1 \end{bmatrix}$,

$\vec{f_2} = \begin{bmatrix} x_2 \\ y_2 \\ z_2 \end{bmatrix}$ and $\vec{f_3} = \begin{bmatrix} x_3 \\ y_3 \\ z_3 \end{bmatrix}$. The matrix notation for expressing this basis as a linear combination of the cartesian coordinate basis is the following:

$\begin{bmatrix} \vec{f_1} \\ f_2 \\ f_3 \end{bmatrix} = \begin{bmatrix} x_1\, y_1\, z_1 \\ x_2\, y_2\, z_2 \\ x_3\, y_3\, z_3 \end{bmatrix} \times \begin{bmatrix} \vec{e_1} \\ \vec{e_2} \\ \vec{e_3} \end{bmatrix}$. Since both sets of vectors are basis, they are linearly independent and the matrix is non-singular. Hence we have the inverse relationship $\begin{bmatrix} \vec{e_1} \\ \vec{e_2} \\ \vec{e_3} \end{bmatrix} = \begin{bmatrix} x_1\, y_1\, z_1 \\ x_2\, y_2\, z_2 \\ x_3\, y_3\, z_3 \end{bmatrix}^{-1} \times \begin{bmatrix} \vec{f_1} \\ f_2 \\ f_3 \end{bmatrix}$. To determine the coordinates of an arbitrary vector $\begin{bmatrix} x \\ y \\ z \end{bmatrix}$, we have the matrix equation

$$\begin{bmatrix} x & y & z \end{bmatrix} \times \begin{bmatrix} \vec{e_1} \\ \vec{e_2} \\ \vec{e_3} \end{bmatrix} = \begin{bmatrix} x & y & z \end{bmatrix} \times \begin{bmatrix} x_1 & y_1 & z_1 \\ x_2 & y_2 & z_2 \\ x_3 & y_3 & z_3 \end{bmatrix}^{-1} \times \begin{bmatrix} \vec{f_1} \\ f_2 \\ f_3 \end{bmatrix}.$$

Therefore we can conclude that the coordinates of a vector in cartesian coordinate system in 3-dimensional space can be converted to an alternative coordinate system using the values given in the matrix product

$$\begin{bmatrix} x & y & z \end{bmatrix} \times \begin{bmatrix} x_1 & y_1 & z_1 \\ x_2 & y_2 & z_2 \\ x_3 & y_3 & z_3 \end{bmatrix}^{-1}.$$

Example

Using the alternative vectors $\begin{bmatrix} 1 \\ 0 \\ 1 \end{bmatrix}$, $\begin{bmatrix} -1 \\ -2 \\ 0 \end{bmatrix}$ and $\begin{bmatrix} 2 \\ 1 \\ 1 \end{bmatrix}$ convert the coordinates of the cartesian vector $\begin{bmatrix} 2 \\ 1 \\ 4 \end{bmatrix}$.

To convert to the alternative coordinates, we first find the inverse of the matrix $\begin{bmatrix} 1 & 0 & 1 \\ -1 & -2 & 0 \\ 2 & 1 & 1 \end{bmatrix}$. The inverse is computed to be $\begin{bmatrix} -2 & 1 & 2 \\ 1 & -1 & -1 \\ 3 & -1 & -2 \end{bmatrix}$. Now we compute the

product $\begin{bmatrix} 2 & 1 & 4 \end{bmatrix} \times \begin{bmatrix} -2 & 1 & 2 \\ 1 & -1 & -1 \\ 3 & -1 & -2 \end{bmatrix}$. The results of the multiplication is $\begin{bmatrix} 9 & -3 & -5 \end{bmatrix}$.

Hence we represent the cartesian vector $\begin{bmatrix} 2 \\ 1 \\ 4 \end{bmatrix}$ as $\begin{bmatrix} 9 \\ -3 \\ -5 \end{bmatrix}$ using the alternative basis

3.5.2 Three-Dimensional Polar Coordinate System

In the discussion of the 2-dimensional vector space $\boldsymbol{R}^2$, we derived the following formulas for the coordinates of the points in the vector space:

$x = \boldsymbol{r} \cdot cos(\phi)$ and $y = \boldsymbol{r} \cdot sin(\phi)$, where the angle ϕ represents the angle between the vector and the x-axis. The value r is the length of the vector. Using the notation of the bases shown in section 3.4, we can express each point as the following:

$\boldsymbol{r} \cdot cos(\phi) \cdot \overrightarrow{\boldsymbol{e}_1} + \boldsymbol{r} \cdot sin(\phi) \cdot \overrightarrow{\boldsymbol{e}_2}$. This representation of the coordinate system is called a polar coordinate system.

For the 3-dimensional space, a third axis (the z-axis) is introduced in the cartesian coordinate system. In the cartesian coordinate system, all points on the z-axis are represented as a multiple of the vector $\overrightarrow{e_3}$. The subspace spanned by the vectors $\overrightarrow{e_1}$ and $\overrightarrow{e_3}$ forms a 2-dimensional vector space of R^3 and is called a plane. A formal definition for a plane is given in Definition 3.5.3.

Definition 3.5.3 A subspace of R^3 having dimension two (2), is called a plane.

The rectangular coordinate for points in R^3 is $x \cdot \overrightarrow{e_1} + y \cdot \overrightarrow{e_2} + z \cdot \overrightarrow{e_3}$. To derive the polar coordinates for a point in R^3, we define the angle of elevation for the point under consideration. The angle of elevation of a vector is measured relative to the vector and the plane spanned by the vectors $\overrightarrow{e_1}$ and $\overrightarrow{e_2}$.

Definition 3.5.4 The angle of elevation from a plane for a point in R^3 is defined as the angle between the vector representing the original point and the vector contained in the plane and representing the point contained in the plane that is closest to the original point.

This definition depends very much on the concept of distance. Using the distance formula, as introduced in basic geometry and algebra, the distance between two points $x_1 \cdot \overrightarrow{e_1} + y_1 \cdot \overrightarrow{e_2} + z_1 \cdot \overrightarrow{e_3}$ and $x_2 \cdot \overrightarrow{e_1} + y_2 \cdot \overrightarrow{e_2} + z_2 \cdot \overrightarrow{e_3}$ in a cartesian coordinate system is defined to be

$dist = \sqrt{(x_2 - x_1)^2 + (y_2 - y_1)^2 + (z_2 - z_1)^2}$. Hence an arbitrary point in the plane spanned by the vectors $\overrightarrow{e_1}$ and $\overrightarrow{e_2}$ has the form $x \cdot \overrightarrow{e_1} + y \cdot \overrightarrow{e_2}$. If we compute the distance of a vector $x_1 \cdot \vec{e_1} + y_2 \cdot \vec{e_2} + z_3 \cdot \vec{e_3}$ with a point contained in this plane then we get $dist = \sqrt{(x-x_1)^2 + (y-y_1)^2 + (z_1)^2}$. To minimize this distance, we choose $x = x_1$ and $y = y_1$. Hence the vector $x_1 \cdot \vec{e_1} + y_2 \cdot \vec{e_2}$ is the vector formed from the closest point in the plane to the vector $x_1 \cdot \vec{e_1} + y_1 \cdot \vec{e_2} + z_1 \cdot \vec{e_3}$. It is also called a projection of the vector $x_1 \cdot \vec{e_1} + y_1 \cdot \vec{e_2} + z_1 \cdot \vec{e_3}$ in the plane.

If the angle θ represents the angle of elevation for the vector from the vector space spanned by the vectors $\vec{e_1}$ and $\vec{e_2}$ then the polar coordinates are given by the following relationships:

$$x = r \cdot cos(\phi) \cdot cos(\theta) \quad ,$$

$$y = r \cdot sin(\phi) \cdot cos(\theta) \quad \text{and}$$

$z = r \cdot sin(\theta)$. In the above relationships, the angle ϕ represents the angle between the vector and the x-axis and r is the length of the vector.

Substituting the values for x, y and z in the rectangular coordinate system, we get the polar coordinate system:

$$r \cdot cos(\phi) \cdot cos(\theta) \cdot \vec{e_1} + r \cdot sin(\phi) \cdot cos(\theta) \cdot \vec{e_2} + r \cdot sin(\theta) \cdot \vec{e_3} \quad .$$

Theorem 3.5.5 Given two vectors represented by polar coordinates in $\mathbf{R}^3$,

$$V_1 = \begin{bmatrix} r_1 \cdot cos(\phi_1) \cdot cos(\theta_1) \\ r_1 \cdot sin(\phi_1) \cdot cos(\theta_1) \\ r_1 \cdot sin(\theta_1) \end{bmatrix}$$

and

$$V_2 = \begin{bmatrix} r_2 \cdot cos(\phi_2) \cdot cos(\theta_2) \\ r_2 \cdot sin(\phi_2) \cdot cos(\theta_2) \\ r_2 \cdot sin(\theta_2) \end{bmatrix}$$

, then the angle between the vectors is designated as δ and the cosine of this angle is given by the following equation:

$$cos(\delta) = cos(\theta_1) \cdot cos(\theta_2) \cdot cos(\phi_2 - \phi_1) - sin(\theta_1) \cdot sin(\theta_2) \ .$$

Proof: The proof of this theorem is left as an exercise for the reader.

Exercises

1. Using the notation shown in theorem 3.5.5, prove that the angle between two vectors V_1 and V_2 is given in the equation shown.

2. Compute the distance between the following two vectors: $-4 \cdot \vec{e_1} + 8 \cdot \vec{e_2} + 11 \cdot \vec{e_3}$ and $2 \cdot \vec{e_1} + 8 \cdot \vec{e_2} + 3 \cdot \vec{e_3}$.

3. Convert to cartesian coordinates when the polar coordinates are the following:

$r = 15$, $\phi = 15°$ and $\theta = 60°$.

4. Convert to cartesian coordinates when the polar coordinates are the following:

$r = 2$, $\phi = 90°$ and $\theta = 45°$.

5. Convert to cartesian coordinates when the polar coordinates are the following:

$r = 20$, $\phi = \pi/5$ and $\theta = \pi/4$.

6. Convert the cartesian vector $\begin{bmatrix} 5 \\ 1 \\ 1 \end{bmatrix}$ to polar coordinates.

7. Using the alternative vectors $\begin{bmatrix} 1 \\ 2 \\ 1 \end{bmatrix}$, $\begin{bmatrix} -1 \\ 3 \\ -5 \end{bmatrix}$ and $\begin{bmatrix} 2 \\ 1 \\ 7 \end{bmatrix}$ convert the coordinates of the cartesian vector $\begin{bmatrix} -1 \\ 1 \\ 2 \end{bmatrix}$ to the alternative coordinate system.

8. Compute the cosine of the angle between the following two vectors specified in polar coordinates: $r = 10$, $\phi = \pi/5$, $\theta = \pi/2$ and $r = 5$, $\phi = \pi/4$ and $\theta = \pi/4$.

3.6 *Physical Properties Of Vectors In* $\mathbf{R}^3$

With the trigonometric and geometric identities established in the previous sections, we are now ready to examine some of the physical properties of the vectors in R^3. This section will introduce such topics as length of vectors, angle between vectors, area of parallelogram formed by two intersecting vectors and the volume of parallelepiped formed by three intersecting vectors.

Definition 3.6.1 Given two vectors $\overrightarrow{V_1}$ and $\overrightarrow{V_2}$, the dot product (or sometimes called inner product) of the two vectors is a binary operation of the vectors resulting in a single real number. If the vectors are expressed in rectangular coordinates as $\overrightarrow{V_1} = x_1 \cdot \overrightarrow{e_1} + y_1 \cdot \overrightarrow{e_2} + z_1 \cdot \overrightarrow{e_3}$ and $\overrightarrow{V_2} = x_2 \cdot \overrightarrow{e_1} + y_2 \cdot \overrightarrow{e_2} + z_2 \cdot \overrightarrow{e_3}$, then the dot product is designated as $\overrightarrow{V_1} \bullet \overrightarrow{V_2} = x_1 \cdot x_2 + y_1 \cdot y_2 + z_1 \cdot z_2$.

If we use polar coordinates to express the dot product we have the results given by Theorem 3.6.2 below.

Theorem 3.6.2 If δ represents the angle between the vectors $\overrightarrow{V_1}$ and $\overrightarrow{V_2}$, then the dot product of the two vectors satisfy the equation $\overrightarrow{V_1} \bullet \overrightarrow{V_2} = r_1 \cdot r_2 \cdot cos(\delta)$, where r_1 and r_2 are the lengths of vectors $\overrightarrow{V_1}$ and $\overrightarrow{V_2}$ respectively.

Proof: Using the polar coordinate representation for each vector and the Definition 3.6.1 we get the following representation for the dot product

$$\vec{V_1} \bullet \vec{V_2} = r_1 \cdot r_2 \cdot (cos(\phi_1) \cdot cos(\theta_1) \cdot cos(\phi_2) \cdot cos(\theta_2) + sin(\phi_1) \cdot cos(\theta_1) \cdot sin(\phi_2) \cdot cos(\theta_2) + sin(\theta_1) \cdot sin(\theta_2))$$

.

By redistributing terms in the above equation the dot product becomes the following:

$$\vec{V_1} \bullet \vec{V_2} = r_1 \cdot r_2 \cdot (cos(\theta_1) \cdot cos(\theta_2) \cdot (cos(\phi_1) \cdot cos(\phi_2) + sin(\phi_1) \cdot sin(\phi_2)) + sin(\theta_1) \cdot sin(\theta_2))$$

.

Using Theorem 3.4.8, we have

$$cos(\phi_2 - \phi_1) = cos(\phi_1) \cdot cos(\phi_2) + sin(\phi_1) \cdot sin(\phi_2) \cdot$$

Therefore, the above equation for the dot product becomes the following:

$$\vec{V_1} \bullet \vec{V_2} = r_1 \cdot r_2 \cdot (cos(\theta_1) \cdot cos(\theta_2) \cdot cos(\phi_2 - \phi_1) + sin(\theta_1) \cdot sin(\theta_2)) .$$

If δ represents the angle between the two vectors we have from Theorem 3.5.5 the relationship

$$cos(\delta) = cos(\theta_1) \cdot cos(\theta_2) \cdot cos(\phi_2 - \phi_1) - sin(\theta_1) \cdot sin(\theta_2)$$

.

Hence the dot product satisfy the relationship

$$\overrightarrow{V_1} \bullet \overrightarrow{V_2} = r_1 \cdot r_2 \cdot cos(\delta) \ .$$

3.6.1 Length of Vectors

Corollary 3.6.3 The square of the length of a vector is equivalent to the dot product of the vector with itself.

Algebraically, this is stated as $\vec{V} \bullet \vec{V} = |\vec{V}|^2$.

Proof: Apply Theorem 3.6.2 when computing the dot product of a matrix with itself.

A new notation is introduced in Corollary 3.6.3; $|\vec{V}|$ is used to indicate the length of a vector.

Theorem 3.6.2 gives a method for deriving the cosine of the angle between two vectors in terms of the dot product between the two vectors. Applying this theorem we derive a theorem that gives the sine of the angle between two vectors.

Theorem 3.6.4 Given vectors $\overrightarrow{V_1}$ and $\overrightarrow{V_2}$ with δ being the angle between the vectors, then using the usual polar coordinates for each vector, the following equation is true:

$$\left|\overrightarrow{V_1}\right|^2 \cdot \left|\overrightarrow{V_2}\right|^2 \cdot sin(\delta)^2 = (y_1 \cdot z_2 - y_2 \cdot z_1)^2 + (x_1 \cdot z_2 - x_2 \cdot z_1)^2 + (x_1 \cdot y_2 - x_2 \cdot y_1)^2$$

Proof: Using the fact that $1 - cos(\delta)^2 = sin(\delta)^2$ and Theorem 3.6.2, we construct the following equation:

$$sin(\delta)^2 = 1 - \left(cos\theta_1 \cdot cos\theta_2 \cdot cos\left(\phi_2 - \phi_1\right) + sin\theta_1 \cdot sin\theta_2\right)^2.$$

$$= 1$$

$$-cos(\theta_1)^2 \cdot cos(\theta_2)^2 \cdot cos(\phi_2 - \phi_1)^2$$

$$-2 \cdot cos(\theta_1) \cdot cos(\theta_2) \cdot cos(\phi_2 - \phi_1) \cdot sin\theta_1 \cdot sin\theta_2$$

$$- sin(\theta_1)^2 \cdot sin(\theta_2)^2$$

Make the substitution

$1 = cos(\theta_1)^2 + sin(\theta_1)^2 \cdot sin(\theta_2)^2 + sin(\theta_1)^2 \cdot cos(\theta_2)^2$ in the above to get the following:

$$= cos(\theta_1)^2 + sin(\theta_1)^2 \cdot sin(\theta_2)^2 + sin(\theta_1)^2 \cdot cos(\theta_2)^2$$
$$-cos(\theta_1)^2 \cdot cos(\theta_2)^2 \cdot cos(\phi_2 - \phi_1)^2$$
$$-2 \cdot cos(\theta_1) \cdot cos(\theta_2) \cdot cos(\phi_2 - \phi_1) \cdot sin\theta_1 \cdot sin\theta_2$$
$$- sin(\theta_1)^2 \cdot sin(\theta_2)^2$$

The $sin(\theta_1)^2 \cdot sin(\theta_2)^2$ terms in the above equation sum to zero, thus

eliminating the terms in the equation. By expanding the $cos(\phi_2 - \phi_1)$ term, the above equation becomes the following:

$$= cos(\theta_1)^2 + sin(\theta_1)^2 \cdot cos(\theta_2)^2$$
$$-2 \cdot cos(\theta_1) \cdot cos(\theta_2) \cdot sin(\theta_1) \cdot sin(\theta_2) \cdot cos(\phi_1) \cdot cos(\phi_2)$$
$$-2 \cdot cos(\theta_1) \cdot cos(\theta_2) \cdot sin(\theta_1) \cdot sin(\theta_2) \cdot sin(\phi_1) \cdot sin(\phi_2)$$
$$- cos(\theta_1)^2 \cdot cos(\theta_2)^2 \cdot cos(\phi_2 - \phi_1)^2$$

Now, let's make the substitution $1 - sin(\phi_2 - \phi_1)^2 = cos(\phi_2 - \phi_1)^2$ to get the following equation:

$$= cos(\theta_1)^2 - cos(\theta_1)^2 \cdot cos(\theta_2)^2 + sin(\theta_1)^2 \cdot cos(\theta_2)^2$$
$$-2 \cdot cos(\theta_1) \cdot cos(\theta_2) \cdot sin(\theta_1) \cdot sin(\theta_2) \cdot cos(\phi_1) \cdot cos(\phi_2)$$
$$-2 \cdot cos(\theta_1) \cdot cos(\theta_2) \cdot sin(\theta_1) \cdot sin(\theta_2) \cdot sin(\phi_1) \cdot sin(\phi_2)$$
$$+ cos(\theta_1)^2 \cdot cos(\theta_2)^2 \cdot sin(\phi_2 - \phi_1)^2$$

Using trigonometric identities that we have proven in the previous sections, we get the following equation:

$$\begin{aligned}
&= \cos(\theta_1)^2 \cdot \sin(\theta_2)^2 \cdot \left[\sin(\phi_1)^2 + \cos(\phi_1)^2\right] \\
&+ \sin(\theta_1)^2 \cdot \cos(\theta_2)^2 \cdot \left[\sin(\phi_2)^2 + \cos(\phi_2)^2\right] \\
&-2 \cdot \cos(\theta_1) \cdot \cos(\theta_2) \cdot \sin(\theta_1) \cdot \sin(\theta_2) \cdot \cos(\phi_1) \cdot \cos(\phi_2) \\
&-2 \cdot \cos(\theta_1) \cdot \cos(\theta_2) \cdot \sin(\theta_1) \cdot \sin(\theta_2) \cdot \sin(\phi_1) \cdot \sin(\phi_2) \\
&+ \cos(\theta_1)^2 \cdot \cos(\theta_2)^2 \cdot \sin(\phi_2 - \phi_1)^2
\end{aligned}$$

As we apply the distributive property of real numbers to the first two lines of the above equation and combine the results with the third and fourth lines above, we get the following equation:

$$\begin{aligned}
&= \cos(\theta_1)^2 \cdot \sin(\theta_2)^2 \cdot \left[\sin(\phi_1)^2 + \cos(\phi_1)^2\right] \\
&+ \sin(\theta_1)^2 \cdot \cos(\theta_2)^2 \cdot \left[\sin(\phi_2)^2 + \cos(\phi_2)^2\right] \\
&-2 \cdot \cos(\theta_1) \cdot \cos(\theta_2) \cdot \sin(\theta_1) \cdot \sin(\theta_2) \cdot \cos(\phi_1) \cdot \cos(\phi_2) \\
&-2 \cdot \cos(\theta_1) \cdot \cos(\theta_2) \cdot \sin(\theta_1) \cdot \sin(\theta_2) \cdot \sin(\phi_1) \cdot \sin(\phi_2) \\
&+ \cos(\theta_1)^2 \cdot \cos(\theta_2)^2 \cdot \sin(\phi_2 - \phi_1)^2
\end{aligned}$$

Combining the first four rows of the above equation give us the following equation:

$$\begin{aligned}
&= \left(\sin(\phi_1) \cdot \cos(\theta_1) \cdot \sin(\theta_2) - \sin(\phi_2) \cdot \cos(\theta_2) \cdot \sin(\theta_1)\right)^2 \\
&+ \left(\cos(\phi_1) \cdot \cos(\theta_1) \cdot \sin(\theta_2) - \cos(\phi_2) \cdot \cos(\theta_2) \cdot \sin(\theta_1)\right)^2 \\
&+ \cos(\theta_1)^2 \cdot \cos(\theta_2)^2 \cdot \sin(\phi_2 - \phi_1)^2
\end{aligned}$$

By multiplying the above by $\left|\overrightarrow{V_1}\right|^2 \cdot \left|\overrightarrow{V_2}\right|^2$ and transforming the polar coordinates to the respective rectangular coordinates we prove the statement of this theorem that $\left|\overrightarrow{V_1}\right|^2 \cdot \left|\overrightarrow{V_2}\right|^2 \cdot sin(\delta)^2 = (y_1 \cdot z_2 - y_2 \cdot z_1)^2 + (x_1 \cdot z_2 - x_2 \cdot z_1)^2 + (x_1 \cdot y_2 - x_2 \cdot y_1)^2$.

3.6.2 Vector Perpendicular To A Plane

Two vectors, like lines in geometry, are perpendicular if the angle between the two vectors has a measure of ninety degrees. If we denote two independent vectors as $\overrightarrow{V_1}$ and $\overrightarrow{V_2}$, then the dot product satisfy the relationship $\overrightarrow{V_1} \bullet \overrightarrow{V_2} = r_1 \cdot r_2 \cdot cos(\delta)$ from theorem 25. These vectors are perpendicular if $cos(\delta) = 0$. In terms of the dot product we define two vectors, $\overrightarrow{V_1}$ and $\overrightarrow{V_2}$, as being perpendicular if $\overrightarrow{V_1} \bullet \overrightarrow{V_2} = 0$. A vector is perpendicular to a plane if it is perpendicular to each vector contained in that plane. A more formal definition for a single vector being perpendicular to a plane is expressed in definition 3.6.2.

Definition 3.6.5 A vector $\vec{P}$ is perpendicular to a plane if it is perpendicular to each vector contained in the plane at its point of intersection. In terms of the dot product, for each vector $\vec{R}$ contained in the plane, we have $\vec{P} \bullet \vec{R} = 0$.

If the two vectors $\overrightarrow{V_1}$ and $\overrightarrow{V_2}$ form a basis for a plane, then it is necessary and sufficient for the vector $\vec{P}$ to be perpendicular to both $\overrightarrow{V_1}$ and $\overrightarrow{V_2}$ for it to be perpendicular to the plane containing the two vectors. Hence we have an alternate and simpler definition for a vector to be perpendicular to a plane. The proof of this statement is left as an exercise for the reader.

With the above working definition for a vector perpendicular to a plane, we can investigate the form of the vector that is perpendicular to a specific plane.

Assuming that the plane is spanned by the two vectors $\overrightarrow{V_1}$ and $\overrightarrow{V_2}$, it is sufficient to find the family of vectors such that each member is perpendicular to both vectors. For a vector to be a member of this family it must satisfy the following two equations $\vec{P} \bullet \overrightarrow{V_1} = 0$ and $\vec{P} \bullet \overrightarrow{V_2} = 0$. Expressing the vectors in rectangular coordinates such that

$\overrightarrow{V_1} = x_1 \cdot \overrightarrow{e_1} + y_1 \cdot \overrightarrow{e_2} + z_1 \cdot \overrightarrow{e_3}$ and

$\overrightarrow{V_2} = x_2 \cdot \overrightarrow{e_1} + y_2 \cdot \overrightarrow{e_2} + z_2 \cdot \overrightarrow{e_3}$, then we can rewrite the above equations as

(1) $a \cdot x_1 + b \cdot y_1 + c \cdot z_1 = 0$ and

(2) $a \cdot x_2 + b \cdot y_2 + c \cdot z_2 = 0$, where $\vec{P} = a \cdot \vec{e_1} + b \cdot \vec{e_2} + c \cdot \vec{e_3}$.

With two equations and three unknowns (a, b, and c), we can solve for two of the unknowns in terms of the third unknown. Let's solve for 'b' and 'c'. We begin by solving for 'c'. We multiply equation (1) by y_2 and equation (2) by y_1 to get equations (1a) and (2a) as follows:

(1a) $a \cdot x_1 \cdot y_2 + b \cdot y_1 \cdot y_2 + c \cdot z_1 \cdot y_2 = 0$ and

(2a) $a \cdot x_2 \cdot y_1 + b \cdot y_2 \cdot y_1 + c \cdot z_2 \cdot y_1 = 0$.

Subtracting equation (1a) from (2a) we get equation (1b) as follows:

(1b) $a \cdot (x_1 \cdot y_2 - x_2 \cdot y_1) + c \cdot (z_1 \cdot y_2 - z_2 \cdot y_1) = 0$.

Solving for the unknown value 'c', we get the following:

$$c = a \cdot (x_1 \cdot y_2 - x_2 \cdot y_1) / (y_1 \cdot z_2 - y_2 \cdot z_1) .$$

To solve for the unknown value 'b', multiply equations (1) and (2) by z_2 and z_1 respectively. Then subtract the first equation from the second equation as in the previous sequence of steps. This sequence of steps results in the following relationship for 'b': $b = -a \cdot (x_1 \cdot z_2 - x_2 \cdot z_1) / (y_1 \cdot z_2 - y_2 \cdot z_1)$.

Since the value of a can be any real number, we specify it to be

$a = \boldsymbol{k} \cdot (y_1 \cdot z_2 - y_2 \cdot z_1)$. In this expression, k is an arbitrary real number. With this substitution, the family of vectors perpendicular to the plane spanned by $\overrightarrow{V_1}$ and $\overrightarrow{V_2}$ is $\vec{P}_{\boldsymbol{k}} =$

$$\left((y_1 \cdot z_2 - y_2 \cdot z_1) \cdot \overrightarrow{e_1} - (x_1 \cdot z_2 - x_2 \cdot z_1) \cdot \overrightarrow{e_2} + (x_1 \cdot y_2 - x_2 \cdot y_1) \cdot \overrightarrow{e_3}\right) \cdot \boldsymbol{k} .$$

The square of the amplitude of this vector is the dot product of the vector with itself. Mathematically, we write this relationship as $\left|\vec{P}_{\boldsymbol{k}}\right|^2 = \vec{P}_{\boldsymbol{k}} \bullet \vec{P}_{\boldsymbol{k}}$ or

$$\left|\vec{P}_{\boldsymbol{k}}\right|^2 = \left((y_1 \cdot z_2 - y_2 \cdot z_1)^2 + (x_1 \cdot z_2 - x_2 \cdot z_1)^2 + (x_1 \cdot y_2 - x_2 \cdot y_1)^2\right) \cdot \boldsymbol{k}^2 .$$

In the special case of $\boldsymbol{k} = 1$, we have $\left|\vec{P}_1\right|^2 = \left|\overrightarrow{V_1}\right|^2 \cdot \left|\overrightarrow{V_2}\right|^2 \cdot sin(\delta)^2$ or

$$\left|\vec{P}_1\right| = \left|\overrightarrow{V_1}\right| \cdot \left|\overrightarrow{V_2}\right| \cdot sin(\delta).$$

The vector $\vec{P}_1$ is formally called the cross product of the vectors $\overrightarrow{V_1}$ and $\overrightarrow{V_2}$. A formal definition of the cross product is given in Definition 3.6.6.

Definition 3.6.6 The cross product of vectors $\overrightarrow{V_1}$ and $\overrightarrow{V_2}$ is defined to be

$$\overrightarrow{V_1} \times \overrightarrow{V_2} = (y_1 \cdot z_2 - y_2 \cdot z_1) \cdot \overrightarrow{e_1} - (x_1 \cdot z_2 - x_2 \cdot z_1) \cdot \overrightarrow{e_2} + (x_1 \cdot y_2 - x_2 \cdot y_1) \cdot \overrightarrow{e_3}$$

.

3.6.3 Area Of Parallelogram

A parallelogram formed by two intersecting vectors (as shown in figure below) has an area equal to $\left|\vec{V_1} \times \vec{V_2}\right|$. This statement is seen in the definition of the cross product. If the angle between the vectors is denoted by δ, then the height of the parallelogram from the base determined by vector $\vec{V_2}$ is defined to be $\left|\vec{V_1}\right| \cdot sin(\delta)$. Since the length of the base is $\left|\vec{V_2}\right|$, the area of the parallelogram is the product of the base and height or

$$\left|\vec{V_1}\right| \cdot \left|\vec{V_2}\right| \cdot sin(\delta) = \left|\vec{V_1} \times \vec{V_2}\right| .$$

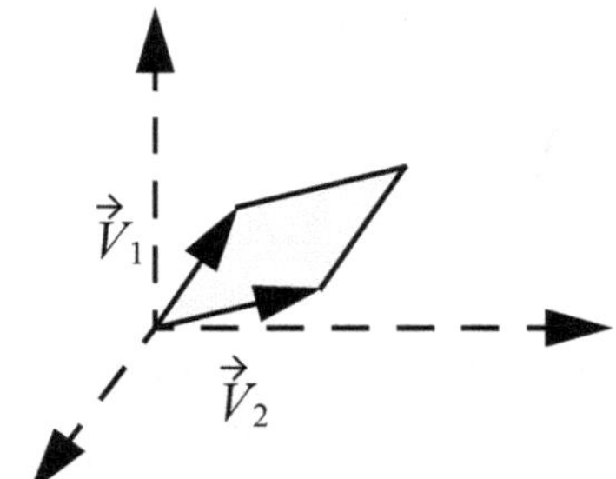

Figure 3.6.2: Parallelogram Formed From Two Vectors

3.6.4 Volume Of Parallelepiped

Having determined the area of a parallelogram, let's move up to three dimensions and determine the volume of a parallelepiped formed using three vectors

$\overrightarrow{V_1}$, $\overrightarrow{V_2}$ and $\overrightarrow{V_3}$. We construct the parallelepiped by first constructing a base formed using the two vectors $\overrightarrow{V_2}$ and $\overrightarrow{V_3}$. Then a second identical parallelogram, with its origin at the end point of vector $\overrightarrow{V_1}$ and lying completely in a plane parallel to the plane containing the base parallelogram, forms the upper surface area of the parallelepiped. This framework is shown in figure 3.6.3 below.

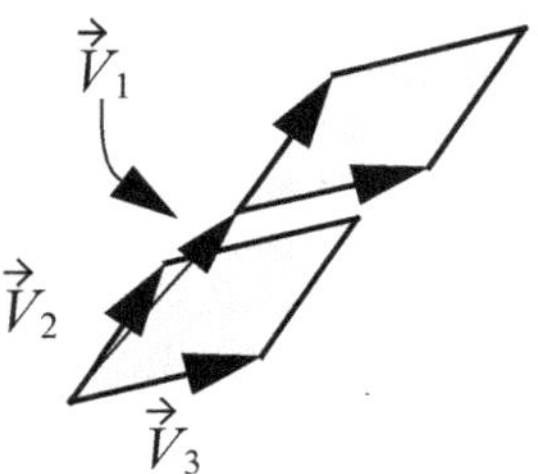

Figure 3.6.3: Parallelepiped Formed From Three Vectors

The volume of the parallelepiped is the area of the base times the height (distance between base and upper area) of the parallelepiped. From section 3.6.4, the area of the base parallelepiped is defined to be $\left|\overrightarrow{V_2} \times \overrightarrow{V_3}\right|$. If the angle between the vector $\overrightarrow{V_1}$ and $\vec{P} = \left(\overrightarrow{V_2} \times \overrightarrow{V_3}\right)$ is θ, then the height of the parallelepiped is computed to be $\left|\overrightarrow{V_1}\right| \cdot cos(\theta)$. Using the definition of the dot product of two vectors we have $\overrightarrow{V_1} \bullet \left(\overrightarrow{V_2} \times \overrightarrow{V_3}\right) = \left(\left|\overrightarrow{V_1}\right| \cdot cos(\theta)\right) \cdot \left|\overrightarrow{V_2} \times \overrightarrow{V_3}\right|$.

Therefore the volume of the parallelepiped is defined to be $\overrightarrow{V_1} \bullet \left(\overrightarrow{V_2} \times \overrightarrow{V_3}\right)$.

Using the cartesian coordinate representation for each vector we have the following formula for the volume:

$$\vec{V_1} \bullet \left(\vec{V_2} \times \vec{V_3}\right) = \begin{matrix} x_1 \cdot (y_2 \cdot z_3 - y_3 \cdot z_2) - y_1 \cdot (x_2 \cdot z_3 - x_3 \cdot z_2) \\ + \quad z_1 \cdot (x_2 \cdot y_3 - x_3 \cdot y_2) \end{matrix}$$

. This representation for the volume will be used in following chapters and will be used to expand additional concepts in the theory of matrices.

Exercises

The following exercises will allow the reader to practise the newly introduced vector operation of dot products. Compute the dot product of each pair of vectors in exercises 1 through 5.

1. $\begin{bmatrix} -1 \\ 5 \\ 2 \end{bmatrix}$ and $\begin{bmatrix} -10 \\ -4 \\ 5 \end{bmatrix}$

2. $\begin{bmatrix} 1 \\ 2 \\ 3 \end{bmatrix}$ and $\begin{bmatrix} 3 \\ 3 \\ 7 \end{bmatrix}$

3. $\begin{bmatrix} 23 \\ 4 \\ -8 \end{bmatrix}$ and $\begin{bmatrix} 1 \\ 0 \\ 4 \end{bmatrix}$

4. $\begin{bmatrix} 9 \\ 5 \\ 3 \end{bmatrix}$ and $\begin{bmatrix} 9 \\ 5 \\ 3 \end{bmatrix}$

5. $\begin{bmatrix} 0 \\ -7 \\ -8 \end{bmatrix}$ and $\begin{bmatrix} 13 \\ -7 \\ 0 \end{bmatrix}$

Exercises

Using the dot product compute the length of each vector and the cosine of the angle between the vectors.

6. $\overrightarrow{V_1} = \begin{bmatrix} 1 \\ 2 \\ -1 \end{bmatrix}$ and $\overrightarrow{V_2} = \begin{bmatrix} 3 \\ 7 \\ 2 \end{bmatrix}$; compute $\left|\overrightarrow{V_1}\right|$, $\left|\overrightarrow{V_2}\right|$ and $cos(\delta)$.

7. $\overrightarrow{V_1} = \begin{bmatrix} 0 \\ -9 \\ 11 \end{bmatrix}$ and $\overrightarrow{V_2} = \begin{bmatrix} 21 \\ 1 \\ 1 \end{bmatrix}$; compute $\left|\overrightarrow{V_1}\right|$, $\left|\overrightarrow{V_2}\right|$ and $cos(\delta)$.

This Page Intentionally Left Blank

CHAPTER 4 Determinants, Matrix Inverses And Partitions

The derivation of physical quantities such as lengths, perpendicular vectors, areas and volumes required some rather involved manipulations of vector coordinates. By generalizing the representation of these vectors in terms of matrices, we get a more elegant and simplistic view of these physical quantities. As we move from the vector representation to the matrix representation, we investigate the physical quantity, the volume of a parallelepiped, and show how a function of a matrix can be used to compute this physical quantity.

We begin by recalling the many computations required for computing the volume when cartesian coordinates are used to represent the vectors. The resultant formula for the volume, as illustrated in chapter 3, is the following:

Volume =

$$x_1 \cdot y_2 \cdot z_3 - x_1 \cdot y_3 \cdot z_2 - x_2 \cdot y_1 \cdot z_3 + x_3 \cdot y_1 \cdot z_2 + x_2 \cdot y_3 \cdot z_1 - x_3 \cdot y_2 \cdot z_1.$$

To move from the realm of vectors to that of matrices; let vector $\vec{V}_1$ represent the first row of the matrix, let vector $\vec{V}_2$ represent the second row of the matrix and let vector $\vec{V}_3$ represent the third row of the matrix. If the resultant matrix is denoted as V, then we have $V = [v_{i,j}]$ where i is an element of $\{1, 2, 3\}$ and j is an element of $\{1, 2, 3\}$. In terms of the elements of the vectors, the following relationships for the elements of the matrix are true:

$$v_{i,1} = x_i \quad \text{(x - coordinate of the i}^{\text{th}} \text{ vector),}$$

$$v_{i,2} = y_i \quad \text{(y - coordinate of the i}^{\text{th}} \text{ vector) and}$$

$$v_{i,3} = z_i \quad \text{(z - coordinate of the i}^{\text{th}} \text{ vector).}$$

Symbolically, we shall represent this composite matrix of the vectors $\vec{V}_1$, $\vec{V}_2$ and $\vec{V}_3$ as

$$V = \begin{bmatrix} \vec{V}_1 \\ \vec{V}_2 \\ \vec{V}_3 \end{bmatrix}.$$

With the above definition of the composite matrix, we are now able to specify the volume of the parallelepiped in terms of the matrix elements. This expression is as follows:

$$\begin{aligned} \text{Volume} = \; & v_{1,1} \cdot v_{2,2} \cdot v_{3,3} - v_{1,1} \cdot v_{3,2} \cdot v_{2,3} - \\ & v_{2,1} \cdot v_{1,2} \cdot v_{3,3} + v_{3,1} \cdot v_{1,2} \cdot v_{2,3} + \\ & v_{2,1} \cdot v_{3,2} \cdot v_{1,3} - v_{3,1} \cdot v_{2,2} \cdot v_{1,3} \quad . \end{aligned}$$

In section 4.1 of this chapter, we discuss and formally define permutations of the set of non repeating integers. In this instance, we are only concerned with the permutations of the set $(1, 2, 3)$. Then the set of all permutations for this set is

$$P = \{(1, 2, 3), (1, 3, 2), (2, 1, 3), (2, 3, 1), (3, 1, 2), (3, 2, 1)\}.$$

With this set notation, the summation notation ($\sum_{(j_1, j_2, j_3) \in P}$) and the product notation $\left(\prod_{k=1}^{3}\right)$, we can write the shorthand formula for the volume of a parallelepiped as the following:

$$\text{Volume} = \sum_{(j_1, j_2, j_3) \in P} \delta(j_1, j_2, j_3) \cdot \prod_{k=1}^{3} v_{j_k, k} .$$

Using the above notation, the function $\delta(j_1, j_2, j_3)$ is a discrete function of the selected permutation of the three discrete integers (j_1, j_2, j_3). Its assigns a value of either -1 or +1 to a selected triplet. The formal specification for this function is given in section 4.1 of this chapter. But for now, it is enough to define the function using the following specification:

$$\delta(1, 2, 3) = 1 ,$$

$$\delta(1, 3, 2) = -1 ,$$

$$\delta(2, 1, 3) = -1 \ ,$$

$$\delta(2, 3, 1) = 1 \ ,$$

$$\delta(3, 1, 2) = 1 \quad \text{and}$$

$$\delta(3, 2, 1) = -1 \ .$$

4.1 Permutations

Before simplification of physical phenomena in terms of matrix operations can be accomplished, some preliminary definitions are required. The first definition is that of a permutation. In the introductory section of this chapter, we discussed the concept of permutations informally. Now we present a more formal definition of a permutation in definition 4.1.1 below.

Definition 4.1.1 **Permutation:** Let $S_n = (1, 2, \ldots, n)$ be the ordered set of positive integers from one (1) to n, then a permutation of S_n is a one to one mapping of S_n to a set $S'_n = (j_1, j_2, \ldots, j_n)$ such that it is a rearrangement of the elements of S_n .

As we observed in the introductory section, there is a function associated with each permutation of an ordered set of integers. This function is indicated using the following terminology: $\delta(j_1, j_2, \ldots, j_n)$. This function assigns either a value of -1 or +1 to a specific permutation. To provide a formal specification for this function, we introduce the concept of an "inversion" in definition 4.1.2 below.

Definition 4.1.2 **Inversion:** The permutation $S'_n = (j_1, j_2, \ldots, j_n)$ of the set $S_n = (1, 2, \ldots, n)$ has an inversion if it has a larger number preceding a smaller number. Mathematically we write

$j_K > j_L$ where $1 \le K < L \le n$.

By counting the number of inversions for a specific permutation, we can derive a discrete function $I(S'_n)$. When a permutation has an odd number of inversions, it is said to be an odd permutation. When it has an even number of permutations it is said to be an even permutation.

Now let's review the inversions for the permutations of the set $(1, 2, 3)$.

TABLE 8. Summary of permutations of $(1, 2, 3)$

Permutation S_3	$\delta(\ldots)$	# Of Inversions $I(S_3)$.	Odd/Even
$(1, 2, 3)$	1	0	Even
$(1, 3, 2)$	-1	1	Odd
$(2, 1, 3)$	-1	1	Odd
$(2, 3, 1)$	1	2	Even
$(3, 1, 2)$	1	2	Even
$(3, 2, 1)$	-1	3	Odd

An immediate observation from this table is that the function $\delta(\ldots)$ is assigned a value of -1 when the permutation is an odd permutation and a value of 1 when it is an even permutation. (We assume that zero inversions is an even permutation.) With this observation, we generalize the definition for the function $\delta(\ldots)$ with the following specification:

$$\delta(S'_n) = (-1)^I \qquad \text{where } I = \boldsymbol{I}(S'_n) \ .$$

Now that the definition of $\delta(\ldots)$ is formalized, it is time to examine some of its properties. As we transform one permutation into another (example exchanging positions of values), we observe the effect on the $\delta(\ldots)$ function. The first and simplest transformation, that we discuss, is the exchanging positions of adjacent values. When adjacent values are exchanged to form a new permutation from the existing permutation, the relationship of the $\delta(\ldots)$ function is stated in corollary 4.1.3 below.

Corollary 4.1.3 Let the permutation $(j_1, \ldots, j_{r+1}, j_r, \ldots, j_n)$ represent the exchanging positions of the adjacent values j_r and j_{r+1} in the permuting $(j_1, \ldots, j_r, j_{r+1}, \ldots, j_n)$, then we have the relationship

$$\delta(j_1, \ldots, j_{r+1}, j_r, \ldots, j_n) = -\delta(j_1, \ldots, j_r, j_{r+1}, \ldots, j_n)$$

.

Proof: To prove this corollary, let $\boldsymbol{I} = \boldsymbol{I}(j_1, \ldots, j_r, j_{r+1}, \ldots, j_n)$, then

$$\delta(j_1, \ldots, j_r, j_{r+1}, \ldots, j_n) = (-1)^{\boldsymbol{I}} .$$

If we exchange the positions of j_r and j_{r+1} when $j_r > j_{r+1}$, then the number of inversions is decreased for the permutation $(j_1, \ldots, j_{r+1}, j_r, \ldots, j_n)$. Hence the function $\delta(\ldots)$ for this new permutation is given by $\delta(j_1, \ldots, j_{r+1}, j_r, \ldots, j_n) = (-1)^{\boldsymbol{I}-1}$. This is rewritten as $\delta(j_1, \ldots, j_{r+1}, j_r, \ldots, j_n) = -\delta(j_1, \ldots, j_r, j_{r+1}, \ldots, j_n)$. So for $j_r > j_{r+1}$, the statement of this corollary is true.

When $j_r < j_{r+1}$, then the number of inversions is increased by one for the permutation $(j_1, \ldots, j_{r+1}, j_r, \ldots, j_n)$. Hence the function $\delta(\ldots)$ for this new permutation is given by $\delta(j_1, \ldots, j_{r+1}, j_r, \ldots, j_n) = (-1)^{\boldsymbol{I}+1}$. This is rewritten as

$\delta(j_1, \ldots, j_{r+1}, j_r, \ldots, j_n) = -\delta(j_1, \ldots, j_r, j_{r+1}, \ldots, j_n)$. So for $j_r < j_{r+1}$, the statement of this corollary is true.

Since the statement is true for all possible relationships between j_r and j_{r+1}, the corollary is proven true.

The next transformation of interest for permutations is the exchanging of values whose positional values differs by a integer value of d, where $1 \le d < n$. This transformation is a generalization of the previous transformation; the exchanging of adjacent values $(d = 1)$. The repeated application of corollary 4.1.3 yields the relationship shown in theorem 4.1.4 below.

Theorem 4.1.4 Let the number d be an integer such that $1 \le d < n$ and let the number k be an integer such that $1 \le k \le n - d$. If the transformation $\boldsymbol{T}_{k,\, d}$ represents a transformation from one permutation to another by exchanging positional values k and $k + d$, then we have the relationship

$\delta(\boldsymbol{T}_{k,\, d}(S'_n)) = -\delta(S'_n)$ where $S'_n = (j_1, j_2, \ldots, j_n)$.

Proof: If we begin with the transformation $\boldsymbol{T}_{k,\, 1}(S'_n)$ which represents the exchange of adjacent positions to form the new permutation then we have $\delta(\boldsymbol{T}_{k,\, 1}(S'_n)) = -\delta(S'_n)$ from corollary 4.1.3. To move the value in position k to position $k + d$, we apply the adjacent exchange transformations as follows

$S^*_n = \boldsymbol{T}_{k+d-1,\, 1}(\boldsymbol{T}_{k+d-2,\, 1}(\ldots \boldsymbol{T}_{k,\, 1}(S'_n)))$. This composite function applies the adjacent exchange transformation a total of d times. Applying corollary 4.1.3 d times, we get

$\delta(S^*_n) = (-1)^d \delta(S'_n)$. After this transformation, the original k^{th} position number is moved to the $(k+d)^{th}$ position and the original $(k+d)^{th}$ position value is now in position $(k + d - 1)$.

To move the value currently in position $(k+d-1)$, we apply the adjacent exchange transformation as follows:

$S^{\tau}{}_n = \boldsymbol{T}_{k,\,1}(\ldots\boldsymbol{T}_{k+d-2,\,1}(S^*{}_n))$. These series effectively moves the original $(k+d)^{th}$ position to the kth position. This composite transformation represents $(d-1)$ adjacent transformation and hence from corollary 4.1.3 we have

$\delta(S^{\tau}{}_n) = (-1)^{(d-1)}\delta(S^*{}_n) = (-1)^{2\cdot d-1}\delta(S'_n)$. Since the quantity $(2\cdot d-1)$ is an odd value, we have the relationship

$\delta(S^{\tau}{}_n) = -\delta(S'_n)$. The permutation $S^{\tau}{}_n$ is the results of the $\boldsymbol{T}_{k,\,d}$ transformation. Hence we have shown that

$\delta(\boldsymbol{T}_{k,\,d}(S'_n)) = -\delta(S'_n)$.

The importance of this theorem will be revealed as we introduce the properties of square matrices and their determinants.

Now we introduce the concept of an inverse transformation. Suppose the permutation $S^T{}_n$ is a result of the transformation $\boldsymbol{T}$ on the zero-inversion permutation $(1, 2, \ldots, n)$. Symbolically, we write this relationship as

$S^T{}_n = \boldsymbol{T}(1, 2, \ldots, n)$. Then the inverse transformation is the transformation $\boldsymbol{T}^*$ such that $\boldsymbol{T}^*(S^T{}_n) = (1, 2, \ldots, n)$. Using the properties of the inverse transformations we are able to prove the relationship of $\delta(\ldots)$ functions of the transformation and its inverse transformation. This relationship is expressed in corollary 4.1.5 below.

Corollary 4.1.5 Let the transformation $\boldsymbol{T}$, map the zero-inversion permutation $(1, 2, \ldots, n)$ into the permutation $S^T{}_n$ (i.e. $S^T{}_n = \boldsymbol{T}(1, 2, \ldots, n)$) Let the transformation $\boldsymbol{T}^*$ be the inverse transformation of $\boldsymbol{T}$ such that $\boldsymbol{T}^*(S^T{}_n) = (1, 2, \ldots, n)$. Then we get the following relationship:
$\delta(\boldsymbol{T}(1, 2, \ldots, n)) = \delta(\boldsymbol{T}^*(1, 2, \ldots, n))$.

Proof: Using the formal definition of the $\delta(\ldots)$ function, we have $\delta(\boldsymbol{T}(1, 2, \ldots, n)) = (-1)^{\boldsymbol{I}}$. The integer 'I' is the number of inversions of the permutation $S^T{}_n$. Then applying the transformation $\boldsymbol{T}^*(S^T{}_n) = (1, 2, \ldots, n)$, we have $\delta(\boldsymbol{T}^*(S^T{}_n)) = \delta(1, 2, \ldots, n)$ or $\delta(\boldsymbol{T}^*(S^T{}_n)) = 1$. Recursively applying theorem 4.1.4, to get the number of exchanges that converts $S^T{}_n$ to $(1, 2, \ldots, n)$, we have $\delta(\boldsymbol{T}^*(S^T{}_n)) = (-1)^M \cdot \delta(S^T{}_n)$; where 'M' is the number of exchanges in the transformation $T^*(\ldots)$. This implies that $(-1)^M \cdot (-1)^I = 1$. For this relationship to hold, the sign of the both terms on the left side of the equation must be the same; functionally we have $(-1)^M = (-1)^I$. Since 'M' and 'I' represent the number of exchanges for the respective transformations we get $\delta(T(1, 2, \ldots, n)) = \delta(T^*(1, 2, \ldots, n))$. Hence the statement of this corollary is true.

Exercises

1. List all permutations of the set $S_4 = (1, 2, 3, 4)$.
2. List all permutations of the set $S_5 = (1, 2, 3, 4, 5)$ that have the value one (1) in the initial position of the permutation (i.e. $(1, j_2, j_3, j_4, j_5)$).
3. Using the set $S_4 = (1, 2, 3, 4)$, construct the summary of permutations table as shown in this section.
4. For problem number 2 above, specify the number of inversions $I(\ )$ for each permutation listed.
5. Evaluate $\delta(1, 3, 4, 2, 5)$.
6. Evaluate $\delta(4, 5, 2, 1, 3)$.
7. Evaluate $\delta(1, 3, 2, 4, 6, 5, 8, 7)$.
8. Evaluate $\delta(10, 9, 8, 7, 6, 5, 4, 3, 2, 1)$.
9. Prove that the number of permutations of S_n is defined to be

   $n! = n \cdot (n-1) \cdot \ldots \cdot 1$.

10. Compute the number of permutations of S_5 .

4.2 Square Matrices and Determinants

Using the concept of the $\delta(\ldots)$ function, introduced in the previous section, we have a formal definition of the volume of a parallelepiped determined by the three vectors $\vec{V}_1$, $\vec{V}_2$ and $\vec{V}_3$ with the associated square matrix

$\boldsymbol{V} = \begin{bmatrix} \vec{V}_1 \\ \vec{V}_2 \\ \vec{V}_3 \end{bmatrix}$ and $\boldsymbol{V} = [v_{i,j}]$. The formalized formula for the volume, in terms

of the elements of the matrix $\boldsymbol{V}$, is Volume = $\sum_{\underset{\sim}{J} \in \boldsymbol{P}} \delta(\underset{\sim}{J}) \cdot \prod_{k=1}^{3} v_{j_k, k}$. By

extending this formula to all square matrices, we get the following definition for determinant of a square matrix.

Definition 4.2.1 Given an **n** by **n** square matrix,

$\boldsymbol{A} = [a_{i,j}] \ (i, j = 1, \ldots, \boldsymbol{n})$, then the determinant of the matrix

$\boldsymbol{A}$ is defined as, $|\boldsymbol{A}| = \sum_{\underset{\sim}{J} \in \boldsymbol{P}} \delta(\underset{\sim}{J}) \cdot \prod_{k=1}^{\boldsymbol{n}} a_{i_k, k}$.

With the above working definition of determinants, we are now ready to derive some properties of determinants of matrices that will be highly useful in developing the theory of matrix algebra. The first property of concern is the determinant of the transpose of a matrix and its relationship to the original matrix. This property is given in corollary 4.2.2 below.

Corollary 4.2.2 If the matrix A^t represents the transpose of matrix $A = [a_{i,j}]$, then we have the relationship $|A^t| = |A|$.

Proof: To begin this proof, we represent the matrix as $A^t = [b_{i,j}]$, where $b_{i,j} = a_{j,i}$. Applying definition 4.2.1 for determinants of square matrices, we get $|A^t| = \sum_{\underset{\sim}{J} \in P} \delta(\underset{\sim}{J}) \cdot \prod_{k=1}^{n} b_{i_k,k}$. Substituting the elements of A into this equation, we have the formula $|A^t| = \sum_{\underset{\sim}{J} \in P} \delta(\underset{\sim}{J}) \cdot \prod_{k=1}^{n} a_{k,i_k}$. Let the transformation T_J, map the permutation $\underset{\sim}{J} = (i_1, i_2, \ldots, i_n)$ into the zero-inversion permutation $(1, 2, \ldots, n)$. Using this transformation we have $|A^t| = \sum_{\underset{\sim}{L} \in P} \delta(\underset{\sim}{J}) \cdot \prod_{k=1}^{n} a_{j_k,k}$ where $\underset{\sim}{L} = T_J(1, \ldots, n)$ and $\underset{\sim}{L} = (j_1, j_2, \ldots, j_n)$. By applying corollary 4.1.5, we have $\delta(\underset{\sim}{L}) = \delta(\underset{\sim}{J})$ and hence we have

$$|A^t| = \sum_{\underset{\sim}{L} \in P} \delta(\underset{\sim}{L}) \cdot \prod_{k=1}^{n} a_{j_k,k}$$ and thus we have shown that $|A^t| = |A|$.

The above corollary illustrates that the determinant function is invariant under the transposing operation of matrices. Other operations of interest are multiplying a single row (or column) by a constant, exchanging a pair of rows (or columns) and finally the addition of a constant value times one row to another row. The determinant of these transformed matrices and their relationship to the original matrix is shown in the following three corollaries.

Corollary 4.2.3 Given the matrix A, let $B = [b_{i,j}]$ representing the matrix such that all elements of matrix are equal to the corresponding elements of the A except for the elements of row 'k' that are a constant value times the elements of corresponding row of the matrix A. Mathematically we write this relationship as $b_{k,j} = c \cdot a_{k,j}$ and $b_{i,j} = a_{i,j};(i \neq k)$. Then the determinant of the matrix B is $|B| = c \cdot |A|$.

Proof: Using the notation $B = [b_{i,j}]$ and applying the definition of determinant, we get, $|B| = \sum_{\underset{\sim}{J} \in P} \delta(\underset{\sim}{J}) \cdot \prod_{j=1}^{n} b_{i_j,j}$. Making the substitution for each element of B in terms of the elements of A we get $|B| = \sum_{\underset{\sim}{J} \in P} \delta(\underset{\sim}{J}) \cdot c \cdot \prod_{j=1}^{n} a_{i_j,j}$. The factor '**c**' is included since one term of the product is multiplied by '**c**' (element $a_{k,j}$). Factoring out the 'c' term in the above equation we get

$|B| = c \cdot \sum_{\underset{\sim}{J} \in P} \delta(\underset{\sim}{J}) \cdot \prod_{j=1}^{n} a_{i_j,j}$. Thus, using the definition of determinant of A we have $|B| = c \cdot \sum_{\underset{\sim}{J} \in P} \delta(\underset{\sim}{J}) \cdot \prod_{j=1}^{n} a_{i_j,j}$ and thus $|B| = c \cdot |A|$.

The recursive application of corollary 4.2.3, when $B = c \cdot A$ and both are **n**-by-**n** square matrices, yields the relationship $|B| = c^n \cdot |A|$. Applying corollary 4.2.2 and 4.2.3 we also get $|B| = c \cdot |A|$ when B is equivalent to the matrix A except for the elements of column 'k' that are a constant value times the elements of the corresponding column of the matrix A. Now we examine the relationship of the resultant matrix that results when two rows (or columns) are exchanged in the original matrix. The relationship of their determinants is specified in corollary 4.2.4 below.

Corollary 4.2.4 Let $\boldsymbol{A}$ and $\boldsymbol{B}$ be **n**-by-**n** square matrices such that for rows **r** and **s** for matrix $\boldsymbol{A}$ the corresponding values of the $\boldsymbol{B}$ matrix are interchanged. Mathematically, this is written as follows:

$$b_{r,j} = a_{s,j} \text{ ,}$$

$$b_{s,j} = a_{r,j} \text{ and}$$

$$b_{i,j} = a_{i,j} \text{ when } i \neq r, s \text{.}$$

Then we have the relationship for the determinants: $|\boldsymbol{B}| = -|\boldsymbol{A}|$. This corollary is also true when the resultant matrix is formed from interchanging columns of the original matrix.

Proof: Using the notation $\boldsymbol{B} = [b_{i,j}]$ and applying the definition of determinant, we get, $|\boldsymbol{B}| = \sum_{\underset{\sim}{J} \in \boldsymbol{P}} \delta(\underset{\sim}{J}) \cdot \prod_{j=1}^{n} b_{i_j,j}$. Now we consider the transformation of one permutation to another. For the permutation $(\ldots, j_i = r, \ldots, j_k = s, \ldots)$,let $\boldsymbol{T}$ be a transformation that exchanges the elements corresponding to the values '**r**' and '**s**'; $\boldsymbol{T}(\ldots, j_i = r, \ldots, j_k = s, \ldots) = (\ldots, j_k = s, \ldots, j_i = r, \ldots)$. For each permutation, denoted as $\underset{\sim}{J}$, we have for each transformation $\boldsymbol{T}$, as described in preceding statement, $\delta(\boldsymbol{T}(\underset{\sim}{J})) = -\delta(\underset{\sim}{J})$ (corollary 4.1.4).

Making the substitution for each element of $\boldsymbol{B}$ in terms of the elements of $\boldsymbol{A}$, the above determinant is $|\boldsymbol{B}| = \sum_{\boldsymbol{T}(\underset{\sim}{J}) \in \boldsymbol{P}} \delta(\boldsymbol{T}(\underset{\sim}{J})) \cdot \prod_{j=1}^{n} a_{i_j,j}$. Since the transformation of the permutation is a one-to-one mapping, we can make the appropriate substitutions to obtain $|B| = \sum_{\underset{\sim}{J} \in P} -\delta(\underset{\sim}{J}) \cdot \prod_{j=1}^{n} a_{i_j, j}$ and hence

$|B| = -|A|$.

Now let us consider the special case in which the '**r**' and '**s**' row (could be column as well) of matrix A are identical. If the matrix B represents the resultant matrix upon exchanging rows '**r**' and '**s**', then $|B| = -|A|$ when applying corollary 4.2.4. Since rows '**r**' and '**s**' are identical we have $B = A$. Hence we have $|A| = -|A|$. This equation can only be true if $|A| = 0$.

The next transformation of interest is that of adding one row to another row to produce the resultant matrix. The corollary (4.2.5) below indicates the relationship of the original matrix to that of the resultant matrix.

Corollary 4.2.5 Let A and B be **n**-by-**n** square matrices such that, row '**r**' is altered by adding '**c**' times row '**s**' in matrix A to form matrix B. Mathematically, this is written as follows:

$$b_{r,j} = a_{r,j} + \mathbf{c} \cdot a_{s,j} \text{ and}$$

$$b_{i,j} = a_{i,j} \text{ when } i \neq r.$$

Then we have the relationship for the determinants: $|B| = |A|$. This corollary is also true when the resultant matrix is formed by adding one column to another column of the original matrix.

Proof: Using the notation $B = [b_{i,j}]$ and apply the definition for determinants, we get $|B| = \sum_{\underset{\sim}{J} \in P} \delta(\underset{\sim}{J}) \cdot \prod_{j=1}^{n} b_{i_j, j}$. If we expand the product term for each permutation of this expression we get $\prod_{j=1}^{n} b_{i_j, j} = b_{i_j, 1} \cdot \ldots \cdot b_{r, k} \cdot \ldots \cdot b_{i_j, n}$. For each term we make the appropriate substitution and we get the following relationship: $\prod_{j=1}^{n} b_{i_j, j} = a_{i_j, 1} \cdot \ldots \cdot (a_{r, k} + c \cdot a_{s, k}) \cdot \ldots \cdot a_{i_j, n}$. Using the distributive property of numbers, the determinant can now be expressed as the following: $|B| = \sum_{\underset{\sim}{J} \in P} \delta(\underset{\sim}{J}) \cdot \prod_{j=1}^{n} a_{i_j, j} + c \cdot \sum_{\underset{\sim}{J} \in P} \delta(\underset{\sim}{J}) \cdot \prod_{j=1}^{n} a^*_{i_j, j}$. By definition of determinant, we can rewrite this expression as $|B| = |A| + c \cdot |A^*|$. The matrix A^* has row '**r**' equivalent to row '**s**'. Hence we have $|A^*| = 0$. Therefore the determinant of the matrix B is $|B| = |A|$.

(Special Matrices)

In this section we will examine the determinants of some special matrices using the definitions, corollaries and theorems. The first matrix that we will consider is the diagonal matrix.

Diagonal Matrix: Define the matrix $D = [d_{i,j}]$ such that $d_{i,j} = 0; i \neq j$. Using the definition of the determinant of a matrix, we have the following expression: $|D| = \sum_{\underset{\sim}{J} \in P} \delta(\underset{\sim}{J}) \cdot \prod_{j=1}^{n} d_{i_j, j}$. Since $d_{i_j, j} = 0$ when $i_j \neq j$, this expression can be reduced to $|D| = \delta(1, \ldots, n) \cdot \prod_{j=1}^{n} d_{j,j}$. Since $(1, \ldots, n)$ has zero inversions, the function $\delta(1, \ldots, n)$ has a value of one (1). Therefore we have $|D| = d_{1,1} \cdot \ldots \cdot d_{n,n}$. Hence the determinant of the diagonal matrix is the product of the diagonal elements.

A special case of the diagonal matrix is the identity matrix (I). Since all diagonal elements are identically equal to one (1), we have $|I| = 1$.

Triangular Matrix (Upper/Lower): Let the upper triangular matrix be represented as $U = [u_{i,j}]$ such that $u_{i,j} = 0;(i<j)$. Using the definition of the determinant of a matrix, we have the following expression: $|U| = \sum_{\underset{\sim}{J} \in P} \delta(\underset{\sim}{J}) \cdot \prod_{j=1}^{n} u_{i_j,j}$. In the product term of this expression, whenever there is an element with its i-subscript is greater than the j-subscript, there must be an element with the i-subscript less than its j-subscript and hence the product is equivalent to zero (0). Hence the only product term that is not identically zero is the product for which $i_j = j$. Hence the determinant of the upper triangular matrix is $|U| = u_{1,1} \cdot \ldots \cdot u_{n,n}$. Hence the determinant of the upper triangular matrix is the product of the diagonal elements. For the lower triangular matrix, its transpose is an upper triangular matrix. Applying corollary 4.2.2 we have that the determinant of the lower triangular matrix is the product of the diagonal elements of the matrix.

Type I Elementary Matrix: Let E_I be an elementary matrix of type I, that exchanges the '**r**' and '**s**' row (or column) when multiplied to another matrix. Upon close inspection of this matrix, it is noticed that it is derived from the identity matrix by exchanging the '**r**' and '**s**' row of the identity matrix. Applying corollary 4.2.3, we have $|E_I| = -|I|$ or $|E_I| = -1$.

Type II Elementary Matrix: Let E_{II} be an elementary matrix of type II, that multiplies row (or column) of a matrix by value '**c**' when multiplied to that matrix. Upon close inspection of this matrix, it is noticed that it is a diagonal matrix with one of the diagonal elements equal to '**c**' and the remaining elements equivalent to one (1). By property of determinant of diagonal matrix, we have $|E_{II}| = c$.

Type III Elementary Matrix: Let E_{III} be an elementary matrix of type III, when multiplied to another matrix, adds to a row of that matrix 'c' times another row of that matrix. Upon close inspection of this matrix, it is noticed that it is a triangular matrix with all diagonal elements equivalent to one (1). By property of determinant of triangular matrix, we have $|E_{III}| = 1$.

Exercises

In the following exercises, compute the determinant of each matrix.

1. $\begin{vmatrix} 1 & 4 \\ 3 & 2 \end{vmatrix}$

2. $\begin{vmatrix} 1 & 2 & -5 \\ 1 & 0 & 1 \\ -2 & -3 & 9 \end{vmatrix}$

3. $\begin{vmatrix} 4 & 3 & -2 \\ 6 & 5 & 0 \\ -3 & -2 & 1 \end{vmatrix}$

4. $\begin{vmatrix} 4 & 1 \\ 8 & 2 \end{vmatrix}$

5. $\begin{vmatrix} 2 & -5 & -6 & 3 \\ 1 & 1 & 7 & 0 \\ 0 & 3 & 0 & -2 \\ 1 & -4 & -3 & 4 \end{vmatrix}$

6. $\begin{vmatrix} 3 & -1 & -1 & 6 \\ -1 & 2 & -4 & -4 \\ -1 & -4 & 3 & 5 \\ 6 & -4 & 5 & 2 \end{vmatrix}$

7. $\begin{vmatrix} 5 & 0 & 4 \\ 1 & 2 & -1 \\ -4 & 3 & -3 \end{vmatrix}$

8. $\begin{vmatrix} 3 & 0 & 0 \\ -1 & 4 & -1 \\ 3 & 4 & 5 \end{vmatrix}$

9. $\begin{vmatrix} 10 & 12 & 14 \\ -7 & 1 & -8 \\ -4 & 14 & -2 \end{vmatrix}$

10. $\begin{vmatrix} 20 & 0 & 0 \\ -4 & 5 & 0 \\ -15 & 3 & 1 \end{vmatrix}$

4.3 Determinants Of Products

As seen in chapter 3, we can repeatedly multiply an **n**-by-**n** matrix by elementary matrices to produce a diagonal matrix with zeros and ones along the diagonal. When the resultant diagonal matrix is the identity matrix, the original matrix is said to be non-singular. If the resultant diagonal matrix has some zeros along the diagonal, the original matrix is singular. Using the properties of determinants of elementary matrices, we can derive the determinant of the product of an elementary matrix with an arbitrary **n**-by-**n** matrix.

Corollary 4.3.1 If E is an **n-by-n** elementary matrix and A is any **n-by-n** arbitrary matrix, then $|E \cdot A| = |A \cdot E| = |E| \cdot |A|$.

Proof: The proof for this corollary will be done for each type of elementary matrix.

(Type I Elementary Matrix) Let $E = E_I$ be a type I elementary matrix. Then the product $E \cdot A$ exchanges two rows of matrix A . Applying corollary 4.2.4 we have $|E \cdot A| = |A \cdot E| = -|A|$. But recall that $|E| = -1$ for type I elementary matrices. Hence we have shown that $|E \cdot A| = |A \cdot E| = |E| \cdot |A|$.

(Type II Elementary Matrix) Let $E = E_{II}$ be a type II elementary matrix. Then the product $E \cdot A$ multiplies a row of matrix A by a constant value '**c**' ($A \cdot E$ multiplies column by '**c**'). Applying corollary 4.2.3 we have $|E \cdot A| = |A \cdot E| = c \cdot |A|$. But recall that $|E| = c$ for type II elementary matrices. Hence we have shown that $|E \cdot A| = |A \cdot E| = |E| \cdot |A|$.

(Type III Elementary Matrix) Let $E = E_{III}$ be a type III elementary matrix. Then the product $E \cdot A$ adds to a row of matrix A a constant times another row of matrix A ($A \cdot E$ adds to a column of matrix A a constant times another column

of matrix A). Applying corollary 4.2.5 we have $|E \cdot A| = |A \cdot E| = |A|$.
But recall that $|E| = 1$ for type III elementary matrices. Hence we have shown that
$|E \cdot A| = |A \cdot E| = |E| \cdot |A|$.

Hence this corollary is true for all elementary matrices.

Matrix With Zero Row Or Column: Using corollary 4.3.1 above for elementary matrices we can show that the matrix with at least one row (or column) having all elements that are equal to zero has a determinant with a value of zero. This relationship is shown in corollary 4.3.2 below.

Corollary 4.3.2 If the matrix A has at least one row or column, containing all zeros, then the determinant of A is zero.

Proof: Assume that row '**r**' of matrix A contains all zero elements. If $E_{c \cdot r}$ is an elementary matrix of type II that multiplies row '**r**' by the arbitrary value '**c**', then $E_{c \cdot r} \cdot A = A$. Hence by corollary 4.3.2 we get $|E_{c \cdot r} \cdot A| = |E_{c \cdot r}| \cdot |A| = |A|$. Since $|E_{c \cdot r}| = c$, we can write $c \cdot |A| = |A|$. Since 'c' is any arbitrary value, we must have $|A| = 0$.

Consider the 3-by-3 matrix $\begin{bmatrix} 1 & 2 & 1 \\ -1 & 1 & 4 \\ 3 & 2 & 1 \end{bmatrix}$. By multiplying a series of elementary matrices, to reduce the matrix to a diagonal matrix we can construct the following relationship:

$$\begin{bmatrix} 1 & 2 & 1 \\ -1 & 1 & 4 \\ 3 & 2 & 1 \end{bmatrix} =$$

$$\begin{bmatrix} 1 & 0 & 0 \\ -1 & 1 & 0 \\ 0 & 0 & 1 \end{bmatrix} \begin{bmatrix} 1 & 0 & 0 \\ 0 & 1 & 0 \\ 3 & 0 & 1 \end{bmatrix} \begin{bmatrix} 1 & \frac{2}{3} & 0 \\ 0 & 1 & 0 \\ 0 & 0 & 1 \end{bmatrix} \begin{bmatrix} 1 & 0 & 0 \\ 0 & 1 & 0 \\ 0 & \frac{-4}{3} & 1 \end{bmatrix} \begin{bmatrix} 1 & 0 & \frac{-1}{2} \\ 0 & 1 & 0 \\ 0 & 0 & 1 \end{bmatrix} \begin{bmatrix} 1 & 0 & 0 \\ 0 & 1 & \frac{15}{14} \\ 0 & 0 & 1 \end{bmatrix} \begin{bmatrix} 1 & 0 & 0 \\ 0 & 3 & 0 \\ 0 & 0 & \frac{14}{3} \end{bmatrix}.$$

This example illustrates that the 3-by-3 matrix can be expressed as a product of elementary matrices and a diagonal matrix. Applying this row reduction methodology to any **n**-by-**n** matrix, allows us to express the **n**-by-**n** matrix as a product of elementary matrices and a diagonal matrix. This approach yields the corollary below.

Corollary 4.3.3 If A is an **n**-by-**n** non-singular matrix and B is an arbitrary **n**-by-**n** matrix, we have the following relationship:

$$|A \cdot B| = |B \cdot A| = |A| \cdot |B|.$$

Proof: From corollary 2.x.x we can express the non-singular matrix A as a finite product of elementary matrices. Hence we have

Corollary 4.3.4 If A is an **n**-by-**n** singular matrix and B is an arbitrary **n**-by-**n** matrix, we have the following relationship:

$|A \cdot B| = |B \cdot A| = |A| \cdot |B|$. Furthermore the resultant product is a singular matrix with

$$|A \cdot B| = 0.$$

Proof: First we shall expand the matrix A as a product of elementary matrices. By definition 2.4.8, the singular matrix A is row equivalent to a matrix that is in row echelon form having at least one row containing all zero elements. Hence we have $A = \left(\prod_{i=1}^{n} E_i\right) \cdot A_0$, where the matrix A_0 is a matrix in row echelon form having at least one row comprised of all zero elements. Using this representation of A and applying corollary 4.3.1 and 4.3.2, we have $|A| = \left(\prod_{i=1}^{n} |E_i|\right) \cdot |A_0| = 0$.

Since the product of the matrix A and B can be expressed as follows: $\left(\prod_{i=1}^{n} E_i\right) \cdot A_0 \cdot B$. Applying the associative law of matrix multiplication we can rewrite this product as $\left(\prod_{i=1}^{n} E_i\right) \cdot (A_0 \cdot B)$.

Note that if row '**r**' of matrix A_0 contains all zero elements, then the product $A_0 \cdot B$ forms a matrix with row '**r**' containing all zero elements. Furthermore this implies that the product $A \cdot B$ is singular, since it is equivalent to a matrix that's in row echelon form with at least one row containing all zero elements. From corollary 4.3.2, we have $|A_0 \cdot B| = 0$ and since $|A_0| = 0$ we have the following relationship: $|A_0 \cdot B| = |A_0| \cdot |B| = 0$. Applying corollary 4.3.1, we can express the determinant of the product as $|A \cdot B| = \left(\prod_{i=1}^{n} |E_i|\right) \cdot |A_0 \cdot B|$. By expanding the right-most determinant we can rewrite this relationship as $|A \cdot B| = \left(\prod_{i=1}^{n} |E_i|\right) \cdot |A_0| \cdot |B| = |A| \cdot |B|$. Since the determinant of A is zero, we have $|A \cdot B| = 0$.

During the proof of corollary 4.3.4, we showed that a singular matrix has a determinant of zero. Conversely, we now ask the question "If a matrix has a determinant of zero, is it a singular matrix?" If a matrix has a determinant of zero, when we express that matrix as the product of elementary matrices and a diagonal matrix, then the determinant of the matrix is the product of the determinants of the elementary matrices and the determinant of the diagonal matrix. The determinants of the elementary matrices are nonzero. (We always select the type II elementary matrix such that $c \neq 0$.) Hence for the determinant of the matrix to be zero, we must have that the determinant of the diagonal matrix is zero. For the determinant of a diagonal matrix to be zero, we must have at least one of the diagonal elements is equal to zero. When one of the diagonal elements of a diagonal entries is zero, that row corresponding to the zero element along the diagonal has all zero elements. Hence we have shown that the matrix with a determinant of zero is equivalent to a matrix in row echelon form containing at least one row with all zero elements. Therefore a matrix with a zero determinant value is a singular matrix.

A property of interest that is an extension of non-singular matrices is that of determinant of an inverse of a matrix. This property is a result of the corollary below.

Corollary 4.3.5 Let $\boldsymbol{A}$ be a non-singular matrix, then

$$|\boldsymbol{A}^{-1}| = 1/|\boldsymbol{A}| \quad .$$

Proof: Since $\boldsymbol{A}$ is non-singular there exists $\boldsymbol{A}^{-1}$ such that $\boldsymbol{A} \cdot \boldsymbol{A}^{-1} = \boldsymbol{I}$. By applying corollary 4.3.3, we get $|\boldsymbol{A} \cdot \boldsymbol{A}^{-1}| = |\boldsymbol{A}| \cdot |\boldsymbol{A}^{-1}| = |\boldsymbol{I}| = 1$. Hence we have $|\boldsymbol{A}^{-1}| = 1/|\boldsymbol{A}|$.

Before completing this section, we introduce another special matrix of interest. This special matrix is best described using an example. Consider the matrix having all elements in the matrix equal to a value of one; $\boldsymbol{A} = [1]$. If $\boldsymbol{A}$ is a 3-by-3 matrix we denote it as $\boldsymbol{A} = \begin{bmatrix} 1 & 1 & 1 \\ 1 & 1 & 1 \\ 1 & 1 & 1 \end{bmatrix}$. If we multiply $\boldsymbol{A}$ times itself, we get $\boldsymbol{A}^2 = \begin{bmatrix} 3 & 3 & 3 \\ 3 & 3 & 3 \\ 3 & 3 & 3 \end{bmatrix}$. Upon closer inspection of this relationship, we can write

$A^2 = 3 \cdot A$. This matrix is called an idempotent matrix. The formal definition for an idempotent matrix is shown below.

Definition 4.3.6 A matrix A is an idempotent matrix if there is a nonzero value '**c**' such that $A^2 = c \cdot A$.

To compute the determinant for an idempotent matrix, we use the property of the idempotent matrix described in the above definition. The derivation is described in the corollary below.

Corollary 4.3.7 Let A be an **n-by-n** idempotent matrix such that $A^2 = c \cdot A$, then we have either $|A| = 0$ (for a singular matrix) or $|A| = c^n$ (for a non-singular matrix).

Proof: Using the relation given in the statement of this corollary and applying corollaries 4.3.1, 4.3.3 and 4.3.4 we get $|A^2| = |c \cdot A|$ which translates to $|A|^2 = c^n \cdot |A|$. Upon subtracting the left term of this equality from both sides of the equal sign, we get $|A|^2 - (c^n \cdot |A|) = 0$. Upon factoring out common terms in the above equation we get the following equation:

$|A| \cdot (|A| - c^n) = 0$. For this equation to be true we must have either $|A| = 0$ or $|A| = c^n$.

Exercises

Using elementary matrices, reduce each matrix in the exercises below to a triangular matrix (All elements above diagonal are equal to zero or all elements below diagonal are equal to zero.). After reducing the matrix to the triangular form, determine the determinant of each matrix.

1. $\begin{vmatrix} 3 & 0 & -1 \\ 6 & 4 & 3 \\ 9 & 7 & 8 \end{vmatrix}$

2. $\begin{vmatrix} 1 & 1 & 9 \\ 0 & 5 & -2 \\ -4 & 6 & -25 \end{vmatrix}$

3. $\begin{vmatrix} 0 & 3 & 9 & 21 \\ 1 & 2 & 4 & 5 \\ 0 & 0 & 0 & 6 \\ 0 & 0 & -4 & 1 \end{vmatrix}$

4. $\begin{vmatrix} 0 & 0 & 4 & 0 \\ 2 & 4 & 0 & 0 \\ 0 & 1 & 1 & 1 \\ 0 & 0 & 3 & 1 \end{vmatrix}$

5. $\begin{vmatrix} 1 & 2 & 4 & 8 & 16 \\ 2 & 1 & 2 & 4 & 8 \\ 4 & 2 & 1 & 2 & 4 \\ 8 & 4 & 2 & 1 & 2 \\ 16 & 8 & 4 & 2 & 1 \end{vmatrix}$

6. $\begin{vmatrix} 5 & -2 & 0 & 1 & -3 \\ 0 & -1 & -2 & 2 & -2 \\ 3 & 3 & 4 & 1 & 3 \\ 1 & -5 & -3 & 2 & 2 \\ 2 & 8 & -4 & 1 & -3 \end{vmatrix}$

7. $\begin{vmatrix} 0 & 9 & 27 & 63 \\ 1 & 2 & 4 & 5 \\ 0 & 0 & 0 & 6 \\ 0 & 0 & -4 & 1 \end{vmatrix}$

8. $\begin{vmatrix} 9 & 4 & 2 \\ 6 & 4 & 3 \\ 15 & 11 & 11 \end{vmatrix}$

9. $\begin{vmatrix} 1 & 3 & 9 & 27 & 81 \\ 3 & 1 & 3 & 9 & 27 \\ 9 & 3 & 1 & 3 & 9 \\ 27 & 9 & 3 & 1 & 3 \\ 81 & 27 & 9 & 3 & 1 \end{vmatrix}$

10. $\begin{vmatrix} 1 & 5 & 0 & 0 \\ 5 & 1 & 5 & 0 \\ 0 & 5 & 1 & 5 \\ 0 & 0 & 5 & 1 \end{vmatrix}$

4.4 Determinants Using Minor Matrices

In this section, we revisit the definition of a determinant of a matrix. The primary goal, of this revisiting, is to simplify the expression for computing determinants. Restating the expression for the determinant of a matrix A, we have $|A| = \sum_{\underset{\sim}{J} \in P} \delta(\underset{\sim}{J}) \cdot \prod_{k=1}^{n} a_{k, i_k}$. If we introduce subsets of '**P**' (we shall call each subset $\mathbf{P_k}$) such that the initial number in the permutation is '**k**', then these subsets are disjoint subsets. If '**P**' represents all permutations of the integers one through '**n**', then we have $P = P_1 \cup \ldots \cup P_n$. Using this notation and the fact that the subsets are disjoint, we can rewrite the equation for the determinant of A as follows: $|A| = \sum_{i=1}^{n} \left(\sum_{\underset{\sim}{J} \in P_i} \delta(\underset{\sim}{J}) \cdot \prod_{k=1}^{n} a_{k, i_k} \right)$. By definition of a permutation in the subset P_i, the product term associated with a permutation of this subset can be expressed in the form $a_{1, i} \cdot \prod_{k=2}^{n} a_{k, i_k}$. Using this expression, we now define the determinant of A to be

$$|A| = \sum_{i=1}^{n} \left(a_{1, i} \cdot \sum_{\underset{\sim}{J} \in P_i} \delta(\underset{\sim}{J}) \cdot \prod_{k=2}^{n} a_{k, i_k} \right).$$

Before we continue simplifying the expression for a determinant, we shall introduce another property of the δ function introduced in the introductory section of this chapter. We begin by introducing a subscript for the function to indicate the number of integers involved in the permutations associated with the function. Specifically, the δ_n function, indicates the sign of the number of inversions associated with the permutation of '**n**' distinct integers. When the permutation $\underset{\sim}{J} \in P_i$ then by definition it has the form $(i, j_2, \ldots, j_n)$. Now we examine the relationship of $\delta_n(i, j_2, \ldots, j_n)$ to $\delta_{n-1}(j_2, \ldots, j_n)$. If there are I_{n-1} inversions associated with the permutation $\underset{\sim}{J} = (j_2, \ldots, j_n)$, then add '**i-1**' inversions by creating the permutation $(i, j_2, \ldots, j_n)$ (by placing the omitted value '**i**' in front). Hence we have $\delta_n(i, j_2, \ldots, j_n) = (-1)^{I_{n-1} + i - 1}$

or $\delta_n(i, j_2, \ldots, j_n) = \left((-1)^{i-1}\right) \cdot \delta_{n-1}(\underset{\sim}{J})$.

Substituting for $\delta_n(i, j_2, \ldots, j_n)$ in terms of $\delta_{n-1}(j_2, \ldots, j_n)$, we rewrite the expression for the determinant of A as

$$|A| = \sum_{i=1}^{n} \left(a_{1,i} \cdot \sum_{\underset{\sim}{J} \in P_i} \left((-1)^{i-1}\right) \cdot \delta_{n-1}(\underset{\sim}{J}) \cdot \prod_{k=2}^{n} a_{k, i_k} \right)$$

. Using the distributive property of numbers, and factoring out the $\left((-1)^{i-1}\right)$ term from within the inner summation, we now have the expression

$$|A| = \sum_{i=1}^{n} \left(\left((-1)^{i-1}\right) \cdot a_{1,i} \cdot \sum_{\underset{\sim}{J} \in P_i} \delta_{n-1}(\underset{\sim}{J}) \cdot \prod_{k=2}^{n} a_{k, i_k} \right).$$

Upon closer inspection of the inner summation, it is determined that it is the sub-matrix within A derived by eliminating the first row and the ith column of A. This matrix is formally defined below as a minor matrix of A.

Definition 4.4.1 If A is an **n-by-n** matrix, then the **(n-1)-by-(n-1)** matrix derived by eliminating the ith row and j^{th} column is called a minor matrix of A and is denoted as $A_{i,j}$.

With this notation for minor matrices, we note that the inner summation is the definition of determinant for the minor matrix $A_{1,\,i}$. Hence the determinant for A is given by the expression $|A| = \sum_{i=1}^{n} \left((-1)^{i-1}\right) \cdot a_{1,\,i} \cdot |A_{1,\,i}|$.

To illustrate an example of this simplified expression for the determinant of a matrix, let's consider the matrix $A = \begin{bmatrix} 1 & 0 & 5 \\ 2 & 3 & 1 \\ 4 & 1 & 7 \end{bmatrix}$. Computing the determinant of A using minor matrices yield the following:

$$\begin{vmatrix} 1 & 0 & 5 \\ 2 & 3 & 1 \\ 4 & 1 & 7 \end{vmatrix} = 1 \cdot \begin{bmatrix} 3 & 1 \\ 1 & 7 \end{bmatrix} - 0 \cdot \begin{bmatrix} 2 & 1 \\ 4 & 7 \end{bmatrix} + 5 \cdot \begin{bmatrix} 2 & 3 \\ 4 & 1 \end{bmatrix} = 1 \cdot (20) - 0 \cdot (10) + 5 \cdot (-10) = -30 .$$

So far we have shown that the determinant of a matrix can be derived from the minor matrices that are formed by eliminating the first row and selected columns of the matrix. Now we prove that the determinant for a matrix can be derived from its minor matrices formed using any arbitrary row of the matrix. The general form of this property is stated in the corollary below.

Corollary 4.4.2 Let A be an **n-by-n** square matrix, then the determinant of the matrix can be expressed as follows: $|A| = \sum_{i=1}^{n} \left((-1)^{i+k}\right) \cdot a_{k,\,i} \cdot |A_{k,\,i}|$.

Proof: To prove the statement of this corollary, we shall transform this matrix A by moving the kth row of this matrix to the first row of the matrix to form a matrix B. To accomplish this transformation we multiply $\boldsymbol{A}$ by the appropriate number of type I matrices. If the matrix $\boldsymbol{E}_{r \leftrightarrow s}$ exchanges rows '**r**' and '**s**' when multiplied to any matrix. Hence if we compute the matrix $\boldsymbol{B} = \boldsymbol{E}_{1 \leftrightarrow 2} \cdot \ldots \cdot \boldsymbol{E}_{k-1 \leftrightarrow k} \cdot \boldsymbol{A}$, it is formed from the original matrix by moving the kth row to row one. This is easily visualize since the **k-1** type I elementary matrices shifts adjacent rows beginning with the kth row until the kth row is in the first row position. Hence if we compute the determinant of the matrix $\boldsymbol{B}$ we have $|\boldsymbol{B}| = \sum_{i=1}^{n} \left((-1)^{i-1}\right) \cdot b_{1,i} \cdot |\boldsymbol{B}_{1,i}|$. Since the matrix $\boldsymbol{B}$ is a transformation of matrix $\boldsymbol{A}$ by moving the kth row to the first row we have $b_{1,i} = a_{k,i}$ and $\boldsymbol{B}_{1,i} = \boldsymbol{A}_{k,i}$. Using these relationships to substitute for the elements of $\boldsymbol{B}$, we can rewrite this determinant as $|\boldsymbol{B}| = \sum_{i=1}^{n} \left((-1)^{i-1}\right) \cdot a_{k,i} \cdot |\boldsymbol{A}_{k,i}|$.

Using corollary 4.3.1 and the fact that the determinant of a type I matrix has a value of -1, we also have $|\boldsymbol{B}| = \left((-1)^{k-1}\right) \cdot |\boldsymbol{A}|$. Using the previous derivation for the determinant of $\boldsymbol{B}$, we have $\left((-1)^{k-1}\right) \cdot |\boldsymbol{A}| = \sum_{i=1}^{n} \left((-1)^{i-1}\right) \cdot a_{k,i} \cdot |\boldsymbol{A}_{k,i}|$. If we solve for the value of $|\boldsymbol{A}|$, we get $|\boldsymbol{A}| = \sum_{i=1}^{n} \left((-1)^{i+k}\right) \cdot a_{k,i} \cdot |\boldsymbol{A}_{k,i}|$.

Hence the determinant of a matrix can be computed by expanding the minor matrices when eliminating the **k**th row.

A similar relationship for the determinant is computed when we use columns instead of the rows of the matrix. The formula when using the **k**th column is the following: $|A| = \sum_{i=1}^{n} \left((-1)^{i+k}\right) \cdot a_{i,k} \cdot |A_{i,k}|$.

Exercises

For this 4-by-4 matrix $A = \begin{bmatrix} 1 & -2 & -4 & 6 \\ -4 & 0 & 2 & 4 \\ 5 & 1 & 0 & 5 \\ 7 & -3 & -6 & 3 \end{bmatrix}$, compute the following listed in the exercises below.

1. Compute $|A_{(1,1)}|$.
2. Compute $|A_{(2,2)}|$.
3. Compute $|A_{(1,2)}|$.
4. Compute $|A_{(1,3)}|$.
5. Compute $|A_{(1,4)}|$.
6. Compute $|A|$.

For this 3-by-3 square matrix $A = \begin{bmatrix} 6 & -1 & 6 \\ 1 & 4 & 0 \\ 2 & 2 & 2 \end{bmatrix}$ compute the following:

7. Compute $|A_{(1,1)}|$.

8. Compute $|A_{(1,2)}|$.

9. Compute $|A_{(1,3)}|$.

10. Compute $|A|$.

4.5 Matrix Inverse Using Minor Matrices/Cramer's Rule

We computed the inverse of a non-singular matrix, in chapter 2, by multiplying the original matrix and the identity matrix by elementary matrices until the original matrix is converted to the identity matrix and the identity matrix is converted to its inverse (Gauss-Jordan Reduction Method). The mathematical expression for this approach is the following: $\left(\prod_{i=1}^{k} \boldsymbol{E}_i\right) \cdot \boldsymbol{A} = \boldsymbol{I}$. In this section, we will use the formulas and properties of determinants to derive an alternate approach for computing the inverse of a matrix. When we apply the Gauss-Jordan Reduction Method to produce the matrix $\boldsymbol{B} = \left(\prod_{i=1}^{k} \boldsymbol{E}_i\right)$, If this matrix is produced using a row reduction technique, will it be equal to the matrix that is produced using a column reduction method? To answer this question, we introduce some corollaries that address properties of non-singular and singular matrices.

According to the definition of an inverse of a matrix the inverse of $\boldsymbol{A}$ is a matrix $\boldsymbol{A}^{-1}$ such that $\boldsymbol{A} \cdot \boldsymbol{A}^{-1} = \boldsymbol{A}^{-1} \cdot \boldsymbol{A} = \boldsymbol{I}$ where $\boldsymbol{I}$ is the identity matrix. In chapter 2, we accepted without proof that if we found a matrix $\boldsymbol{B}$ such that $\boldsymbol{B} \cdot \boldsymbol{A} = \boldsymbol{I}$, then $\boldsymbol{B} = \boldsymbol{A}^{-1}$. So now we ask the question, If we determine an inverse of a matrix through a row elimination method, will the same matrix result if we then determine an inverse through column elimination method? The answer to this question is yes as will be illustrated below. The first corollary to be defined to help support the proof for this statement deals with the determinant of a singular matrix.

Corollary 4.5.1 If $\boldsymbol{A}$ is a matrix such that row '**r**' (denoted as $\underset{\sim}{A}'_{r}$) is a linear combination of other rows of the matrix, $\underset{\sim}{A}'_r = \sum_{i \neq r} c_i \cdot \underset{\sim}{A}'_i$, then A is a singular matrix and $|A| = 0$.

Proof: To begin this proof, we re-introduce the notation of a type III elementary matrix; $E_{c_i \cdot i + r}$ that multiplies the ith row times the value c_i and adds this to row '**r**'. Hence the product, $\prod_{i \neq r} E_{-c_i \cdot i + r}$, when multiplied to the matrix A, subtracts from the **r**th row of this matrix the following: $\sum_{i \neq r} -c_i \cdot \underset{\sim}{A}'_i$. Hence if the **r**th row of the matrix is given by the following relationship $\underset{\sim}{A}'_r = \sum_{i \neq r} c_i \cdot \underset{\sim}{A}'_i$, then the resultant matrix $\left(\prod_{i \neq r} E_{-c_i \cdot i + r}\right) \cdot A$ has an **r**th row that is a zero row vector. Hence we have shown that the matrix A is row equivalent to a matrix containing at least one zero row vector. Applying definition 2.4.8, we conclude that $|A|$ is singular and by corollary 4.3.2, we get $\left|\left(\prod_{i \neq r} E_{-c_i \cdot i + r}\right) \cdot A\right| = 0$. Since the determinant of a type III matrix is one and the sequential application of corollary 4.3.1 we get

$$\left|\left(\prod_{i \neq r} E_{-c_i \cdot i + r}\right) \cdot A\right| = \left|\left(\prod_{i \neq r} E_{-c_i \cdot i + r}\right)\right| \cdot |A| = |A| = 0.$$

The above corollary specifically addresses dependent rows in the matrix. A similar corollary is stated below for dependent columns in the matrix.

Corollary 4.5.2 If A is a matrix such that column '**r**' (denoted as $\underset{\sim}{A}_r$) is a linear combination of other columns of the matrix, $\underset{\sim}{A}_r = \sum_{i \neq r} c_i \cdot \underset{\sim}{A}_i$, then A is a singular matrix and $|A| = 0$.

Proof: The proof for this corollary is similar to the proof for corollary 4.5.1 except that post multiplication by elementary matrices is used in this instance. The details of this proof will be left to the reader as an exercise.

A special case of corollary 4.5.2 (or 4.5.3) is the case in which two distinct rows (or columns) of a matrix are equivalent. From the above corollaries we can conclude that the determinant of such a matrix has a zero value. From corollary 4.4.2 we have the formula $|A| = \sum_{i=1}^{n} \left((-1)^{i+k}\right) \cdot a_{k,i} \cdot |A_{k,i}|$. If we substitute for the $\mathbf{k}^{\text{th}}$ row, the $\mathbf{r}^{\text{th}}$ row where $r \neq k$, then the resultant matrix has the $\mathbf{k}^{\text{th}}$ row and the $\mathbf{r}^{\text{th}}$ row being equivalent. In this case we have

$$\sum_{i=1}^{n} \left((-1)^{i+k}\right) \cdot a_{r,i} \cdot |A_{k,i}| = 0 \text{ for all } r \neq k.$$

The converse of corollary 4.5.1 states that a matrix with a determinant of zero is singular and that at least one of the rows of the matrix is a linear combination of other rows of the matrix. Similarly, the converse of corollary 4.5.2 states that a matrix with a determinant of zero is singular and at least one of its columns is a linear combination of other columns of the matrix.

Corollary 4.5.3 If A is a matrix such that $|A| = 0$, then A is a singular matrix and there is a row $\underset{\sim}{A}'_r$ of the matrix such that it is linear combination of other rows in the matrix. Also there is a is a column $\underset{\sim}{A}_s$ of the matrix such that it is linear combination of other columns in the matrix.

Proof: This corollary will be proved for rows and then columns.

(Dependent Rows) To show that one row is a linear combination of other rows in the matrix, it is sufficient for us to show that there are a sequence of real values c_i such that $\sum_{i=1}^{n} c_i \cdot \underset{\sim}{A}'_i = 0$; where $\underset{\sim}{A}'_i$ represents the $\mathbf{i}^{\text{th}}$ row of the matrix A. We begin this proof by applying the formula for a determinant using minor matrices of A. Using the formula for determinants by expanding the minor matrices along a column we get the following equation $|A| = \sum_{i=1}^{n} \left((-1)^{i+k}\right) \cdot a_{i,k} \cdot |A_{i,k}| = 0$. As shown previously in this section we also have $\sum_{i=1}^{n} \left((-1)^{i+k}\right) \cdot a_{r,k} \cdot |A_{i,k}| = 0$ for all $r \neq k$. If A is not the zero matrix then there is a column of the matrix such that at least one of determinants of the minor matrices is not zero. Assume this column corresponds to the $\mathbf{k}^{\text{th}}$ column of the matrix. Then we have the following sequence of equations:

$$\sum_{i=1}^{n} \left((-1)^{i+k}\right) \cdot a_{i,1} \cdot |A_{i,k}| = 0 \qquad ,$$

$$\sum_{i=1}^{n} \left((-1)^{i+k}\right) \cdot a_{i,2} \cdot |A_{i,k}| = 0 \qquad ,$$

•

•

•

$$\sum_{i=1}^{n} \left((-1)^{i+k}\right) \cdot a_{i,n} \cdot |A_{i,k}| = 0 \quad .$$

In matrix notation, this is written as $\underset{\sim}{B}' \cdot A = \underset{\sim}{0}'$, where $\underset{\sim}{B}'$ is a row matrix such that $\underset{\sim}{B}' = [b_i]$ where $b_i = \left((-1)^{i+k}\right) \cdot |A_{i,k}|$ and i = 1,...,**n**.

For the rows of the matrix A, we have $\sum_{i=1}^{n} b_i \cdot \underset{\sim}{A}_i' = \underset{\sim}{0}'$ and since '**k**' is chosen such that at least one of the coefficients is nonzero, we can write one of the rows as a linear combination of the remaining rows of the matrix.

(Dependent Columns) To show that one column is a linear combination of other columns in the matrix, it is sufficient for us to show that there are a sequence of real values c_i such that $\sum_{i=1}^{n} c_i \cdot A_i = \underset{\sim}{0}$; where A_i represents the **i**th column of the matrix A. We begin this proof by applying the formula for a determinant using minor matrices of A. Using the formula for determinants by expanding the minor matrices along a row we get the following equation $|A| = \sum_{i=1}^{n} \left((-1)^{i+k}\right) \cdot a_{k,i} \cdot |A_{k,i}| = 0$. As shown previously in this section we also have $\sum_{i=1}^{n} \left((-1)^{i+k}\right) \cdot a_{r,i} \cdot |A_{k,i}| = 0$ for all $r \neq k$. If A is not the zero matrix then there is a row of the matrix such that at least one of determinants of the minor matrices is not zero. Assume this row corresponds to the **k**th row of the matrix. Then we have the following sequence of equations:

$$\sum_{i=1}^{n} \left((-1)^{i+k}\right) \cdot a_{1,i} \cdot |A_{k,i}| = 0 \quad ,$$

$$\sum_{i=1}^{n} \left((-1)^{i+k}\right) \cdot a_{2,i} \cdot |A_{k,i}| = 0 \quad ,$$

•

•

•

$$\sum_{i=1}^{n} \left((-1)^{i+k}\right) \cdot a_{n,i} \cdot |A_{k,i}| = 0 \quad .$$

In matrix notation, this is written as $\boldsymbol{A} \cdot \underset{\sim}{\boldsymbol{B}} = \underset{\sim}{\boldsymbol{0}}$, where $\underset{\sim}{\boldsymbol{B}}$ is a row matrix such that $\underset{\sim}{\boldsymbol{B}} = [\boldsymbol{b}_i]$ where $\boldsymbol{b}_i = \left((-1)^{i+k}\right) \cdot |A_{k,i}|$ and i = 1,...,**n**.

For the columns of the matrix $\boldsymbol{A}$, we have $\sum_{i=1}^{n} \boldsymbol{b}_i \cdot \underset{\sim}{\boldsymbol{A}}_i = \underset{\sim}{\boldsymbol{0}}$ and since '**k**' is chosen such that at least one of the coefficients is nonzero, we can write one of the columns as a linear combination of the remaining columns of the matrix.

Having proved that at least one row (or column) is a linear combination of other rows (or columns) of the matrix, we can multiply the matrix by a series of type III elementary matrices to show that the matrix is row (or column) equivalent to a matrix with at least one row (or column) containing all zero elements. Hence the matrix is singular and we have proven the statement of the corollary to be true.

As we proceed with the concept of matrix inverses, we recall the definition of an inverse of the matrix $\boldsymbol{A}$ is a matrix denoted as $\boldsymbol{A}^{-1}$ such that $\boldsymbol{A}^{-1} \cdot \boldsymbol{A} = \boldsymbol{A} \cdot \boldsymbol{A}^{-1} = \boldsymbol{I}$. The question arising from this definition is the following: If we find a matrix $\boldsymbol{M}$ such that $\boldsymbol{M} \cdot \boldsymbol{A} = \boldsymbol{I}$, is this sufficient to declare that it is an inverse of the matrix $\boldsymbol{A}$? The answer to this question is "yes". Using corollaries 4.5.1, 4.5.2 and 4.5.3 we prove corollary 4.5.4 below.

Corollary 4.5.4 If $\boldsymbol{M}$ is a non-singular matrix such that $\boldsymbol{M} \cdot \boldsymbol{A} = \boldsymbol{I}$ then it is the unique inverse of the matrix $\boldsymbol{A}$. Mathematically we state

$$\boldsymbol{A}^{-1} = \boldsymbol{M}.$$

Proof: The proof for this corollary follows from the following steps:

1. Let $\boldsymbol{M}$ be a non-singular matrix such that $\boldsymbol{M} \cdot \boldsymbol{A} = \boldsymbol{I}$. This is a restatement of the statement of this corollary.

2. If we multiply both sides of the equality given in step 1 by the matrix $\boldsymbol{A}$, we get the following relationship $\boldsymbol{A} \cdot \boldsymbol{M} \cdot \boldsymbol{A} = \boldsymbol{A}$ or $(\boldsymbol{A} \cdot \boldsymbol{M} - \boldsymbol{I}) \cdot \boldsymbol{A} = \boldsymbol{0}$.

3. If the matrix $(\boldsymbol{A} \cdot \boldsymbol{M} - \boldsymbol{I})$ is not the zero matrix, then one column of the matrix $\boldsymbol{A}$ is a linear combination of other columns of the matrix. By corollary 4.5.2 the matrix $\boldsymbol{A}$ is singular and $|\boldsymbol{A}| = 0$. This creates a contradiction since $|\boldsymbol{M}| \cdot |\boldsymbol{A}| = 1$. Therefore we must have $\boldsymbol{A} \cdot \boldsymbol{M} - \boldsymbol{I} = 0$ and hence $\boldsymbol{A} \cdot \boldsymbol{M} = \boldsymbol{I}$.

4. Since we have a matrix $\boldsymbol{M}$ such that $\boldsymbol{M} \cdot \boldsymbol{A} = \boldsymbol{A} \cdot \boldsymbol{M} = \boldsymbol{I}$ then $\boldsymbol{M}$ is the inverse matrix of $\boldsymbol{A}$.

Some terms that will prove useful in the remainder of this section are coveters, cofactor matrix and adjoint matrix. The formal definition of these terms are given below.

Definition 4.5.5 A cofactor of the matrix $\boldsymbol{A}$ is the term

$$c_{i,j} = \left((-1)^{i+j}\right) \cdot |A_{i,j}| \quad .$$

Definition 4.5.6 The cofactor matrix of the matrix $\boldsymbol{A}$ is the matrix with elements corresponding to the cofactors of $\boldsymbol{A}$. Functionally we write the cofactor matrix to be $\boldsymbol{C}$ such that $\boldsymbol{C} = [c_{i,j}]$.

Definition 4.5.7 The Adjoint matrix of $\boldsymbol{A}$ is denoted as $Adj(\boldsymbol{A})$ and is defined to be the transpose of the cofactor matrix of $\boldsymbol{A}$. In matrix notation we write $Adj(\boldsymbol{A}) = \boldsymbol{C}'$.

As stated in the beginning of this section, we will now use the properties of determinants to show an alternate method of computing matrix inverses using minor matrices. Let the matrix $\boldsymbol{A}$ be a non-singular matrix; then we have $|A| = \sum_{i=1}^{n} \left((-1)^{i+k}\right) \cdot a_{i,k} \cdot |A_{i,k}|$. Now we create a series of matrices $\boldsymbol{A}_{k=j}$, in which the **k**th column is replaced with the **j**th column of the matrix. This implies that $\boldsymbol{A}_{k=k} = \boldsymbol{A}$. When $k \neq j$, we get $|\boldsymbol{A}_{k=j}| = 0$ using corollary 4.5.2. This is true since, in this instance, column '**k**' can be considered a linear combination of row '**j**'. When $k = j$ we have $|\boldsymbol{A}_{k=k}| = |\boldsymbol{A}|$. If we define a matrix

$\boldsymbol{M}$ where $\boldsymbol{M} = [m_{i,j}]$ and $m_{i,j} = ((-1)^{i+j} \cdot |A_{j,i}|) / |\boldsymbol{A}|$, then in matrix notation we have that $\boldsymbol{M} \cdot \boldsymbol{A} = \boldsymbol{I}$. Applying corollary 4.5.4, the matrix $\boldsymbol{M}$ is an inverse of the matrix $\boldsymbol{A}$.

We formally state this as the following theorem.

Theorem 4.5.8 Let $\boldsymbol{A}$ be a non-singular matrix then the inverse of the matrix can be expressed in terms of the determinant of the matrix and the determinants of the minor matrices of the matrix. Functionally the inverse of the matrix is defined to be $\boldsymbol{A}^{-1} = [((-1)^{i+j} \cdot |A_{j,i}|) / |\boldsymbol{A}|]$, **i** represents the row index and **j** represents the column index. In terms of the adjoint matrix of $\boldsymbol{A}$ the inverse is written as follows:

$$\boldsymbol{A}^{-1} = Adj(\boldsymbol{A}) / |\boldsymbol{A}|.$$

Examples of computing matrix inverses using minor matrices is shown below.

With two approaches for computing the inverse of a non-singular matrix (gaussian elimination and determinants of minor matrices) we revisit the task of solving linear independent equations. We shall investigate the use of determinants in solving **n** independent linear equations with "**n**" unknown variables.

We start with the matrix notation in the form $\boldsymbol{A} \cdot \underset{\sim}{X} = \underset{\sim}{b}$. For this to be **n** independent equations, the matrix $\boldsymbol{A} = [a_{i,j}]$ must be an **n-by-n** non-singular matrix. The column matrix $\underset{\sim}{\boldsymbol{X}} = \begin{bmatrix} x_1 \\ x_2 \\ \circ \\ \circ \\ x_n \end{bmatrix}$ has **n** unknown variables and is an element of an n-dimensional vector space. The matrix $\underset{\sim}{\boldsymbol{b}} = \begin{bmatrix} b_1 \\ b_2 \\ \circ \\ \circ \\ b_n \end{bmatrix}$ is a column matrix of **n** known constant values.

Traditionally such an equation would be written in the following form:

$$a_{1,1} \cdot x_1 + a_{1,2} \cdot x_2 + \ldots + a_{1,n} \cdot x_n = b_1$$
$$a_{2,1} \cdot x_1 + a_{2,2} \cdot x_2 + \ldots + a_{2,n} \cdot x_n = b_2$$
$$\circ$$
$$\circ$$
$$a_{n,1} \cdot x_1 + a_{n,2} \cdot x_2 + \ldots + a_{n,n} \cdot x_n = b_n$$

The application of the adjoint matrix approach for computing the inverse of a matrix leads to the theorem below. This theorem is well known throughout the study of matrix algebra and is referred to as Cramer's Rule.

Theorem 4.5.9 (Cramer's Rule) Let the equation denoted by the matrix equation $A \cdot \underset{\sim}{X} = \underset{\sim}{b}$ be a system of "**n**" independent linear equations. Then A is non-singular and the solution of the "**n**" unknown variables of the column matrix $\underset{\sim}{X}$ is given using the following "**n**" determinants:

$$x_1 = \begin{bmatrix} b_1 & a_{1,2} & \cdots & a_{1,n} \\ b_2 & a_{2,2} & \cdots & a_{2,n} \\ \triangledown & \triangledown & \cdots & \triangledown \\ b_n & a_{n,2} & \cdots & a_{n,n} \end{bmatrix} / |A| \ ,$$

$$x_2 = \begin{bmatrix} a_{1,1} & b_1 & \cdots & a_{1,n} \\ a_{2,1} & b_2 & \cdots & a_{2,n} \\ \triangledown & \triangledown & \cdots & \triangledown \\ a_{n,1} & b_n & \cdots & a_{n,n} \end{bmatrix} / |A| \ ,$$

$$\cdots\cdots \ x_n = \begin{bmatrix} a_{1,1} & a_{1,2} & \cdots & b_1 \\ a_{2,1} & a_{2,2} & \cdots & b_2 \\ \triangledown & \triangledown & \cdots & \triangledown \\ a_{n,1} & a_{n,2} & \cdots & b_n \end{bmatrix} / |A| \ .$$

Proof: We begin with a system of "**n**" independent linear equations, for the variables $\underset{\sim}{X} = \begin{bmatrix} x_1 \\ x_2 \\ \circ \\ \circ \\ x_n \end{bmatrix}$, denoted by the linear equation

$A \cdot \underset{\sim}{X} = \underset{\sim}{b}$ where A is a non-singular matrix. The solution for the $\underset{\sim}{X}$ column matrix is defined to be $\underset{\sim}{X} = A^{-1} \cdot \underset{\sim}{b}$. From corollary 4.5.8, we can express the inverse in terms of the adjoint matrix of A. Using this corollary, we rewrite the solution of the equation as $\underset{\sim}{X} = (A_{dj}(A) \cdot \underset{\sim}{b}) / |A|$. If we expand the right side of this equation, using definition of adjoint matrix, we have the following family of equations for the elements of $\underset{\sim}{X}$:

$$x_k = \sum_{i=1}^{n} (-1)^{i+k} \cdot b_i \cdot |A_{i,k}| / |A| \quad \text{for k} = 1, \ldots, \mathbf{n}.$$

For the kth variable in the above equation, the right side of equation is the determinant of the matrix that is derived from the matrix A by replacing the elements of the kth column with the corresponding elements of the constant matrix $\underset{\sim}{b}$. Hence we get the following expression: $x_k = \begin{bmatrix} a_{1,1} & a_{1,2} & \cdots & b_1 & \cdots & a_{1,n} \\ a_{2,1} & a_{2,2} & \cdots & b_2 & \cdots & a_{2,n} \\ \nabla & \nabla & \cdots & \nabla & \cdots & \nabla \\ a_{n,1} & a_{n,2} & \cdots & b_n & \cdots & a_{n,n} \end{bmatrix} / |A|$

and the statement of the theorem is proved true.

(Example - Applying Cramer's Rule)

Given the following three equations: $\left\{\begin{matrix} 2\cdot x+4\cdot y+z=11 \\ 5\cdot x+2\cdot y-z=8 \\ 3\cdot x+2\cdot y+4\cdot z=11 \end{matrix}\right\}$. Representing these equations using matrix notation we get $\begin{bmatrix} 2 & 4 & 1 \\ 5 & 2 & -1 \\ 3 & 2 & 4 \end{bmatrix} \times \begin{bmatrix} x \\ y \\ z \end{bmatrix} = \begin{bmatrix} 11 \\ 8 \\ 11 \end{bmatrix}$.

Applying Cramer's Rule we get

$$x = \left|\begin{matrix} 11 & 4 & 1 \\ 8 & 2 & -1 \\ 11 & 2 & 4 \end{matrix}\right| / \left|\begin{matrix} 2 & 4 & 1 \\ 5 & 2 & -1 \\ 3 & 2 & 4 \end{matrix}\right| = (-68)/(-68) = 1,$$

$$y = \left|\begin{matrix} 2 & 11 & 1 \\ 5 & 8 & -1 \\ 3 & 11 & 4 \end{matrix}\right| / \left|\begin{matrix} 2 & 4 & 1 \\ 5 & 2 & -1 \\ 3 & 2 & 4 \end{matrix}\right| = (-136)/(-68) = 2 \text{ and}$$

$$z = \left|\begin{matrix} 11 & 4 & 1 \\ 8 & 2 & -1 \\ 11 & 2 & 4 \end{matrix}\right| / \left|\begin{matrix} 2 & 4 & 1 \\ 5 & 2 & -1 \\ 3 & 2 & 4 \end{matrix}\right| = (-68)/(-68) = 1.$$

Exercises

For the following matrix $A = \begin{bmatrix} 1 & 2 \\ 4 & 5 \end{bmatrix}$, compute the quantities in the exercises below.

1. Compute the quantities $c_{1,1}$, $c_{1,2}$, $c_{2,1}$ and $c_{2,2}$.
2. Compute the matrix $Adj(A)$.
3. Compute $|A|$.
4. Compute the matrix A^{-1} .

For the matrix $A = \begin{bmatrix} -5 & 0 & 5 \\ 1 & 2 & 3 \\ -2 & 1 & -1 \end{bmatrix}$, compute the quantities in the exercises below.

5. Compute the quantities $c_{1,1}$, $c_{1,2}$ and $c_{1,3}$.
6. Compute the quantities $c_{2,1}$, $c_{2,2}$ and $c_{2,3}$.
7. Compute the quantities $c_{3,1}$, $c_{3,2}$ and $c_{3,3}$.
8. Compute the matrix $Adj(A)$.
9. Compute $|A|$.
10. Compute the matrix A^{-1} .

(Exercises using Cramer's Rule)

Using Cramer's Rule, solve for the unknowns in the linear equations in the exercises below.

11. Solve for **x**, **y** and **z** for equation $\begin{bmatrix} 17 \\ 20 \\ -18 \end{bmatrix} = \begin{bmatrix} 6 & 3 & -1 \\ 4 & -2 & 0 \\ -3 & 4 & 2 \end{bmatrix} \times \begin{bmatrix} x \\ y \\ z \end{bmatrix}$.

12. Solve for **x** and **y** for equation $\begin{bmatrix} 36 \\ 19 \end{bmatrix} = \begin{bmatrix} 8 & 3 \\ 9 & -2 \end{bmatrix} \times \begin{bmatrix} x \\ y \end{bmatrix}$.

13. Solve for **x**, **y** and **z** for equation $\begin{bmatrix} 8 \\ 2 \\ 3 \end{bmatrix} = \begin{bmatrix} 4 & 0 & 4 \\ 2 & 1 & 0 \\ 2 & 5 & 1 \end{bmatrix} \times \begin{bmatrix} x \\ y \\ z \end{bmatrix}$.

14. Solve for **w**, **x**, **y** and **z** for equation $\begin{bmatrix} 12 \\ -9 \\ -9 \\ -18 \end{bmatrix} = \begin{bmatrix} 1 & 2 & 4 & 8 \\ 2 & 1 & 2 & 4 \\ 4 & 2 & 1 & 2 \\ 8 & 4 & 2 & 1 \end{bmatrix} \times \begin{bmatrix} w \\ x \\ y \\ z \end{bmatrix}$.

15. Solve for **x**, **y** and **z** for equation $\begin{bmatrix} 9 \\ -12 \\ 18 \end{bmatrix} = \begin{bmatrix} 7 & 10 & 10 \\ 3 & -5 & 7 \\ 1 & 9 & 3 \end{bmatrix} \times \begin{bmatrix} x \\ y \\ z \end{bmatrix}$.

4.6 *Inner Products Using Determinants*

Using the concept of inner products (dot products and cross products) and the notation introduced in this chapter we write the expression for the volume of a parallelepiped as the following:

$$\text{volume} = \vec{V}_1 \bullet \vec{V}_2 \times \vec{V}_3 = |V| \quad \text{or}$$

$$\text{volume} = v_{1,1} \cdot |V_{1,1}| - v_{1,2} \cdot |V_{1,2}| + v_{1,3} \cdot |V_{1,3}| .$$

Recall that we define $V = [v_{i,j}]$ such that vector $\vec{V}_1$ represents the first row of the matrix, vector $\vec{V}_2$ represents the second row of the matrix and vector $\vec{V}_3$ represents the third row of the matrix.

Now we examine the cross product of the vectors and show how it can also be defined in terms of a determinant. Using the cartesian unit vectors $\vec{\boldsymbol{E}}_1$, $\vec{\boldsymbol{E}}_2$ and $\vec{\boldsymbol{E}}_3$ we can express the cross product as $\vec{V}_2 \times \vec{V}_3 =$ $|V_{1,1}| \cdot \vec{\boldsymbol{E}}_1 - |V_{1,2}| \cdot \vec{\boldsymbol{E}}_2 + |V_{1,3}| \cdot \vec{\boldsymbol{E}}_3$. Using determinant notation, we adopt the nomenclature $\vec{V}_2 \times \vec{V}_3 = \begin{vmatrix} \vec{\boldsymbol{E}}_1 & \vec{\boldsymbol{E}}_2 & \vec{\boldsymbol{E}}_3 \\ v_{2,1} & v_{2,2} & v_{2,3} \\ v_{3,1} & v_{3,2} & v_{3,3} \end{vmatrix}$ in which the resultant is a vector. Using this notation it is easily seen that

$\vec{V}_3 \times \vec{V}_2 = |\boldsymbol{E}_{2 \leftrightarrow 3}| \cdot \vec{V}_2 \times \vec{V}_3$. From the properties of type I matrices in section 4.2, we get $|\boldsymbol{E}_{2 \leftrightarrow 3}| = -1$. Hence we get the corollary below.

Corollary 4.6.1 Given two distinct vectors $\vec{V}_2$ and $\vec{V}_3$, then we have the following relationship

$$\vec{V}_3 \times \vec{V}_2 = -\vec{V}_2 \times \vec{V}_3 \quad .$$

A special case of the cross product occurs when one vector is a constant value times the other vector. If we have the relationship $\vec{V}_3 = c \cdot \vec{V}_2$, then we have $\vec{V}_3 \times \vec{V}_2 = -\vec{V}_2 \times \vec{V}_3$ from corollary 4.6.1 and hence

$(c \cdot \vec{V}_2) \times \vec{V}_2 = -\vec{V}_2 \times (c \cdot \vec{V}_2)$. If we designate $\vec{\kappa} = \vec{V}_2 \times \vec{V}_2$ then we have $c \cdot \vec{\kappa} = -c \cdot \vec{\kappa}$ or $\vec{\kappa} = \vec{0}$. This property of vectors will be useful a we discuss other properties of inner products.

In contrast to the above (two dependent vectors), when the vectors $\vec{V}_2$ and $\vec{V}_3$ are independent vectors, they span the plane containing them. This property is shown to be true in chapter 3. By definition every vector contained in the plane spanned by these two vectors are expressed as a linear combination of the two base vectors $\vec{V}_2$ and $\vec{V}_3$ (basis). Mathematically we state that for any real values $a \; and \; b$, the vector $a \cdot \vec{V}_2 + b \cdot \vec{V}_3$ is contained in the plane spanned by the vectors. This property leads to the theorem below.

Theorem 4.6.2 Let $\vec{V}_2$ and $\vec{V}_3$ be independent vectors, then the cross product $\vec{V}_2 \times \vec{V}_3$ is perpendicular to all vectors contained in the plane which is spanned by the two independent vectors. Conversely, if a vector is perpendicular to the cross product $\vec{V}_2 \times \vec{V}_3$,then it is a linear combination of the two vectors and hence it is a vector in the plane spanned by the two independent vectors.

Proof: Select an arbitrary vector contained in the plane spanned by the arbitrary vectors $\vec{V}_2$ and $\vec{V}_3$. We will denote this vector as $\vec{V} = a \cdot \vec{V}_2 + b \cdot \vec{V}_3$. In chapter 3, we showed that the dot product of two vectors is equivalent to the product of the length of each vector and the cosine of the angle between the two vectors. Hence, if the dot product is equal to zero of two non zero vectors, we conclude that the vectors are perpendicular.

Using the expression for $\vec{V}$, the dot product for $\vec{V}$ and $\vec{V}_2 \times \vec{V}_3$ is $\vec{V} \bullet \vec{V}_2 \times \vec{V}_3 = (a \cdot \vec{V}_2 + b \cdot \vec{V}_3) \bullet (\vec{V}_2 \times \vec{V}_3)$. From the determinant representation of dot product we get the following equation:

$$\vec{V} \bullet \vec{V}_2 \times \vec{V}_3 = \begin{vmatrix} a \cdot x_2 + b \cdot x_3 & a \cdot y_2 + b \cdot y_3 & a \cdot z_2 + b \cdot z_3 \\ x_2 & y_2 & z_2 \\ x_3 & y_3 & z_3 \end{vmatrix}$$

. As we examine the determinant in this expression, we observe that row one is a linear combination of rows two and three for the resultant matrix. Applying corollary 4.5.1, we can conclude that $\vec{V} \bullet \vec{V}_2 \times \vec{V}_3 = 0$ and hence $\vec{V}$ and $\vec{V}_2 \times \vec{V}_3$ are perpendicular. Since $\vec{V}$ is an arbitrary vector contained in the plane spanned by $\vec{V}_2$ and $\vec{V}_3$, we can conclude that $\vec{V}_2 \times \vec{V}_3$ is perpendicular to all vectors contained in the plane.

Conversely, if $\vec{V} \bullet \vec{V}_2 \times \vec{V}_3 = 0$, then by corollary 4.5.3 the first row of the matrix formed in the determinant expression is a linear combination of rows two and three. Rows two and three are independent since they correspond to vectors $\vec{V}_2$ and $\vec{V}_3$ respectively which are independent. Hence we must have the relationship that $\vec{V} = a \cdot \vec{V}_2 + b \cdot \vec{V}_3$ for which at least one of the coefficients a *or* b is not equal to zero.

The next corollary shows that a basis can be derived from two vectors $\vec{V}_2$ and $\vec{V}_3$ for the three dimensional vector space $\mathbf{R}^3$ when they are independent.

Corollary 4.6.3 Let $\vec{V}_2$ and $\vec{V}_3$ be two independent vectors, then the vector $\vec{V}_2 \times \vec{V}_3$ and the two original vectors form a basis for the three dimensional vector space $\mathbf{R}^3$.

Proof: To prove this corollary it is sufficient to show that matrix formed by the vectors $\vec{V}_2$, $\vec{V}_3$ and $\vec{V}_2 \times \vec{V}_3$ is non-singular and any vector in $\mathbf{R}^3$can be expressed as a linear combination of these three vectors. We begin by assuming that the cross product is nonzero and $\vec{V}_2 \times \vec{V}_3 = a \cdot \vec{V}_2 + b \cdot \vec{V}_3$. If we compute the dot product of the cross product we can write $\left(\vec{V}_2 \times \vec{V}_3\right) \bullet \left(\vec{V}_2 \times \vec{V}_3\right) = \left(a \cdot \vec{V}_2 + b \cdot \vec{V}_3\right) \bullet \left(\vec{V}_2 \times \vec{V}_3\right)$. Since both $\vec{V}_2$ and $\vec{V}_3$ are perpendicular to the cross product, the right side of this equation is equal to zero;

$(\vec{V}_2 \times \vec{V}_3) \bullet (\vec{V}_2 \times \vec{V}_3) = 0$. This can only be true if the cross product is equal to the zero matrix. Since the two original vectors are independent, this is not the case. Hence by contradiction, we have that the three vectors are independent. Since the vectors are independent, the matrix $A = \begin{bmatrix} \vec{V}_2 \times \vec{V}_3 \\ \vec{V}_2 \\ \vec{V}_3 \end{bmatrix}$ is non-singular. Furthermore, if an arbitrary vector $\vec{V} = (x, y, z)$ is contained in $\mathbf{R}^3$, then the coefficients derived as follows $\begin{bmatrix} a \\ b \\ c \end{bmatrix} = A^{-1} \cdot \begin{bmatrix} x \\ y \\ z \end{bmatrix}$, derives the coefficients such that

$\vec{V} = a \cdot (\vec{V}_2 \times \vec{V}_3) + b \cdot \vec{V}_2 + c \cdot \vec{V}_3$. Hence these three vectors span the vector space $\mathbf{R}^3$.

Now we examine the cross product of $\vec{V}_1 \times (\vec{V}_2 \times \vec{V}_3)$. It turns out that this cross product can be written as a linear combination of two of the vectors. The theorem below gives the explicit form of this relationship.

Theorem 4.6.4 Given the vectors $\vec{V}_1$, $\vec{V}_3$ and $\vec{V}_3$ then we have the relationship

$$\vec{V}_1 \times (\vec{V}_2 \times \vec{V}_3) = (\vec{V}_1 \bullet \vec{V}_3) \cdot \vec{V}_2 - (\vec{V}_1 \bullet \vec{V}_2) \cdot \vec{V}_3 \quad .$$

Proof: We begin this proof by defining the vector $\vec{V}$ to be $\vec{V} = \vec{V}_1 \times (\vec{V}_2 \times \vec{V}_3)$. Applying corollary 4.6.2, the resultant vector $\vec{V}$ is perpendicular to both of matrices in the cross product. Specifically it is perpendicular to $\vec{V}_1$ and to $(\vec{V}_2 \times \vec{V}_3)$. Since $\vec{V}$ is perpendicular to $(\vec{V}_2 \times \vec{V}_3)$, corollary 4.6.2 implies that it is a linear combination of the vectors $\vec{V}_2$ and $\vec{V}_3$. Hence we can write $\vec{V} = a \cdot \vec{V}_2 + b \cdot \vec{V}_3$. Since this vector is also perpendicular to the vector $\vec{V}_1$,we have $\vec{V}_1 \bullet \vec{V} = 0$. Using this equation we have the following relationship for the coefficients of $\vec{V}$: $(\vec{V}_1 \bullet \vec{V}_2) \cdot a + (\vec{V}_1 \bullet \vec{V}_3) \cdot b = 0.$

In the above equation, we have one equation and two unknowns. Hence if we make the substitution $a = c \cdot (\vec{V}_1 \bullet \vec{V}_3)$, we get $b = -c \cdot (\vec{V}_1 \bullet \vec{V}_2)$. Now it is sufficient for us to determine the value of 'c' for any three vectors contained in a cross product such as this. The new representation of the vector $\vec{V}$ is $\vec{V} = c \cdot ((\vec{V}_1 \bullet \vec{V}_3) \cdot \vec{V}_2 - (\vec{V}_1 \bullet \vec{V}_2) \cdot \vec{V}_3)$. To compute the value of 'c', let's examine a couple of special cases.

<u>**Case 1:** $\vec{V}_1 = \vec{V}_2$</u> For this case we get

$\vec{V}_2 \times (\vec{V}_2 \times \vec{V}_3) = c \cdot ((\vec{V}_2 \bullet \vec{V}_3) \cdot \vec{V}_2 - (\vec{V}_2 \bullet \vec{V}_2) \cdot \vec{V}_3)$. By computing the dot product of both sides of this equation with $\vec{V}_2 \times (\vec{V}_2 \times \vec{V}_3)$ we have the following:

$$\left|\vec{V}_2\right|^4 \cdot \left|\vec{V}_3\right|^2 \cdot sin(\theta_{2,3})^2 = c(\vec{V}_2 \bullet \vec{V}_3) \cdot 0 - c \cdot (\vec{V}_2 \bullet \vec{V}_2) \cdot (\vec{V}_3 \bullet \vec{V}_2 \times (\vec{V}_2 \times \vec{V}_3)) \quad \text{or}$$

$$\left|\vec{V}_2\right|^4 \cdot \left|\vec{V}_3\right|^2 \cdot sin(\theta_{2,3})^2 = c \cdot \left|\vec{V}_2\right|^2 \cdot (\vec{V}_2 \times \vec{V}_3) \bullet (\vec{V}_2 \times \vec{V}_3)$$

.This relationship is derived using the determinant representation for the triple inner product. Also note that the angle $\vec{V}_3$ is the angle between $\vec{V}_2$ and $\vec{V}_3$. Noting that the dot product of the cross product of $\vec{V}_2$ and $\vec{V}_3$ with itself is simply the square of the amplitude of the cross product. Since the amplitude of the cross product is the amplitude of each vector times the sine of the angle between them we get the following equation:

$$\left|\vec{V}_2\right|^4 \cdot \left|\vec{V}_3\right|^2 \cdot sin(\theta_{2,3})^2 = c \cdot \left|\vec{V}_2\right|^4 \cdot \left|\vec{V}_3\right|^2 \bullet sin(\theta_{2,3})^2$$

. Dividing by all like terms in the above, we get the results $c = 1$. Hence for this case we have

$$\vec{V}_2 \times (\vec{V}_2 \times \vec{V}_3) = (\vec{V}_2 \bullet \vec{V}_3) \cdot \vec{V}_2 - (\vec{V}_2 \bullet \vec{V}_2) \cdot \vec{V}_3.$$

<u>**Case 2:** $\vec{V}_1 = \vec{V}_3$</u> For this case we get

$\vec{V}_3 \times (\vec{V}_2 \times \vec{V}_3) = (\vec{V}_3 \bullet \vec{V}_3) \cdot \vec{V}_2 - (\vec{V}_3 \bullet \vec{V}_2) \cdot \vec{V}_3$. The proof for this statement is similar to that for case 1.

<u>**Case 3:** $\vec{V}_1 \in \boldsymbol{R}^3$</u> From corollary 4.6.3 the vectors $\vec{V}_2$, $\vec{V}_3$ and $(\vec{V}_2 \times \vec{V}_3)$ forms basis for the vector space $\boldsymbol{R}^3$. Hence we can express $\vec{V}_1$ as a linear combination of these vectors such as

$\vec{V}_1 = \alpha \cdot \vec{V}_2 + \beta \cdot \vec{V}_3 + \upsilon \cdot (\vec{V}_2 \times \vec{V}_3)$. Since the cross product of a vector with itself is the zero vector, we get the following relationship for the vector $\vec{V}$:

$\vec{V} = \alpha \cdot \vec{V}_2 \times (\vec{V}_2 \times \vec{V}_3) + \beta \cdot \vec{V}_3 \times (\vec{V}_2 \times \vec{V}_3)$.Using the relationships established in cases one and two above, this relationship converts to the following: $\vec{V} = \alpha \cdot ((\vec{V}_2 \bullet \vec{V}_3) \cdot \vec{V}_2 - (\vec{V}_2 \bullet \vec{V}_2) \cdot \vec{V}_3) +$. By

$$\beta \cdot ((\vec{V}_3 \bullet \vec{V}_3) \cdot \vec{V}_2 - (\vec{V}_3 \bullet \vec{V}_2) \cdot \vec{V}_3)$$

combining the coefficients of each vector in the above and the fact that the dot product of $(\vec{V}_2 \times \vec{V}_3)$ with either $\vec{V}_2$ or $\vec{V}_3$ is zero then we have the following specification:

$$\vec{V}_1 \times (\vec{V}_2 \times \vec{V}_3) = (\vec{V}_1 \bullet \vec{V}_3) \cdot \vec{V}_2 - (\vec{V}_1 \bullet \vec{V}_2) \cdot \vec{V}_3 .$$

With the results in case three we have shown that for independent vectors $\vec{V}_2$ and $\vec{V}_3$,we get the results given in the statement of the proof. So now we ask ourselves the question what about the instance in which the vectors are not independent, or stated mathematically $\vec{V}_3 = a \cdot \vec{V}_2$ .Under this condition, the cross product $\left(\vec{V}_2 \times \vec{V}_3\right)$ is the zero vector and hence we examine the quantity $(\vec{V}_1 \bullet \vec{V}_3) \cdot \vec{V}_2 - (\vec{V}_1 \bullet \vec{V}_2) \cdot \vec{V}_3$,it reduces to $(\vec{V}_1 \bullet (a \cdot \vec{V}_2)) \cdot \vec{V}_2 - (\vec{V}_1 \bullet \vec{V}_2) \cdot (a \cdot \vec{V}_2) = \vec{0}$, Since quantity $\vec{V}_1 \times \left(\vec{V}_2 \times \vec{V}_3\right) = \vec{0}$, we conclude in this instance that the relationship $\vec{V}_1 \times \left(\vec{V}_2 \times \vec{V}_3\right) = (\vec{V}_1 \bullet \vec{V}_3) \cdot \vec{V}_2 - (\vec{V}_1 \bullet \vec{V}_2) \cdot \vec{V}_3$ is true. Hence we have shown that the state of the theorem is true for both independent and dependent vectors $\vec{V}_2$ and $\vec{V}_3$.

Before we leave this section, we examine the concept of orthogonal vectors using the concept of cross products. Two vectors are considered orthogonal if the angle between the vector is 90 degrees (right angle). In terms of the dot product, two vectors are orthogonal if the dot product of the two matrices is zero. Whenever, we construct a basis for a vector space, it is advantageous to construct an orthogonal basis; that is each pair of vectors are orthogonal.

From corollary 4.6.3, we can construct a basis for $\mathbf{R}^3$ using the following three vectors: $\vec{V}_2$, $\vec{V}_3$ and $\left(\vec{V}_2 \times \vec{V}_3\right)$. By chance these three vectors may form an orthogonal vector space. To insure an orthogonal vector space, we replace the vector $\vec{V}_3$ with the vector $\vec{V}_2 \times \left(\vec{V}_2 \times \vec{V}_3\right)$. From theorem 4.6.4 we have $\vec{V}_2 \times \left(\vec{V}_2 \times \vec{V}_3\right) = (\vec{V}_2 \bullet \vec{V}_3) \cdot \vec{V}_2 - (\vec{V}_2 \bullet \vec{V}_2) \cdot \vec{V}_3$. Hence the vector triplet: $\vec{V}_2$, $\left(\vec{V}_2 \times \vec{V}_3\right)$ and $(\vec{V}_2 \bullet \vec{V}_3) \cdot \vec{V}_2 - (\vec{V}_2 \bullet \vec{V}_2) \cdot \vec{V}_3$ form an orthogonal basis for the vector space $\mathbf{R}^3$. The use of orthogonal vectors to represent a vector space, simplifies operations such dot products and cross products.

Exercises

From theorem 4.6.2 the vector $\vec{V}_1 \times \vec{V}_2$ is perpendicular to all vectors contained in the plane spanned by the distinct vectors $\vec{V}_1$ and $\vec{V}_2$. This suggests the equation $(\vec{V}_1 \times \vec{V}_2) \bullet \vec{V} = 0$ for all vectors $\vec{V}$ contained in the plane. In scalar notation, when $\vec{V}_1 \times \vec{V}_2 = \begin{bmatrix} a & b & c \end{bmatrix}$ and $\vec{V} = \begin{bmatrix} x & y & z \end{bmatrix}$, the equation is written as $a \cdot x + b \cdot y + c \cdot z = 0$. For exercises 1 through 5 below, specify the equation of the plane in scalar notation for the plane spanned the specified distinct vector pair.

1. $\vec{V}_1 = \begin{bmatrix} 1 & 1 & 1 \end{bmatrix}$ and $\vec{V}_2 = \begin{bmatrix} -1 & 0 & 1 \end{bmatrix}$.
2. $\vec{V}_1 = \begin{bmatrix} 2 & 1 & 1 \end{bmatrix}$ and $\vec{V}_2 = \begin{bmatrix} 3 & -1 & -2 \end{bmatrix}$.
3. $\vec{V}_1 = \begin{bmatrix} 3 & 4 & 5 \end{bmatrix}$ and $\vec{V}_2 = \begin{bmatrix} 5 & 4 & 3 \end{bmatrix}$.
4. $\vec{V}_1 = \begin{bmatrix} 1 & 2 & 4 \end{bmatrix}$ and $\vec{V}_2 = \begin{bmatrix} 1 & 1 & 0 \end{bmatrix}$.
5. $\vec{V}_1 = \begin{bmatrix} 6 & 7 & 2 \end{bmatrix}$ and $\vec{V}_2 = \begin{bmatrix} -2 & -2 & 1 \end{bmatrix}$.

(Exercises - Orthogonal Basis)

Beginning with vectors $\vec{V}_1$ and $\vec{V}_2$, construct an orthogonal basis for R^3 with $\vec{V}_1$ as one of the orthogonal vectors.

6. $\vec{V}_1 = \begin{bmatrix} 2 & 1 & 2 \end{bmatrix}$ and $\vec{V}_2 = \begin{bmatrix} -1 & 4 & 3 \end{bmatrix}$.
7. $\vec{V}_1 = \begin{bmatrix} -1 & -2 & 4 \end{bmatrix}$ and $\vec{V}_2 = \begin{bmatrix} 1 & 2 & 4 \end{bmatrix}$.
8. $\vec{V}_1 = \begin{bmatrix} 4 & -1 & 5 \end{bmatrix}$ and $\vec{V}_2 = \begin{bmatrix} 0 & 2 & -4 \end{bmatrix}$.
9. $\vec{V}_1 = \begin{bmatrix} -6 & 2 & 1 \end{bmatrix}$ and $\vec{V}_2 = \begin{bmatrix} 3 & -2 & -2 \end{bmatrix}$.
10. $\vec{V}_1 = \begin{bmatrix} 3 & 9 & 1 \end{bmatrix}$ and $\vec{V}_2 = \begin{bmatrix} 1 & 4 & 16 \end{bmatrix}$.

4.7 Matrix Partitions

Having digressed from the general topic of matrices to discuss vectors in $\mathbf{R}^3$,we can now return to the general topic of matrices. Another approach to examining the properties of matrices is to divide the matrix into partitions. In some applications, it is advantageous to subdivide the matrix into many sub-matrices. The appearance of such partitions may be as follows:

$$\boldsymbol{A} = \begin{bmatrix} A_{1,1} & A_{1,2} & \dots & A_{1,k} \\ A_{2,1} & A_{2,2} & \dots & A_{2,k} \\ \dots & \dots & \dots & \dots \\ A_{k,1} & A_{k,2} & \dots & A_{k,k} \end{bmatrix}$$

. By subdividing matrices in this manner we are able to examine such properties as determinants and inverses in terms of the sub-matrices.

(Matrix Inverse Using Partitions)

The first partition that we will consider is a two-by-two partition. The original matrix will be partitioned such that it consist of two rows of sub-matrices and two columns of sub-matrices. Furthermore, each sub-matrix in a given column will have an equal number of columns and each matrix in a given row will have an equal number of rows. An **n-by-n** matrix A , that has a two-by-two partition will have the following appearance: $\boldsymbol{A} = \begin{bmatrix} A_{1,1} & A_{1,2} \\ A_{2,1} & A_{2,2} \end{bmatrix}$. The sub-matrix $A_{1,1}$ has **k** rows and **k** columns. The sub-matrix $A_{1,2}$ has **k** rows and **m** columns. The sub-matrix $A_{2,1}$ has **m** rows and **k** columns. The sub-matrix $A_{2,2}$ has **m** rows and **m** columns. In this notation, we must have $n = k+m$.

If A is non-singular, then there is an **n-by-n** matrix B such that $A \cdot B = I_n$. In this notation, the matrix I_n is the **n-by-n** identity matrix. For compatibility in multiplying the matrices, the matrix B will also be partitioned with a two-by-two partition. Each sub-matrix of this matrix will have the following dimensions: the sub-matrix $B_{1,1}$ has **k** rows and **k** columns, the sub-matrix $B_{1,2}$ has **k** rows and **m** columns, the sub-matrix $B_{2,1}$ has **m** rows and **k** columns and the sub-matrix $B_{2,2}$ has **m** rows and **m** columns. From corollary 4.5.4, the matrix B is a unique inverse of A .

Applying the commutative principles of real numbers, it is easily shown that the product of $A \cdot B$ can be expanded to the product:

$$A \cdot B = \begin{bmatrix} A_{1,1} \cdot B_{1,1} + A_{1,2} \cdot B_{2,1} & A_{1,1} \cdot B_{1,2} + A_{1,2} \cdot B_{2,2} \\ A_{2,1} \cdot B_{1,1} + A_{2,2} \cdot B_{2,1} & A_{2,1} \cdot B_{1,2} + A_{2,2} \cdot B_{2,2} \end{bmatrix}$$

. Since the product is equivalent to the identity matrix, we have four equations shown in the list below.

1. $A_{1,1} \cdot B_{1,1} + A_{1,2} \cdot B_{2,1} = I_k$,
2. $A_{1,1} \cdot B_{1,2} + A_{1,2} \cdot B_{2,2} = \mathbf{0}$,
3. $A_{2,1} \cdot B_{1,1} + A_{2,2} \cdot B_{2,1} = \mathbf{0}$ and
4. $A_{2,1} \cdot B_{1,2} + A_{2,2} \cdot B_{2,2} = I_m$.

Before we proceed further, we must make an assumption concerning the sub-matrices along the diagonal of matrix A . The assumption is that the sub-matrix $A_{1,1}$ is non-singular. (If it is singular, we multiply $\boldsymbol{A}$ by the appropriate type I elementary matrices until it is non-singular. If we cannot find at least **k** rows such that this sub-matrix is non-singular, then we can conclude that the matrix $\boldsymbol{A}$ is singular. this is a contradiction.) With this assumption, we multiply equation 2 by $-A_{2,1} \cdot A^{-1}{}_{1,1}$ and add the resulting equation to equation 4. This results in a new equation number 4 such as

4a. $(A_{2,2} - A_{2,1} \cdot A^{-1}{}_{1,1} \cdot A_{1,2}) \cdot B_{2,2} = \boldsymbol{I}_m$.

This equation indicates that the sub-matrix $B_{2,2}$ is non-singular and that it is equivalent to $(A_{2,2} - A_{2,1} \cdot A^{-1}{}_{1,1} \cdot A_{1,2})^{-1}$. If we now multiply equation 2 by $A^{-1}{}_{1,1}$, we get an expression for $B_{1,2}$ and the following equations:

1. $A_{1,1} \cdot B_{1,1} + A_{1,2} \cdot B_{2,1} = \boldsymbol{I}_k$,

2a. $B_{1,2} = -A^{-1}{}_{1,1} \cdot A_{1,2} \cdot (A_{2,2} - A_{2,1} \cdot A^{-1}{}_{1,1} \cdot A_{1,2})^{-1}$,

3. $A_{2,1} \cdot B_{1,1} + A_{2,2} \cdot B_{2,1} = \boldsymbol{0}$ and

4a. $B_{2,2} = (A_{2,2} - A_{2,1} \cdot A^{-1}{}_{1,1} \cdot A_{1,2})^{-1}$.

By multiplying equation 1 by $-A_{2,1} \cdot A^{-1}{}_{1,1}$ and adding equation 3 to the resultant number 1, then we can replace equation 3 above with this equation:

3a. $(A_{2,2} - A_{2,1} \cdot A^{-1}{}_{1,1} \cdot A_{1,2}) \cdot B_{2,1} = -A_{2,1} \cdot A^{-1}{}_{1,1}$.

Use this equation to solve for $B_{2,1}$ and then substitute the results into equation 1 to get a solution for $B_{1,1}$. With this sequence of algebraic steps we now have solutions for all sub-matrices contained in matrix $\boldsymbol{B}$. The solutions are shown in the list below.

1a. $B_{1,1} = A^{-1}_{1,1} \cdot (\boldsymbol{I}_k - A_{1,2} \cdot B_{2,1})$,

2a. $B_{1,2} = -A^{-1}_{1,1} \cdot A_{1,2} \cdot \left(A_{2,2} - A_{2,1} \cdot A^{-1}_{1,1} \cdot A_{1,2}\right)^{-1}$,

3a. $B_{2,1} = -\left(A_{2,2} - A_{2,1} \cdot A^{-1}_{1,1} \cdot A_{1,2}\right)^{-1} \cdot A_{2,1} \cdot A^{-1}_{1,1}$ and

4a. $B_{2,2} = \left(A_{2,2} - A_{2,1} \cdot A^{-1}_{1,1} \cdot A_{1,2}\right)^{-1}$.

For large matrices (dimension greater than 6), this approach may not yield a significantly easier approach to computing the inverse of a matrix. However, for a **4-by-4**, a **5-by-5** or a **6-by-6**, this approach may ease the pain in computing the inverse of a matrix. To use this new approach for computing inverses, see the following examples.

(Example)

For the matrix $\boldsymbol{A} = \begin{bmatrix} 1 & 0 & 0 & 2 \\ 1 & 2 & 3 & 4 \\ -1 & 4 & 2 & 5 \\ -3 & 2 & 0 & 1 \end{bmatrix}$ apply the matrix partitioning techniques learned in this section to partition this 4-by-4 matrix so that two 2-by-2 matrices lie along the diagonal. The partitioning of this matrix will be

$$A = \begin{bmatrix} \begin{bmatrix} 1 & 0 \\ 1 & 2 \end{bmatrix} & \begin{bmatrix} 0 & 2 \\ 3 & 4 \end{bmatrix} \\ \begin{bmatrix} -1 & 4 \\ -3 & 2 \end{bmatrix} & \begin{bmatrix} 2 & 5 \\ 0 & 1 \end{bmatrix} \end{bmatrix} = \begin{bmatrix} A_{1,1} & A_{1,2} \\ A_{2,1} & A_{2,2} \end{bmatrix}.$$

Letting $B = A^{-1}$ and applying equations 1a through 4a above, we have

$$B_{2,2} = \begin{bmatrix} 5 & -3 \\ 3 & -4 \end{bmatrix} / (-11), \quad B_{2,1} = \begin{bmatrix} -3 & 7 \\ 7 & 2 \end{bmatrix} / 11 \quad ,$$

$$B_{1,2} = \begin{bmatrix} 6 & -8 \\ \frac{21}{2} & -\frac{17}{2} \end{bmatrix} / 11 \text{ and } B_{1,1} = \begin{bmatrix} -3 & -4 \\ -8 & -7 \end{bmatrix} / 11.$$ Hence we have

$$A^{-1} = \begin{bmatrix} -3 & -4 & 6 & -8 \\ -8 & -7 & \frac{21}{2} & -\frac{17}{2} \\ -3 & 7 & -5 & 3 \\ 7 & 2 & -3 & 4 \end{bmatrix} / 11 .$$

Special Case: k=1

In this special case with **k** equal to one, the first diagonal element of the partition of A is not a sub-matrix; rather it is a real valued number (scalar). The partition in this special instance has the following appearance:

$A = \begin{bmatrix} a_{1,1} & A_{1,2} \\ A_{2,1} & A_{2,2} \end{bmatrix}$. The sub-matrix $A_{1,2}$ is row matrix with **n-1** elements

and $A_{2,1}$ is a column matrix with **n-1** elements. To solve for the inverse matrix B, use the four solution equations derived above. In this instance, the requirement is that $a_{1,1} \neq 0$.

1a. $b_{1,1} = (1 - A_{1,2} \cdot B_{2,1})/a_{1,1}$,

2a. $B_{1,2} = -A_{1,2} \cdot (A_{2,2} - A_{2,1} \cdot A_{1,2}/a_{1,1})^{-1}/a_{1,1}$,

3a. $B_{2,1} = -(A_{2,2} - A_{2,1} \cdot A_{1,2}/a_{1,1})^{-1} \cdot A_{2,1}/a_{1,1}$ and

4a. $B_{2,2} = (A_{2,2} - A_{2,1} \cdot A_{1,2}/a_{1,1})^{-1}$.

Expanding equation 1a fully in terms of elements of the matrix A gives us the expression

$b_{1,1} = (1 + A_{1,2} \cdot (A_{2,2} - A_{2,1} \cdot A_{1,2}/a_{1,1})^{-1} \cdot A_{2,1}/a_{1,1})/a_{1,1}$.
But recall that in section 4.5 the element in row one and column one of the inverse matrix is computed to be $b_{1,1} = |A_{2,2}|/|A|$. In this particular partition, $A_{2,2}$ is the minor matrix that corresponds to eliminating row one and column one. Hence we have a relationship that relates the ratio of these determinants to the inverse of a matrix:

$|A_{2,2}|/|A| = (1 + A_{1,2} \cdot (A_{2,2} - A_{2,1} \cdot A_{1,2}/a_{1,1})^{-1} \cdot A_{2,1}/a_{1,1})/a$

or

$a_{1,1} \cdot |A_{2,2}|/|A| - 1 = A_{1,2} \cdot (A_{2,2} - A_{2,1} \cdot A_{1,2}/a_{1,1})^{-1} \cdot A_{2,1}/a_{1,1}$
. In the field of statistics, the left side of the equation is often used in regression

analysis. The use of determinants gives us an approach to simplifying the computations required for this expression.

Another equation for the ratio of these determinants can be computed if we assume the sub-matrix $A_{2,2}$ is non-singular. We use the same approach that was used when we assumed that $A_{1,1}$ was non-singular. If we follow similar algebraic steps as above, time we get the relationship: $a_{1,1} \cdot |A_{2,2}| / |A| = 1 / (a_{1,1} - A_{1,2} \cdot A^{-1}{}_{2,2} \cdot A_{2,1})$. Below are some examples of applying these equations to simplify calculations.

(Example)

For the matrix $A = \begin{bmatrix} 3 & 1 & -1 \\ 0 & 1 & 2 \\ 4 & -1 & 5 \end{bmatrix}$, partition the matrix so that the diagonal contains a scalar and a 2-by-2 matrix and then compute the inverse of the matrix.

We begin with $A = \begin{bmatrix} 3 & \begin{bmatrix} 1 & -1 \end{bmatrix} \\ \begin{bmatrix} 0 \\ 4 \end{bmatrix} & \begin{bmatrix} 1 & 2 \\ -1 & 5 \end{bmatrix} \end{bmatrix} = \begin{bmatrix} a_{1,1} & A_{1,2} \\ A_{2,1} & A_{2,2} \end{bmatrix}$ and from the techniques learned in this section we get

$$B_{2,2} = \left(\begin{bmatrix} 1 & 2 \\ -1 & 5 \end{bmatrix} - \begin{bmatrix} 0 \\ 4 \end{bmatrix} \cdot \begin{bmatrix} 1 & -1 \end{bmatrix} / 3 \right)^{-1} = \begin{bmatrix} \frac{19}{33} & \frac{7}{33} \\ -\frac{2}{11} & \frac{1}{11} \end{bmatrix},$$

$$B_{1,2} = -\begin{bmatrix} 1 & -1 \end{bmatrix} \cdot \begin{bmatrix} \frac{19}{33} & \frac{7}{33} \\ -\frac{2}{11} & \frac{1}{11} \end{bmatrix} /3 = \begin{bmatrix} -\frac{25}{99} & -\frac{4}{99} \end{bmatrix},$$

$$B_{2,1} = -\begin{bmatrix} \frac{19}{33} & \frac{7}{33} \\ -\frac{2}{11} & \frac{1}{11} \end{bmatrix} \cdot \begin{bmatrix} 0 \\ 4 \end{bmatrix} /3 = \begin{bmatrix} -\frac{28}{99} \\ -\frac{4}{33} \end{bmatrix} \text{ and}$$

$$b_{1,1} = \left(1 - \begin{bmatrix} 1 & -1 \end{bmatrix} \cdot \begin{bmatrix} -\frac{28}{99} \\ -\frac{4}{33} \end{bmatrix} \right) /3 = \frac{116}{297}$$

. Hence the inverse of this matrix is

$$A^{-1} = \begin{bmatrix} \frac{116}{297} & -\frac{25}{99} & -\frac{4}{99} \\ -\frac{28}{99} & \frac{19}{33} & \frac{7}{33} \\ -\frac{4}{33} & -\frac{2}{11} & \frac{1}{11} \end{bmatrix}.$$

Partition Matrices Into Columns

Another partition of interests for an **n-by-n** matrix is the partition that has "**n**" column matrices. Using a row and column gaussian reduction approach, we can express any n-by-n matrix as the product of three matrices in which one of the matrices is a diagonal matrix. Such a representation for the matrix $\boldsymbol{A}$ would be the following: $\boldsymbol{A} = \boldsymbol{P} \cdot \boldsymbol{D} \cdot \boldsymbol{Q}'$. The matrix $\boldsymbol{D}$ represents the diagonal matrix and it has the form $\boldsymbol{D} = \begin{bmatrix} d_1 & 0 & \dots & 0 \\ 0 & d_2 & \dots & 0 \\ \dots & \dots & \dots & \dots \\ 0 & 0 & & d_n \end{bmatrix}$. Now we partition both $\boldsymbol{P}$ and $\boldsymbol{Q}$ into column matrices. Thus we can represent each as $\boldsymbol{P} = \begin{bmatrix} P_1 \ P_2 \ \dots \ P_n \end{bmatrix}$ and $\boldsymbol{Q} = \begin{bmatrix} Q_1 \ Q_2 \ \dots \ Q_n \end{bmatrix}$. With this partition and the notation for the diagonal matrix we have a new representation for the original matrix; $\boldsymbol{A} = \boldsymbol{P} \cdot \boldsymbol{D} \cdot \boldsymbol{Q}' = \sum_{i=1}^{n} d_i \cdot (P_i \cdot Q'_i)$. If we multiply the matrix $P_i \cdot Q'_i$ times itself we get a multiple of the matrix. Hence the product $P_i \cdot Q'_i$ is an idempotent matrix. If we can find the sequence of matrices such that $(P_i \cdot Q'_i) \cdot (P_j \cdot Q'_j) = 0$ when $i \neq j$,then we have an orthogonal representation of the matrix. This representation will be discussed in more details in the next chapter.

(Example)

For this example, we use the matrix $\boldsymbol{A} = \begin{bmatrix} 1 & 2 & 3 \\ -1 & 0 & 2 \\ -4 & -8 & 1 \end{bmatrix}$ to represent the matrix in the notation $\boldsymbol{A} = \boldsymbol{P} \cdot \boldsymbol{D} \cdot \boldsymbol{Q}' = \sum_{i=1}^{n} d_i \cdot (P_i \cdot Q'_i)$. To begin this conversion we pre-multiply the matrix A and the identity matrix by the appropriate elementary matrices to obtain an upper triangular matrix.

$$E_{1R1+R2} \times E_{4R1+R3} \times \begin{bmatrix} 1 & 2 & 3 \\ -1 & 0 & 2 \\ -4 & -8 & 1 \end{bmatrix} \rightarrow \begin{bmatrix} 1 & 0 & 0 \\ 1 & 1 & 0 \\ 4 & 0 & 1 \end{bmatrix} \text{ or } \begin{bmatrix} 1 & 2 & 3 \\ 0 & 2 & 5 \\ 0 & 0 & 13 \end{bmatrix} \rightarrow \begin{bmatrix} 1 & 0 & 0 \\ 1 & 1 & 0 \\ 4 & 0 & 1 \end{bmatrix}.$$

Now that we have converted the original matrix to an upper triangular matrix, we can not post-multiply the upper triangular matrix to obtain a diagonal matrix.

$$\begin{bmatrix} 1 & 2 & 3 \\ 0 & 2 & 5 \\ 0 & 0 & 13 \end{bmatrix} \times E_{-2C1+C2} \times E_{-3C1+C3} \times E_{\left(-\frac{5}{2}\right)C2+C3} \rightarrow \begin{bmatrix} 1 & -2 & 2 \\ 0 & 1 & -\frac{5}{2} \\ 0 & 0 & 1 \end{bmatrix} \text{ or}$$

$$\begin{bmatrix} 1 & 0 & 0 \\ 0 & 2 & 0 \\ 0 & 0 & 13 \end{bmatrix} \rightarrow \begin{bmatrix} 1 & -2 & 2 \\ 0 & 1 & -\frac{5}{2} \\ 0 & 0 & 1 \end{bmatrix}.$$

(Exercises)

For the exercises one through fourteen below, use partitions to compute the inverse of each matrix.

1. $\begin{bmatrix} 89 & -32 & 20 \\ 78 & -27 & 18 \\ -258 & 96 & -57 \end{bmatrix}$

2. $\begin{bmatrix} -206 & -148 & -120 \\ 702 & 504 & 408 \\ -496 & -356 & -288 \end{bmatrix}$

3. $\begin{bmatrix} -148 & 202 & -58 \\ 70 & -100 & 28 \\ 645 & -897 & 255 \end{bmatrix}$

4. $\begin{bmatrix} 56 & -27 & -6 \\ 102 & -45 & -4 \\ -39 & 11 & -9 \end{bmatrix}$

5. $\begin{bmatrix} 40 & 39 & 20 \\ 114 & 103 & 54 \\ -300 & -276 & -144 \end{bmatrix}$

6. $\begin{bmatrix} 54 & 32 & -28 \\ -24 & -22 & 8 \\ 64 & 32 & -38 \end{bmatrix}$

7. $\begin{bmatrix} -23 & 12 & 10 \\ 26 & -33 & -13 \\ -110 & 102 & 52 \end{bmatrix}$

8. $\begin{bmatrix} 4 & -4 & -2 \\ 2 & 19 & 7 \\ -4 & -26 & -8 \end{bmatrix}$

9. $\begin{bmatrix} 62 & 92 & 24 \\ -42 & -64 & -16 \\ -6 & -4 & -4 \end{bmatrix}$

10. $\begin{bmatrix} 70 & -240 & 64 \\ 0 & -10 & 0 \\ -80 & 240 & -74 \end{bmatrix}$

11. $\begin{bmatrix} 16 & -13 & -11 & 11 \\ -22 & 61 & 43 & -43 \\ 36 & 90 & 44 & -50 \\ 0 & 162 & 96 & -102 \end{bmatrix}$

12. $\begin{bmatrix} -16 & 31 & 40 & 14 \\ -2 & -97 & -98 & -30 \\ 20 & 77 & 69 & 19 \\ -54 & 75 & 102 & 38 \end{bmatrix}$

13. $\begin{bmatrix} 138 & 0 & 32 & 48 \\ -43 & -2 & -11 & -16 \\ 39 & 6 & 13 & 18 \\ -410 & -4 & -98 & -146 \end{bmatrix}$

14. $\begin{bmatrix} -104 & 2 & -68 & -32 \\ 241 & -10 & 166 & 79 \\ 151 & -4 & 100 & 49 \\ 37 & 2 & 22 & 7 \end{bmatrix}$

This Page is Intentionally Left Blank

CHAPTER 5

Characteristic Roots, Equations And Vectors

In chapter 4, we introduced the **n-by-n** nonzero idempotent matrix A such that $A^2 = c \cdot A$. Corollary 4.3.7 shows that either $|A| = 0$ or $|A| = c^n$. In addition to the property shown in corollary 4.3.7, it is readily shown that $|A - c \cdot I| = 0$. This property is illustrated by showing that the matrix $(A - c \cdot I)$ is singular. If this matrix is non-singular, then there is a non-singular matrix M that is an inverse of the matrix. This condition implies the matrix equation $M \cdot (A - c \cdot I) = I$. By rearranging terms, we can rewrite the equation to have $M \cdot A = I + c \cdot M$. Multiplying both sides of the equation by the matrix $(A - c \cdot I)$ results in the equation $M \cdot A \cdot (A - c \cdot I) = (I + c \cdot M) \cdot (A - c \cdot I)$. Using the properties of the idempotent matrix and the properties of inverses this equation becomes $\mathbf{0} = ((\boldsymbol{A} - \boldsymbol{c} \cdot \boldsymbol{I}) + \boldsymbol{c} \cdot \boldsymbol{I})$ or $\boldsymbol{A} = \mathbf{0}$. This presents a contradiction since $\boldsymbol{A}$ is a nonzero matrix. Hence $(\boldsymbol{A} - \boldsymbol{c} \cdot \boldsymbol{I})$ is singular and

$|\boldsymbol{A} - \boldsymbol{c} \cdot \boldsymbol{I}| = 0$. In this chapter, we investigate the function $|\boldsymbol{A} - x \cdot \boldsymbol{I}|$ for the arbitrary **n-by-n** matrix.

5.1 Characteristic Roots(Eigenvalues) And Equations

Eigenvalues or characteristic roots of a square matrix $\boldsymbol{A}$ are the values of **x** that are solutions to the following equation: $|\boldsymbol{A} - x \cdot \boldsymbol{I}| = 0$. This equation is traditionally called the characteristic equation for the matrix $\boldsymbol{A}$. If we apply definition 4.2.1 for the determinant of a matrix then the expression $|\boldsymbol{A} - x \cdot \boldsymbol{I}|$ is a polynomial function of **x**. This section will investigate some of the properties of the polynomial equations of the characteristic roots. In many instances, a real solution may not exist for the polynomial function. Before we progress to the point of determining the solutions of the equation, our first step will be that of constructing the polynomial equation.

To simplify the notation for the characteristic equation of a matrix $\boldsymbol{A}$, we introduce the short-hand notation for the characteristic function $\rho_A(x) = |\boldsymbol{A} - x \cdot \boldsymbol{I}|$. This notation will be used throughout the remainder of the book whenever we reference the characteristic equation or characteristic function. If we expand the characteristic function as a polynomial of degree **M**, then we have $\rho_A(x) = c_M \cdot x^M + c_{M-1} \cdot x^{M-1} + \ldots + c_0$. To this point all coefficients and the degree **M** are unknown values for the characteristic polynomial. The determination of the value of **M**, the degree of the polynomial, may be determined by finding the largest positive integer **m** such that $\lim_{x \to \infty} (\rho_A(x) / x^m) \neq 0$ and $\left|\lim_{x \to \infty} (\rho_A(x) / x^m)\right| < \infty$. If $\boldsymbol{A}$ is an **n-by-n** matrix, we have the relationship $\boldsymbol{A} - x \cdot \boldsymbol{I} = x \cdot (\boldsymbol{A} / x - \boldsymbol{I})$. Computing the determinant of both sides of this relationship and using corollary 4.2.3 recursively, we can rewrite the relationship as $\rho_A(x) = x^n \cdot |\boldsymbol{A} / x - \boldsymbol{I}|$. Hence if **m** is less than **n** then we have $n - m > 0$ and hence

$\lim_{x \to \infty} (\rho_A(x) / x^m) = \lim_{x \to \infty} (x^{n-m} \cdot |A / x - I|) = \infty$. When **m** is greater than **n** then we have $n - m < 0$ and hence $\lim_{x \to \infty} (\rho_A(x) / x^m) = \lim_{x \to \infty} (x^{n-m} \cdot |A / x - I|) = 0$. When **m** is equal to **n** then we have $n - m = 0$ and hence $\lim_{x \to \infty} (\rho_A(x) / x^m) = \lim_{x \to \infty} (|A / x - I|) = (-1)^n$. Therefore we have shown that the polynomial $\rho_A(x) = |A - x \cdot I|$ is an **n**-degree polynomial.

Not only have we shown that the characteristic polynomial is an **n**-degree polynomial, we have also derived the **n**th coefficient of the polynomial to be (-1)**n**. If we evaluate the characteristic function at zero, we get the following equation: $\rho_A(0) = |A| = c_0$. Hence the 0th coefficient of the characteristic polynomial is equal to the determinant of the matrix A. Another coefficient that is easily defined is the (**n-1**)th coefficient. Consider the basic definition for computing the determinant of a matrix. If the matrix B is defined to be $A - x \cdot I$, then we compute the determinant as $|B| = \sum_{\underset{\sim}{J} \in P} \delta(\underset{\sim}{J}) \cdot \prod_{k=1}^{n} b_{i_k, k}$. The only terms of this matrix that includes the **x-terms** are those along the diagonal. The only product in the definition to contain **n-1** or more "**x-terms**" is the product of the diagonal elements. Upon diagnosis of the product $(a_{1,1} - x) \cdot (a_{2,2} - x) \cdot \ldots \cdot (a_{n,n} - x)$, we discover the coefficient of the x^{n-1} term is observed to be

$(-1)^{n-1} \cdot (a_{1,1} + a_{2,2} + \ldots + a_{n,n})$. The sum of the diagonal elements along a square matrix is called the trace of the matrix and is denoted as $Tr(A)$. Therefore, the "n-1" coefficient of the polynomial is determined to be $c_{n-1} = (-1)^{n-1} \cdot Tr(A)$. Summarizing the three coefficients that we have determined so far for the characteristic polynomial, we present the table below.

Coefficient	c_0	c_{n-1}	c_n
Value	$\lvert A \rvert$	$(-1)^{n-1} \cdot Tr(A)$	$(-1)^n$

(2-By-2 Matrix)

For a **2-by-2** matrix, all coefficients of the characteristic function can be derived using the relationships derived above. We begin by showing a few examples for the two dimensional matrix.

Example 1 Given the matrix $\begin{bmatrix} 2 & 5 \\ 1 & -1 \end{bmatrix}$, derive the characteristic equation associated with this matrix and derive the characteristic roots.

For a 2-by-2 matrix we have $c_2 = 1$, $c_1 = -(2-1) = -1$, and $c_0 = -2-5 = -7$. Hence the characteristic equation associated with this matrix is $\rho_A(x) = x^2 - x - 7$. The solution for the equation $x^2 - x - 7 = 0$ is obtained using the quadratic equation. From the quadratic equation we get the solutions $x = \frac{1-\sqrt{29}}{2}$ and $x = \frac{1+\sqrt{29}}{2}$.

Example 2

Given the matrix $\begin{bmatrix} 3 & 2 \\ 1 & 2 \end{bmatrix}$, derive the characteristic equation associated with this matrix and derive the characteristic roots.

For a 2-by-2 matrix we have $c_2 = 1$, $c_1 = -(3+2) = -5$, and $c_0 = (6-2) = 4$. Hence the characteristic equation associated with this matrix is $\rho_A(x) = x^2 - 5 \cdot x + 4$. The solution for the equation $x^2 - 5 \cdot x + 4 = 0$ is obtained using the quadratic equation. From the quadratic equation we get the solutions $x = 1$ and $x = 4$.

The derivation for the 2-by-2 matrix characteristic equation could have been as easily accomplished by evaluating the determinant of the matrix $\boldsymbol{A} - x \cdot \boldsymbol{I}$. But for higher dimension matrices this is not a reasonable alternative.

(3-By-3 Matrix)

Moving up one dimension to three dimension matrices, we have three known coefficients; $c_0, c_2 \, and \, c_3$. By evaluating the characteristic function at $x = 1$, we have an equation for determining the value of c_1. The resultant solution for this coefficient is $c_1 = \rho_A(1) - c_3 - c_2 - c_0$, We summarize the coefficients for the **3-by-3** matrix in the table below.

Coefficient	Value
c_0	$\lvert A \rvert$
c_1	$\rho_A(1) - c_3 - c_2 - c_0$
c_2	$Tr(A)$
c_3	-1

Example 3 Given the matrix $\begin{bmatrix} 1 & 0 & 1 \\ 2 & 4 & 1 \\ 3 & -1 & -2 \end{bmatrix}$, derive the characteristic equation associated with this matrix.

For a 3-by-3 matrix we have $c_3 = -1$, $c_2 = 3$ and $c_0 = (-8+1)+(-2-12) = -21$. The determinant of the matrix derived by subtracting the identity matrix from the original matrix is the quantity $\rho_A(1) = -11$. Using the formula in the table above we have $c_1 = -11+1-3+21 = 8$. Hence the characteristic equation associated with this matrix is $\rho_A(x) = -x^3 + 3 \cdot x^2 + 8 \cdot x - 21$. As you may observe, the higher the dimension of the matrix, the more complex are the calculations for determining the coefficients of the characteristic equation and computing the roots of the equation becomes a major exercise even for a **3-by-3** matrix. Fortunately, computers are well designed to handle the number of iterations required to solve polynomial equations of practically any order. The solution of the polynomial equations will be left as a computer exercises for the reader.

(4-By-4 Matrix)

For the 4-by-4 matrix, two coefficients of the characteristic equation are unknown. The unknown coefficients are c_1 and c_2. To assist in deriving these coefficients we introduce another function involving the characteristic function and the known coefficients c_4, c_3 and c_0. The new function is denoted as $G_A(x)$ and is defined as

$G_A(x) = \rho_A(x) - c_4 \cdot x^4 - c_3 \cdot x^3 - c_0$. This function is a polynomial function and it has the following equivalence: $G_A(x) = c_2 \cdot x^2 + c_1 \cdot x$. Upon evaluating this function for the values of -1 and 1, we get the following equations:

5. $$G_A(-1) = c_2 - c_1 \quad \text{and}$$

6. $$G_A(1) = c_2 + c_1 \quad .$$

Using these equations we get the following solutions for c_1 and c_2 :

7. $$c_1 = \frac{G_A(1) - G_A(-1)}{2} \quad \text{and}$$

8. $$c_2 = \frac{G_A(1) + G_A(-1)}{2} \quad .$$

(N-Dimensional Square Matrix)

As the dimension of the square matrix increases above the value of four (4), it becomes increasingly difficult to estimate all the parameters of the characteristic function associated with the matrix. However, there are numerical approaches available for computing the coefficients. Using an approach that is similar to the approach used when $n = 4$, we define the general case of the function $G_A(x)$.

The new function is defined as $G_A(x) = \rho_A(x) - c_n \cdot x^n - c_{n-1} \cdot x^{n-1} - c_0$. The polynomial representation for this function is $G_A(x) = c_{n-2} \cdot x^{n-2} + \ldots + c_1 \cdot x$. By evaluating this function at "**n-2**" distinct values, we can construct a system of linear equations to solve for the "**n-2**" coefficients. Choosing the distinct values $(x_{n-2}, x_{n-3}, \ldots, x_1)$, the linear system of equations is the following:

$$\begin{bmatrix} x_{n-2}^{n-2} & x_{n-2}^{n-3} & \cdots & x_{n-2} \\ x_{n-3}^{n-2} & x_{n-3}^{n-3} & \cdots & x_{n-3} \\ \nabla & \nabla & \cdots & \nabla \\ x_1^{n-2} & x_1^{n-3} & \cdots & x_1 \end{bmatrix} \cdot \begin{bmatrix} c_{n-2} \\ c_{n-3} \\ \nabla \\ c_1 \end{bmatrix} = \begin{bmatrix} G_A(x_{n-2}) \\ G_A(x_{n-3}) \\ \nabla \\ G_A(x_1) \end{bmatrix}$$

. Applying Cramer's Rule to this system of linear equations we get the following "n-2" solutions for the remaining coefficients for the characteristic function:

$$c_{n-2} = \begin{vmatrix} G_A(x_{n-2}) & x_{n-2}^{n-3} & \cdots & x_{n-2} \\ G_A(x_{n-3}) & x_{n-3}^{n-3} & \cdots & x_{n-3} \\ \nabla & \nabla & \cdots & \nabla \\ G_A(x_1) & x_1^{n-3} & \cdots & x_1 \end{vmatrix} \div |A|,$$

$$\nabla \quad \nabla \quad \nabla \quad \nabla$$

$$c_1 = \begin{vmatrix} x_{n-2}^{n-2} & x_{n-2}^{n-3} & \cdots & G_A(x_{n-2}) \\ x_{n-3}^{n-2} & x_{n-3}^{n-3} & \cdots & G_A(x_{n-3}) \\ \nabla & \nabla & \cdots & \nabla \\ x_1^{n-2} & x_1^{n-3} & \cdots & G_A(x_1) \end{vmatrix} \div |A| .$$

With these equations, all coefficients are now computable and the characteristic equation is representable as known polynomial equation. Now the determination of the characteristic roots is an exercise in solving polynomial equations.

Exercises

For the following matrices, derive the characteristics equation for each matrix. Then solve each characteristic equation for the characteristic roots of the matrix.

1. $\begin{bmatrix} -80 & 34 \\ -170 & 73 \end{bmatrix}$

2. $\begin{bmatrix} -99 & 36 \\ -288 & 105 \end{bmatrix}$

3. $\begin{bmatrix} 56 & -27 & -6 \\ 102 & -45 & -4 \\ -39 & 11 & -9 \end{bmatrix}$

4. $\begin{bmatrix} 40 & 39 & 20 \\ 114 & 103 & 54 \\ -300 & -276 & -144 \end{bmatrix}$

5. $\begin{bmatrix} 54 & 32 & -28 \\ -24 & -22 & 8 \\ 64 & 32 & -38 \end{bmatrix}$

6. $\begin{bmatrix} -57 & 76 & 22 \\ 94 & -135 & -38 \\ -444 & 624 & 177 \end{bmatrix}$

7. $\begin{bmatrix} -23 & 12 & 10 \\ 26 & -33 & -13 \\ -110 & 102 & 52 \end{bmatrix}$

8. $\begin{bmatrix} 4 & -4 & -2 \\ 2 & 19 & 7 \\ -4 & -26 & -8 \end{bmatrix}$

9. $\begin{bmatrix} 62 & 92 & 24 \\ -42 & -64 & -16 \\ -6 & -4 & -4 \end{bmatrix}$

10. $\begin{bmatrix} 70 & -240 & 64 \\ 0 & -10 & 0 \\ -80 & 240 & -74 \end{bmatrix}$

11. $\begin{bmatrix} 16 & -13 & -11 & 11 \\ -22 & 61 & 43 & -43 \\ 36 & 90 & 44 & -50 \\ 0 & 162 & 96 & -102 \end{bmatrix}$

12. $\begin{bmatrix} -16 & 31 & 40 & 14 \\ -2 & -97 & -98 & -30 \\ 20 & 77 & 69 & 19 \\ -54 & 75 & 102 & 38 \end{bmatrix}$

13. $\begin{bmatrix} 138 & 0 & 32 & 48 \\ -43 & -2 & -11 & -16 \\ 39 & 6 & 13 & 18 \\ -410 & -4 & -98 & -146 \end{bmatrix}$

14. $\begin{bmatrix} -104 & 2 & -68 & -32 \\ 241 & -10 & 166 & 79 \\ 151 & -4 & 100 & 49 \\ 37 & 2 & 22 & 7 \end{bmatrix}$

5.2 Characteristic Vector - Spectral Decomposition

For each characteristic root "x" of the matrix $\boldsymbol{A}$, there is a nonzero vector $\underset{\sim}{\boldsymbol{P}}'_x$ such that $\underset{\sim}{\boldsymbol{P}}'_x \cdot (\boldsymbol{A} - x \cdot \boldsymbol{I}) = \underset{\sim}{0}'$ .These vectors are called eigenvectors or characteristic vectors.

When the matrix is a diagonal matrix, for example $\boldsymbol{A} = \begin{bmatrix} d_1 & 0 & \dots & 0 \\ 0 & d_2 & \dots & 0 \\ \dots & \dots & \dots & \dots \\ 0 & 0 & \dots & d_n \end{bmatrix}$, then the following relationship $\underset{\sim}{\boldsymbol{E}}'_k \cdot (\boldsymbol{A} - x \cdot \boldsymbol{I}) = \underset{\sim}{0}'$ is true. In this expression the vector $\underset{\sim}{\boldsymbol{E}}'_k$ is the n-dimensional cartesian row vector in which the $\mathbf{k}^{\text{th}}$ element of the vector has a value of one and all other elements have a value of zero. Further investigation by the observant eye shows that we can express the matrix $\boldsymbol{A}$ as the following summation of vector products:

$$\boldsymbol{A} = \sum_{k=1}^{n} d_i \cdot (\underset{\sim}{\boldsymbol{E}}_k \cdot \underset{\sim}{\boldsymbol{E}}'_k) .$$

Now let's suppose that we have a matrix $\boldsymbol{A}$ that can be expressed as the following product: $\boldsymbol{A} = \boldsymbol{P}^{-1} \cdot \boldsymbol{D} \cdot \boldsymbol{P}$ (spectral decomposition), where the matrix $\boldsymbol{D}$ is a diagonal matrix; $\boldsymbol{D} = \begin{bmatrix} d_1 & 0 & \dots & 0 \\ 0 & d_2 & \dots & 0 \\ \dots & \dots & \dots & \dots \\ 0 & 0 & \dots & d_n \end{bmatrix}$. With this representation, we have $(\mathbf{A} - x \cdot \mathbf{I}) = (\mathbf{P}^{-1} \cdot \mathbf{D} \cdot \mathbf{P} - x \cdot \mathbf{I})$. This relationship can be written as $(\mathbf{A} - x \cdot \mathbf{I}) = \mathbf{P}^{-1} \cdot (\mathbf{D} - x \cdot \mathbf{I}) \cdot \mathbf{P}$. Hence if $x = d_i$, then we have for $i \in \{1, 2, \ldots, n\}$ that $|\mathbf{A} - d_i \cdot \mathbf{I}| = 0$. Since the determinant of the product of matrices is the product of the determinants, we can express the characteristic polynomial function of such a matrix as the product $\rho_A(x) = \prod_{i=1}^{n} (d_i - x)$. Furthermore, if we express the matrix $\mathbf{P}$ as a matrix partitioned into row vectors $\mathbf{P} = \begin{bmatrix} \underset{\sim}{\mathbf{P}}'_1 \\ \underset{\sim}{\mathbf{P}}'_2 \\ \\ \underset{\sim}{\mathbf{P}}'_n \end{bmatrix}$ and the inverse $\mathbf{P}^{-1}$ as a matrix partitioned into column vectors $\mathbf{P}^{-1} = \begin{bmatrix} \underset{\sim}{\mathbf{Q}}_1 & \underset{\sim}{\mathbf{Q}}_2 & \cdots & \underset{\sim}{\mathbf{Q}}_n \end{bmatrix}$. Using this notation and implementing the multiplication of the matrices using partitions, we get the following expression:

$\mathbf{A} = \sum_{k=1}^{n} d_k \cdot (\underset{\sim}{\mathbf{Q}}_k \cdot \underset{\sim}{\mathbf{P}}'_k)$. Note that the matrices in this linear expression are pairwise orthogonal and hence this is defined to be a spectral decomposition.

Now that we have explored the simple cases described above, we will now examine the general case of the matrix $\mathbf{A}$. It turns out that any square matrix can be described as a product of a diagonal matrix between a non-singular matrix and its inverse. After stating a few definitions and proving a few preliminary corollaries, we will prove the theorem that states this fact. Reviewing the concept of characteristic roots, we examine the possibility of a single root having multiple independent characteristic vectors associated with the matrix. This concept leads to definition 5.2.1 below.

Definition 5.2.1 Let x_1 be a characteristic root of matrix $\boldsymbol{A}$. If there are "**m**" independent characteristic vectors associated with the root x_1 such that $\underset{\sim}{\boldsymbol{P}}'_i \cdot (\boldsymbol{A} - x_1 \cdot \boldsymbol{I}) = \underset{\sim}{0}'$ where $i \in \{1, 2, \ldots, m\}$ then the root x_1 is of multiplicity "**m**" for matrix $\boldsymbol{A}$.

With the definition of multiplicity of a root of a matrix, several questions immediately come to mind.

1). How is the multiplicity of a root related to the characteristic polynomial function associated with the matrix?
2). Can the multiplicity of the root be determined from the characteristic polynomial function associated with the matrix?
3). What is the value of the sum of the multiplicities of all roots of the characteristic equation associated with the matrix?

To answer the above questions concerning the multiplicity of roots, the following corollary will prove useful.

Corollary 5.2.2 If there are **k** characteristic roots $\{x_i\}$ and **k** associated independent characteristic vectors

$\{\underset{\sim}{P'}_i\}$, then there are **n** independent vectors $\{\underset{\sim}{P'}_i\}$ such that

$$\begin{bmatrix} \underset{\sim}{P'}_1 \\ \dots \\ \underset{\sim}{P'}_k \\ \dots \\ \underset{\sim}{P'}_n \end{bmatrix} \cdot A = \begin{bmatrix} x_1 & \dots & 0 & \dots & 0 \\ \dots & \dots & \dots & \dots & \dots \\ 0 & \dots & x_k & \dots & 0 \\ \dots & \dots & \dots & \dots & \dots \\ \alpha_{n,1} & \dots & \alpha_{n,k} & \dots & \alpha_{n,k} \end{bmatrix} \cdot \begin{bmatrix} \underset{\sim}{P'}_1 \\ \dots \\ \underset{\sim}{P'}_k \\ \dots \\ \underset{\sim}{P'}_n \end{bmatrix} .$$

Proof: For the **k** characteristic roots $\{x_i\}$ and **k** associated independent characteristic vectors $\{\underset{\sim}{P'}_i\}$. We have **k** vector equations such that $\underset{\sim}{P'}_i \cdot A = x_i \cdot \underset{\sim}{P'}_i$. Since the order of the characteristic polynomial has the value of **n**, we have $k \le n$. When the value of **k** is less than the value of **n**, we can find **n-k** independent vectors that are independent of **k** characteristic vectors. For ease of computation we will use the notation $\{\underset{\sim}{P'}_{k+1}, \dots, \underset{\sim}{P'}_n\}$ to represent the independent vectors not in the original characteristic vectors. Since all **n** vectors are independent they form a basis for the **n**-dimensional space $\boldsymbol{R}^n$. Hence for $i > k$ we have $\underset{\sim}{P'}_i \cdot A = \alpha_{i,1} \cdot \underset{\sim}{P'}_1 + \dots + \alpha_{i,n} \cdot \underset{\sim}{P'}_n$. Placing these equations in matrix notation, we show that the statement of the corollary is true; that is

$$\begin{bmatrix} \underset{\sim}{P}'_1 \\ \cdots \\ \underset{\sim}{P}'_k \\ \cdots \\ \underset{\sim}{P}'_n \end{bmatrix} \cdot A = \begin{bmatrix} x_1 & \cdots & 0 & \cdots & 0 \\ \cdots & \cdots & \cdots & \cdots & \cdots \\ 0 & \cdots & x_k & \cdots & 0 \\ \cdots & \cdots & \cdots & \cdots & \cdots \\ \alpha_{n,1} & \cdots & \alpha_{n,k} & \cdots & \alpha_{n,k} \end{bmatrix} \cdot \begin{bmatrix} \underset{\sim}{P}'_1 \\ \cdots \\ \underset{\sim}{P}'_k \\ \cdots \\ \underset{\sim}{P}'_n \end{bmatrix} \cdot$$

Additional corollaries that will prove useful are expressed below. If we know the values of characteristic roots for a matrix can we derive a relationship between the matrix and the characteristic roots?

Corollary 5.2.3 Let x_1 be a characteristic root of matrix A with a multiplicity value of **m**. Then the characteristic polynomial of the matrix is of the form $\rho_A(x) = (x_1 - x)^m \cdot \rho_1(x)$ and the polynomial $\rho_1(x)$ is of order **n-m**.

Proof: The details of this proof is left to the reader as an exercise.

Now let's examine the converse of corollary 5.2.3. If the characteristic polynomial can be expressed as two polynomials such that it can be expressed as $\rho_A(x) = (x_1 - x)^m \cdot \rho_1(x)$, where the polynomial $\rho_1(x)$ is of order **n-m** and x_1 is not a root of it. Is the multiplicity of the characteristic root x_1 equal to **m**? The answer to this question is yes. The proof is shown in corollary 5.2.4 below.

Corollary 5.2.4 If the value specified by x_1 is a characteristic root of the "**n-by-n**" matrix $\boldsymbol{A}$ and the characteristic function polynomial of the matrix is of the form $\rho_A(x) = (x_1 - x)^m \cdot \rho_1(x)$ and the polynomial $\rho_1(x)$ is of order "**n-m**" and does not have x_1 as a root, then there is a non-singular matrix $\boldsymbol{P}$ such that

$$\boldsymbol{A} = \boldsymbol{P}^{-1} \cdot \begin{bmatrix} x_1 \cdot \boldsymbol{I}_{m \times m} & \varnothing \\ \varnothing & A_{2,2} \end{bmatrix} \cdot \boldsymbol{P} .$$

Proof: We begin this proof by defining the matrix $\boldsymbol{B} = (\boldsymbol{A} - x_1 \cdot \boldsymbol{I})$.

Since x_1 is a characteristic root of the "**n-by-n**" matrix $\boldsymbol{A}$, the matrix $\boldsymbol{B}$ is singular and there are a sequence of elementary matrices that reduces it to canonical form. The matrix $\boldsymbol{B}$ is reduced to canonical form through the use of elementary matrices. By having the elementary matrices operate on row one only of the original matrix (when it is a pre-multiplier - multiplies from left), we get a product series of elementary matrices $\boldsymbol{P}_1 = \prod_j \boldsymbol{E}_j$ such that

$$\boldsymbol{P}_1 \cdot \boldsymbol{B} = \begin{bmatrix} \boldsymbol{0} & \underset{\sim}{\boldsymbol{0}'} \\ \underset{\sim}{\boldsymbol{B}_1} & \boldsymbol{B}_{2,2} \end{bmatrix}$$

. Since the matrix $\boldsymbol{P}_1$ is a product of elementary matrices, it is a non-singular matrix and hence $P^{-1}{}_1$ exist and is the product of elementary matrices that affects only column 1 of the matrix (when it is a post-multiplier - multiplies from right). Hence the resultant matrix $P_0 \cdot B \cdot P_0^{-1} = \begin{bmatrix} \mathbf{0} & \underset{\sim}{\mathbf{0}'} \\ \underset{\sim}{B_2} & B_{2,2} \end{bmatrix}$

has all zero elements in the first row. If the column vector denoted by $\underset{\sim}{B_2}$ is not a zero vector, then we can pre-multiply the resultant matrix above, using an elementary matrix, to place a zero in the top position of the vector. This would be row 2 and column 1 of the matrix. We shall denote this elementary matrix as E . It will either exchange row two of the matrix with another row (greater than 2) having a zero in position 1 or it will subtract a multiple of another row not having zero in position 1. Since E is an elementary matrix it has an inverse. Also the inverse of the elementary matrix affects only column 2 of the matrix when post-multiplying to the matrix above. Therefore we get the following relationship:

$$E \cdot \left(P_0 \cdot B \cdot P_0^{-1} \right) \cdot E^{-1} = \begin{bmatrix} \mathbf{0} & \underset{\sim}{\mathbf{0}'} \\ \underset{\sim}{\begin{pmatrix} 0 \\ B_3 \end{pmatrix}} & C_{2,2} \end{bmatrix}$$

. If we let

$P_1 = E \cdot P_0$ then we write the equation as

$$P_1 \cdot B \cdot P_1^{-1} = \begin{bmatrix} \mathbf{0} & \underset{\sim}{\mathbf{0}'} \\ \underset{\sim}{\begin{pmatrix} 0 \\ B_3 \end{pmatrix}} & C_{2,2} \end{bmatrix}$$

. Since

$\rho_A(x) = (x_1 - x)^m \cdot \rho_1(x)$, the next "**m-1**' resulting diagonal matrices will be singular and we can apply this approach to the sub-matrix $C_{2,\,2}$ and the other "**m-2**" resulting diagonal matrices. Recursively applying this approach "**m-1**" times, we get the following: $P_m \cdot B \cdot P_m^{-1} = \begin{bmatrix} \varnothing_1 & \varnothing_2 \\ \underset{\sim}{B}_{m+1} & C_{m,\,m} \end{bmatrix}$ where $\varnothing_1$ is an "**m-by-m**" zero matrix and $\varnothing_2$ is an "**m-by-(n-m)**" zero matrix.

Since the characteristic equation has "**m**" roots with a value equal to a value of x_1 , the sub-matrix $C_{m,\,m}$ is non-singular. Hence its inverse $C^{-1}{}_{m,\,m}$ exists. Hence we are able to define the non-singular matrix $M = \begin{bmatrix} I_{m \times m} & \varnothing_2 \\ C^{-1}{}_{m,\,m} \cdot B_{m+1} & C_{m,\,m} \end{bmatrix}$ where the inverse is computed to be $M^{-1} = \begin{bmatrix} I_{m \times m} & \varnothing_2 \\ -(C^{-1}{}_{m,\,m} \cdot B_{m+1}) & C_{m,\,m} \end{bmatrix}$. This choice of M gives us the following relationship:

$M \cdot \left(P_m \cdot B \cdot P_m^{-1} \right) \cdot M^{-1} = \begin{bmatrix} \varnothing_1 & \varnothing_2 \\ \varnothing & C_{m,\,m} \end{bmatrix}$. If we define $P = M \cdot P_m$ resulting in expression $P^{-1} = P_m^{-1} \cdot M^{-1}$, then we can write $\boldsymbol{P} \cdot (\boldsymbol{A} - x_1 \cdot \boldsymbol{I}) \cdot \bar{\boldsymbol{P}}^1 = \begin{bmatrix} \varnothing_1 & \varnothing_2 \\ \varnothing & \boldsymbol{C}_{m,m} \end{bmatrix}$. By

adding the matrix quantity $x_1 \cdot \boldsymbol{I}_{m \times m}$ to both sides of the equation

and letting $A_{2,2} = \boldsymbol{C}_{m,m} + x_1 \cdot \boldsymbol{I}_{m \times m}$, we get

$\boldsymbol{P} \cdot \boldsymbol{A} \cdot \bar{\boldsymbol{P}}^1 = \begin{bmatrix} x_1 \cdot \boldsymbol{I}_{m \times m} & \varnothing_2 \\ \varnothing & A_{2,2} \end{bmatrix}$ and hence we can write

$\boldsymbol{A} = \boldsymbol{P}^{-1} \cdot \begin{bmatrix} x_1 \cdot \boldsymbol{I}_{m \times m} & \varnothing \\ \varnothing & A_{2,2} \end{bmatrix} \cdot \boldsymbol{P}$. With this result, we have

proven the statement of the corollary when the characteristic equation of matrix is of the form $\rho_A(x) = (x_1 - x)^m \cdot \rho_1(x)$ and x_1 is a not a root of the polynomial $\rho_1(x)$.

Corollary 5.2.4, when stated in literal terms, states that the multiplicity of the characteristic root x_1 is "**m**" when the characteristic equation of matrix is of the form $\rho_A(x) = (x_1 - x)^m \cdot \rho_1(x)$ and x_1 is a not a root of the polynomial $\rho_1(x)$. Corollaries 5.2.3 and 5.24 allow us to define the multiplicity of roots in terms of the number of times the root appears in the characteristic equation. Also, corollary 5.2.4 implies that the sum of the multiplicity of all the characteristic roots of a matrix is equal to the number of columns (or rows) in the matrix. A repeated application of corollary 5.2.4 to a matrix and its subsequent diagonal sub-matrices yields the theorem below.

Theorem 5.2.5 If the "**n-by-n**" matrix A has the following characteristic roots $\{x_1, x_2, \ldots, x_n\}$,then there is a non-singular matrix

$$P \text{ such that } A = P^{-1} \cdot \begin{bmatrix} x_1 & 0 & \ldots & 0 \\ 0 & x_2 & \ldots & 0 \\ \ldots & \ldots & \ldots & \ldots \\ 0 & 0 & \ldots & x_n \end{bmatrix} \cdot P.$$

Proof: To prove this theorem, apply corollaries 5.2.3 and 5.2.4 to the matrix A and all the sub-matrices of derived along the diagonal as you apply corollary 5.2.4. I leave the details of this proof to the reader.

Now that we have shown that any arbitrary matrix A can be expressed as the product $A = P^{-1} \cdot \begin{bmatrix} x_1 & 0 & \ldots & 0 \\ 0 & x_2 & \ldots & 0 \\ \ldots & \ldots & \ldots & \ldots \\ 0 & 0 & \ldots & x_n \end{bmatrix} \cdot P$, we can express it in terms of the linear sum of idempotent matrices. In particular, we have the relationship

$A = \sum_{k=1}^{n} x_k \cdot (\underset{\sim}{Q}_k \cdot \underset{\sim}{P'}_k)$. In this relationship the row vectors $\underset{\sim}{P'}_k$ are the row vectors that make up the matrix P . The column vectors $\underset{\sim}{Q}_k$ are the column vectors that make up the matrix P^{-1} . We have the relationships $\underset{\sim}{P'}_k \cdot \underset{\sim}{Q}_k = 1$ and $\underset{\sim}{P'}_j \cdot \underset{\sim}{Q}_k = 0 \text{ for } j \neq k$. This representation of the matrix

$\boldsymbol{A}$ will prove useful as we discuss properties of matrices and properties of the characteristic vectors in the following sections.

(Special Case $x_k = 1 \; for\, all \;\; k$**)**

For the special case when all characteristic roots are equal to the same value, namely one (1), then the diagonal matrix in theorem 5.2.4 is the identity matrix. Hence in this instance we have $\boldsymbol{A} = \boldsymbol{P}^{-1} \cdot \boldsymbol{I} \cdot \boldsymbol{P}$. Using the commutative property of multiplication for matrices we have $\boldsymbol{A} = \boldsymbol{I}$. Using the expansion of $\boldsymbol{A}$ as a sum of idempotent matrices, we have the following formula:

$\boldsymbol{I} = \sum_{k=1}^{n} \underset{\sim}{\boldsymbol{Q}}_k \cdot \underset{\sim}{\boldsymbol{P}}'_k$. Hence for any non-singular matrix and its inverse, we have this relationship of the respective row vectors and column vectors.

(Special Case Symmetric Matrix)

When the matrix $\boldsymbol{A}$ is symmetric, by definition of symmetric, we have $\boldsymbol{A}' = \boldsymbol{A}$. Expanding both in terms of the idempotent matrices, we get

$\boldsymbol{A} = \sum_{k=1}^{n} x_k \cdot (\underset{\sim}{\boldsymbol{Q}}_k \cdot \underset{\sim}{\boldsymbol{P}}'_k)$ and $\boldsymbol{A}' = \sum_{k=1}^{n} x_k \cdot (\underset{\sim}{\boldsymbol{P}}_k \cdot \underset{\sim}{\boldsymbol{Q}}'_k)$. From the definition of a characteristic vector for $\boldsymbol{A}$, we have $\underset{\sim}{\boldsymbol{P}}'_k \cdot \boldsymbol{A} = x_k \cdot \underset{\sim}{\boldsymbol{P}}'_k$ and

$\underset{\sim}{\boldsymbol{Q}}'_k \cdot \boldsymbol{A} = x_k \cdot \underset{\sim}{\boldsymbol{Q}}'_k$. If we add the diagonal matrix $x \cdot \boldsymbol{I}$ to $\boldsymbol{A}$, we get similar relationships, namely $(\boldsymbol{A} + x \cdot \boldsymbol{I})' = \boldsymbol{A} + x \cdot \boldsymbol{I}$,

$$(\boldsymbol{A} + x \cdot \boldsymbol{I})' = \sum_{k=1}^{n} (x_k + x) \cdot (\underset{\sim}{\boldsymbol{P}}_k \cdot \underset{\sim}{\boldsymbol{Q}}'_k) \text{ and}$$

$$\underset{\sim}{\boldsymbol{P}}'_k \cdot (\boldsymbol{A} + x \cdot \boldsymbol{I}) = (x_k + x) \cdot \underset{\sim}{\boldsymbol{P}}'_k .$$

Since the sequence of row vectors are independent and form a basis for the n-dimensional vector space $\boldsymbol{R}_n$, we can express each row vector $\underset{\sim}{\boldsymbol{P}}'_k$ as the linear combination of the independent row vectors $\underset{\sim}{\boldsymbol{Q}}'_k$; $\underset{\sim}{\boldsymbol{P}}'_k = \sum_{k=1}^{n} a_{k,j} \cdot \underset{\sim}{\boldsymbol{Q}}'_j$. With this representation, we use the definition of a characteristic vector to form the following relationship: $\underset{\sim}{\boldsymbol{P}}'_k \cdot (\boldsymbol{A} + x \cdot \boldsymbol{I}) = (x_k + x) \cdot \left(\sum_{k=1}^{n} a_{k,j} \cdot \underset{\sim}{\boldsymbol{Q}}'_j \right)$.

Since the matrix $(\boldsymbol{A} + x \cdot \boldsymbol{I})$ is symmetric we can replace the expression for the transpose $(\boldsymbol{A} + x \cdot \boldsymbol{I})'$ in this expression to yield the results $\underset{\sim}{\boldsymbol{P}}'_k \cdot \sum_{j=1}^{n} (x_j + x) \cdot (\underset{\sim}{\boldsymbol{P}}_j \cdot \underset{\sim}{\boldsymbol{Q}}'_j) = (x_k + x) \cdot \left(\sum_{j=1}^{n} a_{k,j} \cdot \underset{\sim}{\boldsymbol{Q}}'_j \right)$. Carrying through this vector and scalar multiplication, we rewrite the expression as

$\underset{\sim}{\varnothing}' = \sum_{j=1}^{n} ((x_j + x) \cdot (\underset{\sim}{\boldsymbol{P}}'_k \cdot \underset{\sim}{\boldsymbol{P}}_j) - a_{k,j} \cdot (x_k + x)) \cdot \underset{\sim}{\boldsymbol{Q}}'_j$. But recall that the row vectors $\underset{\sim}{\boldsymbol{Q}}'_k$ are independent vectors. Hence, for this equation to be true we must have for each value of **j** and **k** the following equation:

$(x_j + x) \cdot (\underset{\sim}{\boldsymbol{P}}'_k \cdot \underset{\sim}{\boldsymbol{P}}_j) - a_{k,j} \cdot (x_k + x) = 0$. By rearranging the terms in this expression we get

$(x_j \cdot (\underset{\sim}{P}'_k \cdot \underset{\sim}{P}_j) - a_{k,j} \cdot x_k) + x \cdot ((\underset{\sim}{P}'_k \cdot \underset{\sim}{P}_j) - a_{k,j}) = 0$. With x being an arbitrary value, we must have that the coefficient term and the additive term must both be zero for this equation to be true for all x values. Therefore we have the following two equations:

(1). $(\underset{\sim}{P}'_k \cdot \underset{\sim}{P}_j) - a_{k,j} = 0$ and

(2). $x_j \cdot (\underset{\sim}{P}'_k \cdot \underset{\sim}{P}_j) - a_{k,j} \cdot x_k = 0$.

From equation (1) above, the solution for the coefficients are $a_{k,j} = (\underset{\sim}{P}'_k \cdot \underset{\sim}{P}_j)$. Substituting this value for the coefficients in equation (2) above, We have either $(\underset{\sim}{P}'_k \cdot \underset{\sim}{P}_j) = 0$ or $x_j = x_k$. When $x_j \neq x_k$ then we must have for a symmetric matrix that the row vectors $\underset{\sim}{P}'_k$ and $\underset{\sim}{P}_j'$ of the matrix P are orthogonal to one another. When the multiplicity of the characteristic root x_k is greater than one, then we may have independent characteristic vectors associated with the root x_k that are not orthogonal. If the multiplicity of root x_k is **m**, then we have $\underset{\sim}{P}'_k = \sum_{k=1}^{m} (\underset{\sim}{P}'_k \cdot \underset{\sim}{P}_j) \cdot \underset{\sim}{Q}'_j$. If we designate the matrix $M_{(k)}$ as $M_{(k)} = [\underset{\sim}{P}'_i \cdot \underset{\sim}{P}_j]$, then, using matrix notation we can write the following relationship: $P = \begin{bmatrix} M_{(1)} & \dots & 0 \\ \dots & \dots & \dots \\ 0 & \dots & M_{(r)} \end{bmatrix} \cdot (P^{-1})^T$.

Each of the sub-matrices along the diagonal has a rank equal to the multiplicity of the characteristic root that it is associated with as described in the previous statements. When multiplying this matrix by the diagonal matrix, we can illustrate that the commutative property holds;

$D \cdot \begin{bmatrix} M_{(1)} & \dots & 0 \\ \dots & \dots & \dots \\ 0 & \dots & M_{(r)} \end{bmatrix} = \begin{bmatrix} M_{(1)} & \dots & 0 \\ \dots & \dots & \dots \\ 0 & \dots & M_{(r)} \end{bmatrix} \cdot D$. This is true since the diagonal

matrix has the form $D = \begin{bmatrix} x_k \cdot I_{(1)} & \dots & 0 \\ \dots & \dots & \dots \\ 0 & \dots & x_k \cdot I_{(r)} \end{bmatrix}$. For ease of computation, we will refer to this matrix as N. These relationships leads to the following corollary for symmetric matrices.

Corollary 5.2.6 If the matrix $A = P^{-1} \cdot D \cdot P$ is symmetric the we have $P = N \cdot (P^{-1})^T$ where N satisfies the relationship $D \cdot N = N \cdot D$. Furthermore, if all characteristic roots are of multiplicity one, then the matrix N is the identity matrix and we have that $P = (P^{-1})^T$ or $P^{-1} = P^T$.

Proof: The details of this proof are left as an exercise for the reader.

Exercises

1. Prove corollary 5.2.3, that is if x_1 is a characteristic root of matrix $\boldsymbol{A}$ with a multiplicity value of **m**. Then the characteristic polynomial of the matrix is of the form

   $\rho_A(x) = (x_1 - x)^m \cdot \rho_1(x)$ and the polynomial $\rho_1(x)$ is of order **n-m**.

2. Prove that the following relationship is true for every **n-by-n** matrix, partitioned as follows

   $$\boldsymbol{A} = \begin{bmatrix} a_{1,1} & \underset{\sim}{A}'_{1,2} \\ \underset{\sim}{A}_{2,1} & \boldsymbol{A}_{2,2} \end{bmatrix} : |\boldsymbol{A}| = a_{1,1} \cdot |\boldsymbol{A}_{2,2}| \cdot \left| \boldsymbol{I} - \frac{(\underset{\sim}{A}_{2,1} \cdot \boldsymbol{A}_{2,2}^{-1} \cdot \underset{\sim}{A}'_{1,2})}{a_{1,1}} \right| \text{ when}$$

   $a_{1,1} \neq 0$ and the sub-matrix $\boldsymbol{A}_{2,2}$ is non-singular. (Hint: Represent the matrix $\boldsymbol{A}$ as the product of two matrices.)

3. Prove that the following relationship is true for every **n-by-n** matrix, partitioned as follows

   $$\boldsymbol{A} = \begin{bmatrix} a_{1,1} & \underset{\sim}{A}'_{1,2} \\ \underset{\sim}{A}'_{2,1} & \boldsymbol{A}_{2,2} \end{bmatrix} : \quad |\boldsymbol{A}| = a_{1,1} \cdot |\boldsymbol{A}_{2,2}| \cdot \left(1 - \frac{(\underset{\sim}{A}'_{1,2} \cdot \boldsymbol{A}_{2,2} \cdot \underset{\sim}{A}_{2,1})}{a_{1,1}}\right)$$

   when $a_{1,1} \neq 0$ and the sub-matrix $\boldsymbol{A}_{2,2}$ is non-singular. (Hint: Represent the matrix $\boldsymbol{A}$ as the product of two matrices.)

4. In this section, we introduced the concept of a characteristic vector associated with a characteristic root of a matrix. To construct an explicit formula for computing the associated characteristic vectors, show that if $(x, \underset{\sim}{X}')$ represents the associated characteristic vector

   of the characteristic root z and matrix $\boldsymbol{A} = \begin{bmatrix} a_{1,1} & \underset{\sim}{A}'_{1,2} \\ \underset{\sim}{A}'_{2,1} & \boldsymbol{A}_{2,2} \end{bmatrix}$ then we have

   $$\underset{\sim}{X}' = x \cdot \underset{\sim}{A}'_{1,2} \cdot (z \cdot \boldsymbol{I} - \boldsymbol{A}_{2,2})^{-1} \text{ when } (z \cdot \boldsymbol{I} - \boldsymbol{A}_{2,2})^{-1} \text{ exists.}$$

Exercises - Computing Characteristic Vectors

For exercises five through ten below, compute the characteristic vector of the matrix specified as A in each exercise and the associated characteristic root specified as z . Use the approach described in exercise 4 above with $x = 1$.

5. $A = \begin{bmatrix} -148 & 202 & -58 \\ 70 & 100 & 28 \\ 645 & -897 & 255 \end{bmatrix}$ and $z = -2$

6. $A = \begin{bmatrix} 56 & -27 & -6 \\ 102 & -45 & -4 \\ -39 & 11 & -9 \end{bmatrix}$ and $z = 1$

7. $A = \begin{bmatrix} -22 & 9 & -8 & 3 \\ -26 & 9 & -10 & 1 \\ -2 & 1 & -4 & -5 \\ -16 & 8 & -8 & -8 \end{bmatrix}$ and $z = -10$

8. $A = \begin{bmatrix} -80 & -15 & -43 & -1 \\ 268 & 21 & 149 & 23 \\ 71 & 15 & 34 & 1 \\ 339 & 45 & 192 & 15 \end{bmatrix}$ and $z = 12$

9. $A = \begin{bmatrix} 61 & -130 & -72 & -260 & -74 \\ -51 & 104 & 78 & 213 & 63 \\ -223 & 448 & 198 & 877 & 245 \\ -405 & 808 & 326 & 1577 & 437 \\ 1777 & -3558 & -1532 & -6970 & -1942 \end{bmatrix}$ and $z = 5$

10. $A = \begin{bmatrix} -88 & -48 & 37 & 62 & 10 \\ -2 & 22 & -4 & -16 & -11 \\ -90 & 30 & 21 & 0 & -30 \\ -85 & -90 & 45 & 103 & 36 \\ 170 & 180 & -90 & -200 & -69 \end{bmatrix}$ and $z = -2$

5.3 Determination of Characteristic Roots

This section represents a deviation from matrix theory. We discuss some approaches for solving polynomial equations and introduce a numerical algorithm that is readily implemented using high speed computers.

As stated in the section 5.1, the characteristic equation, $\rho_A(x) = 0$, has "**n**" roots when the matrix A is an "**n-by-n**" square matrix. But all roots may not necessarily be in the realm of real numbers. For example, if the characteristic equation of a "**2-by-2**" matrix is $x^2 + 1 = 0$, then there is no real solution to this equation. But if we extend the solution set to that of complex numbers (imaginary numbers); we have two complex number solutions. In this section we will look for the real number solutions and leave the complex number solutions for a more advanced treatise of matrix algebra.

5.3.1 Solving for Roots of a "2-by-2" matrix

The simplest case for computing the characteristic roots is that of the "**2-by-2**" matrix. Unlike the higher ordered matrices, an explicit formula is easily obtained for this matrix. Let the matrix A be denoted as $A = \begin{bmatrix} a_{1,1} & a_{1,2} \\ a_{2,1} & a_{2,2} \end{bmatrix}$. For this matrix, the characteristic equation is

$$\rho_A(x) = x^2 - (a_{1,1} + a_{2,2}) \cdot x + (a_{1,1} \cdot a_{2,2} - a_{1,2} \cdot a_{2,1}) = 0 .$$

Applying the quadratic formula to this 2nd degree polynomial we get the following solution

$$x = \frac{(a_{1,1} + a_{2,2}) \pm \sqrt{(a_{1,1} + a_{2,2})^2 + 4 \cdot (a_{1,1} \cdot a_{2,2} - a_{1,2} \cdot a_{2,1})}}{2}$$

.

5.3.2 Solving for Roots of a "3-by-3" Matrix

Unlike the **2-by-2** square matrix, the roots of the **3-by-3** square matrix are the solution of a three degree polynomial equation with solutions that are not consistently expressed as an explicit equation. To consistently solve a three degree polynomial, we develop a numerical algorithm to compute the real roots of the polynomial equation. To develop an algorithm, to determine roots of the three degree polynomial, we examine the general form of the three degree polynomial in graphical form. To simplify, the graphing of the polynomial, we will assume that it is of the form $-\rho_A(x) = x^3 + a_2 \cdot x^2 + a_1 \cdot x + a_0$. (The coefficient of the highest term is positive one.)

The graph to the right represents the third order polynomial with three real roots. The points marked with solid filled circles represent the local minimum and local maximum of the polynomial function.

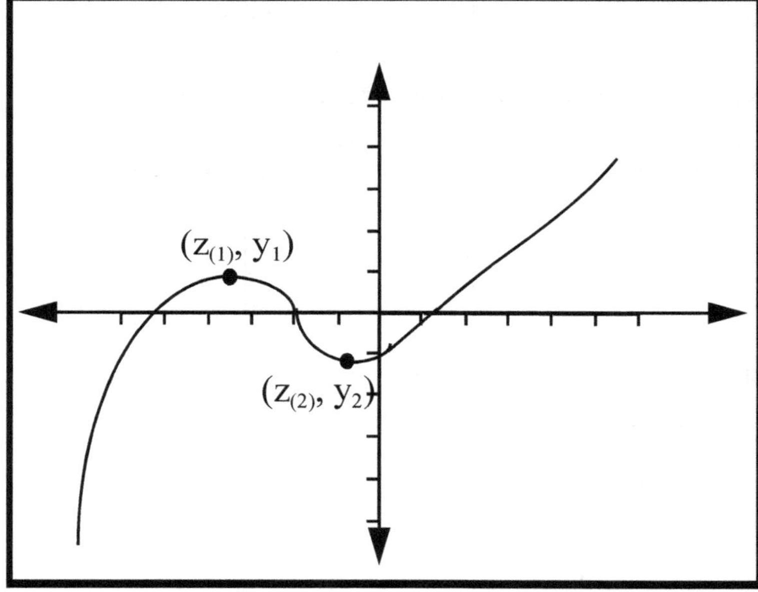

Using the knowledge that as the value of **x** approaches negative infinity, the value of the polynomial approaches negative infinity. Also as the value of **x** approaches positive infinity, the value of the polynomial approaches positive infinity. Hence, we find a value $\mathbf{x}_0$ such that $x_0 < z_{(1)}$, and $-\rho_A(x_0) < 0$. The first zero of the polynomial is found by performing a search between x_0 *and* $z_{(1)}$. This algorithm requires the knowledge of the local maxima and minima of the polynomial function. We turn to the principles of differential calculus and locate the zero values of the derivative of the polynomial relative to the variable "**x**". In this case the derivative of the third degree polynomial, shown above, is $\frac{d}{dx}(-\rho_A(x)) = 3 \cdot x^2 + 2 \cdot a_2 \cdot x + a_1$. The numerical search algorithm for finding the smallest zero value of the third degree polynomial is described in the table below.

TABLE 9. Numerical Solution of Smallest Root of 3rd Degree Polynomial

Step	Parameter	Description of Calculations
1	Compute $\mathbf{Z}_{(1)}$ and $\mathbf{Z}_{(2)}$ such that $\mathbf{Z}_{(1)} < \mathbf{Z}_{(2)}$.	The values of $\mathbf{Z}_{(1)}$ and $\mathbf{Z}_{(2)}$ are the zeros of the derivative of the third degree polynomial.
2	Find a value $X_0 < \mathbf{Z}_{(1)}$.	Evaluate the function $(-\mathbf{p}_A(x))$ until it has a negative value. This will give us an initial value $\mathbf{X}_0$.
3	Compute initial value of **L.** This value will serve as a limiting value of the search algorithm.	The initial value of **L** is computed as follows: $$\mathbf{L} = \left\{ \begin{array}{c} Z_{(1)} when (-p_A(Z_{(1)}) \geq 0) \\ Z_{(2)} when (-p_A(Z_{(2)}) \geq 0) \\ U \; such that \; (-p_A(U) \geq 0) \end{array} \right\}$$
4	Obtain better estimate of smallest zero of polynomial; X_k.	$$X_k = X_{k-1} - \frac{p_A(X_{k-1}) \cdot (X_{k-1} - \boldsymbol{L})}{(p_A(X_{k-1}) - p_A(\boldsymbol{L}))}$$

TABLE 9. Numerical Solution of Smallest Root of 3rd Degree Polynomial

Step	Parameter	Description of Calculations
5	Update value of **L**.	$\mathbf{L} = \begin{Bmatrix} X_{k-1}\ if(sgn(p_A(X_{k-1})) \neq sgn(p_A(X_k))) \\ \boldsymbol{L}\ if\ (sgn(p_A(\boldsymbol{L})) \neq sgn(p_A(X_k))) \end{Bmatrix}$
6	Repeat steps 4 and 5 if $\mathbf{D} > \mathbf{d_0}$	$\boldsymbol{D} = \lvert p_A(X_k) \rvert$
7	Upon computing the lowest root of the equation use synthetic division to reduce polynomial to a second degree polynomial.	
8	Use quadratic equation to obtain remaining roots.	

An example of finding roots of a third degree polynomial is shown in the following example.

(Example)

To find the roots of the characteristic equation $-\rho_A(x) = x^3 - x^2 - 9 \cdot x + 9$, we implement the steps shown in the table above.

1). The derivative of the function

$-\rho_{A}(x) = x^3 - x^2 - 9 \cdot x + 9$ is the 2nd order polynomial $-\rho'_{A}(x) = 3 \cdot x^2 - 2 \cdot x - 9$. The roots of the derivative are computed using the quadratic equation and are determined to be $Z_{(1)} = \dfrac{1 - 2 \cdot \sqrt{7}}{3} \approx -1.4305009$ and

$$Z_{(2)} = \frac{1 + 2 \cdot \sqrt{7}}{3} \approx 2.0971675 \quad .$$

2). In this step we search for a value $X_0 \leq -1.4305009$ such that $-\rho_{A}(X_0) \leq 0$. A first guest would be $X_0 = -5$. Upon evaluating this value, we get $-\rho_{A}(-5) = -96$ which is less than zero.

3). Since $-\rho_{A}(Z_{(1)}) = -\rho_{A}(-1.4305) = -96$, we have L = -1.4305.

4). Using the equations $X_k = X_{k-1} - \frac{p_A(X_{k-1}) \cdot (X_{k-1} - \boldsymbol{L})}{(p_A(X_{k-1}) - p_A(\boldsymbol{L}))}$ and

$$\boldsymbol{L} = \left\{ \begin{array}{c} X_{k-1} \, if(sgn(p_A(X_{k-1})) \neq sgn(p_A(X_k))) \\ \boldsymbol{L} \text{ } if \text{ } (sgn(p_A(\boldsymbol{L})) \neq sgn(p_A(X_k))) \end{array} \right\}$$, we get

the sequence of estimates:

$$X_1 = -1.9648, (-\rho_{\boldsymbol{A}}(X_1) = 15.2375)$$

$$X_5 = -2.9059, (-\rho_{\boldsymbol{A}}(X_5) = 2.1698)$$

$$X_{10} = -2.9970, (-\rho_{\boldsymbol{A}}(X_{10}) = 0.0726) \cdot$$

$$X_{16} = -3.0000, (-\rho_{\boldsymbol{A}}(X_{16}) = 0)$$

5). From the sequence of steps above, we conclude that -3 is the smallest root of the polynomial equation

$-\rho_{\boldsymbol{A}}(x) = x^3 - x^2 - 9 \cdot x + 9 = 0$. Hence if we divide the

polynomial by (x+3) we get $x^2 - 4 \cdot x + 3 = 0$. To determine the remaining roots, it is sufficient to solve for the roots of this quadratic equation.

6). Using the quadratic formula to solve the quadratic equation in step 5

above, we get $x = 1$ and $x = 3$.

7). Hence the roots of the equation

$-\rho_{\boldsymbol{A}}(x) = x^3 - x^2 - 9 \cdot x + 9 = 0$ is determined to be -3, 1 and 3.

5.3.3 Generalizing Algorithm for Solving Roots of an n-by-n Matrix

The general solution for calculating roots of the **n-by-n** square matrix is an algorithm that's a generalization of the **3**-degree polynomial. For this case, the polynomial is an **n**-degree polynomial. We insure that the coefficient of the highest power is one by multiplying the characteristic equation by the constant minus one raise to the power of **n**. The resultant polynomial is

$(-1)^n \cdot \rho_A(x) = \sum_{k=0}^{n} a_k \cdot x^k$. The nth coefficient, a_n, is equivalent to positive one in this equation. If we are to use the approach, established in the previous section, we need to compute the derivative of this polynomial and determine the zero roots of the resultant polynomial. The derivative of this polynomial function is expressed as follows:

$$\frac{d}{dx}((-1)^n \cdot \rho_A(x)) = \sum_{k=0}^{n} k \cdot a_k \cdot x^{k-1}.$$

To find the root of the characteristic equation, we use the algorithm described in the following list.

1). If the value of **n** is less than or equal to two, use the explicit formulas to obtain the real roots of the polynomial equations.

2). If the value of **n** is equal to three, obtain the 2nd degree polynomial that represents the derivative function of the 3rd degree characteristic polynomial function. Find the real roots of the derivative function using step 1.

3). Once the real roots of the derivative function are obtained, determine the two extreme points of the characteristic function that enclose the roots of the derivative function of the polynomial.

4). Using the set of values determined in step 3 above, use a binary search algorithm to determine the roots between adjacent values.

5). If the value of "n" is greater than 3, obtain the $(n-1)^{th}$ degree polynomial that represents the derivative function of the nth degree characteristic polynomial function. Determine the roots of the $(n-1)^{th}$ polynomial by recursively constructing derivatives until the derivative function is a 3rd degree polynomial. When the 3rd degree polynomial is reached, use steps 2 and 3 to obtain the roots of the 3rd degree polynomial. If the recursion for the derivatives have been done appropriately, a recursive application of steps 3 and 4 for each higher order polynomial (derivatives) will allow the computation of the real roots of the nth degree characteristic polynomial function.

The algorithm described in the list above is complex to implement. If you intend to perform the calculations without the use of a high speed computer, good luck, you will need plenty of luck and patience. In this era of high speed computing, the smart approach is to develop a computer program that will implement each step of the above algorithm. Using Visual Basic, I put together a few modules that can be implemented in an electronic spreadsheet by importing the code into the spreadsheet. Refer to appendix B for a listing of the code to determine real zeros of a polynomial function using spreadsheets. For computing the roots polynomial using the Linear Algebra LMS tool, see appendix B that discusses the use of the LMS.

(Exercises: Roots for Two-by-Two Matrices)

For exercises one through ten below, derive the characteristic roots of the two-by-two matrices.

1. $\begin{bmatrix} -6 & -9 \\ 0 & 3 \end{bmatrix}$

2. $\begin{bmatrix} -19 & -12 \\ 40 & 25 \end{bmatrix}$

3. $\begin{bmatrix} -12 & -2 \\ 1 & -9 \end{bmatrix}$

4. $\begin{bmatrix} -46 & -20 \\ 100 & 44 \end{bmatrix}$

5. $\begin{bmatrix} 13 & 7 \\ -42 & -22 \end{bmatrix}$

6. $\begin{bmatrix} 3 & -13 \\ 0 & -10 \end{bmatrix}$

7. $\begin{bmatrix} 10 & 0 \\ 10 & 0 \end{bmatrix}$

8. $\begin{bmatrix} 21 & 9 \\ -72 & -30 \end{bmatrix}$

9. $\begin{bmatrix} 19 & -6 \\ 45 & -14 \end{bmatrix}$

10. $\begin{bmatrix} -52 & -18 \\ 135 & 47 \end{bmatrix}$

(Exercises: Roots of 3rd Degree Polynomials)

Using the sequential approach for computing the smallest root of a 3rd degree polynomial, compute the roots of the polynomials in exercises 11 through 16. See root computation program in the learning management system (LMS) for a computer algorithm to help with the computation of roots.

11. $x^3 - 15 \cdot x^2 + 74 \cdot x - 120$

12. $x^3 + 6 \cdot x^2 - 16 \cdot x - 96$

13. $x^3 - 15 \cdot x^2 + 12 \cdot x + 352$

14. $x^3 - x^2 - 9 \cdot x + 9$

15. $x^3 + 13 \cdot x^2 - 58 \cdot x - 880$

16. $x^3 - 9 \cdot x^2 - 40 \cdot x + 300$

(Exercises: Roots for Polynomials)

To generalize the approach discussed in section 5.3.2, an algorithm is discussed in the appendix and provided with the linear algebra learning management system (LMS). Use the subroutines in the LMS to compute the real roots of the following polynomials.

1. $x^4 - 19 \cdot x^3 + 28 \cdot x^2 + 804 \cdot x - 1584$

2. $x^4 + 6 \cdot x^3 - 41 \cdot x^2 - 150 \cdot x + 400$

3. $x^4 - 3 \cdot x^3 - 64 \cdot x^2 - 60 \cdot x$

4. $x^4 + 13 \cdot x^3 - 21 \cdot x^2 - 433 \cdot x + 440$

5. $x^5 - 6 \cdot x^4 - 117 \cdot x^3 + 598 \cdot x^2 + 1284 \cdot x - 3960$

6. $x^5 + 8 \cdot x^4 - 115 \cdot x^3 - 1250 \cdot x^2 - 2856 \cdot x - 1728$

7. $x^5 - x^4 - 109 \cdot x^3 - 47 \cdot x^2 + 2748 \cdot x + 6048$

8. $x^6 - 17 \cdot x^5 - 40 \cdot x^4 + 1820 \cdot x^3 - 6816 \cdot x^2 - 13248 \cdot x + 69120$

5.4 Cayley-Hamilton Theorem

In this section, we investigate the behavior of the matrix function as A is placed in a linear combination of powers of the matrix A. This linear combination of powers of matrices is an extention of the real number concept of polynomials to the realm of square matrices, which we will reference as matrix polynomials. A well known theorem associated with the matrix polynomials is the Cayley-Hamilton theorem. The Caley-Hamilton theorem relates the behavior of the matrix A when it is evaluated in its associated characteristic polynomial. Before proving the Cayley-Hamilton theorem, some preliminary properties are established. These properties are stated in the corrollaries shown below.

We begin our discussion of matrix polynomials with a discussion of the polynomial of rank two (2). For the polynomial $p_2(x) = c_2 \cdot x^2 + c_1 \cdot x + c_0$, if the roots of the polynomail are $\{r_1, r_2\}$ then we can express polynomial as the product rank one polynomials. Hence we have the equation $c_2 \cdot x^2 + c_1 \cdot x + c_0 = (r_1 - x) \cdot (r_2 - x)$. The matrix polynomial for the matrix A using the polynomial $p_2(x)$ is defined to be $p_2(A) = c_2 \cdot A^2 + c_1 \cdot A + c_0$. Using the roots of the polynomial, the matrix polynomial can be expressed in terms of polynomial roots as the following products of matrices $p_2(A) = (r_1 \cdot I - A) \cdot (r_2 \cdot I - A) = (r_2 \cdot I - A) \cdot (r_1 \cdot I - A)$. Hence we can state for a polynomial of rank two, the matrix product $(r_1 \cdot I - A) \cdot (r_2 \cdot I - A)$, for any arbitrary matrix, satisfy the commutative property of binary matrix multiplication. We extend this to polynomials of rank n polynomial.(See corollary 5.4.1 below).

Corollary 5.4.1 If the matrix A is a square matrix and $p_n(x)$ *is a polynomial of rank n, then the matrix* *polynomial $p_n(A)$ is represented as a product of n matrices which satisfy the commutative property of matrix multiplication. We write*

$\prod_{K}(r_k \cdot \boldsymbol{I} - \boldsymbol{A}) = \prod_{L}(r_l \cdot \boldsymbol{I} - \boldsymbol{A})$, where K and L are different permutations of the integers from 1 to n.

Proof: We have shown, for a polynomial of rank 2, the matrix polynomial is represented as a commutative product of two polynomials of rank one each. From mathematical induction, we conclude that for a polynomial of rank n, the matrix polynomial can be expressed as a commutative product of n polynomials of rank one each.

As promised earlier, we now investigate the behavior of the square matrix as an argument of the characteristic polynomial of the matrix.

Theorem 5.4.2 The **Cayley-Hamilton Theorem** states that if the polynomial $p_A(x)$ is the characteristic polynomial of the matrix $\boldsymbol{A}$, then the matrix polynomial $p_A(\boldsymbol{A})$ satisfy the relationship $p_A(\boldsymbol{A}) = \varnothing$.

Proof: The proof of the Cayley-Hamilton theorem is shown in the table

TABLE 10. Proof of Cayley-Hamilton Theorem (5.4.2)

Step	Statement	Rationale for Statement
1	$p_A(A) = \prod_K (r_k \cdot \boldsymbol{I} - \boldsymbol{A})$	Since the characteristic polynomial of the matrix n-by-n matrix $\boldsymbol{A}$ has n roots, we can express it as a product of n polynomials of rank one (see corollary 5.4.1).
2	There are n independent row vectors $\underset{\sim}{\boldsymbol{P}}_k'$ such that $\underset{\sim}{\boldsymbol{P}}_k' \cdot (r_k \cdot \boldsymbol{I} - \boldsymbol{A}) = \varnothing$.	This relationship is obtained using the definition of a characteristic root of the matrix $\boldsymbol{A}$.
3	For each of the n row vectors we get $\underset{\sim}{\boldsymbol{P}}_k' \cdot p_A(A) = \varnothing$.	Apply commutative property of the matrix polynomial $p_A(A) = \prod_K (r_k \cdot \boldsymbol{I} - \boldsymbol{A})$.
4	There is a nonsingular/nonzero matrix $\boldsymbol{B}$ such that $\boldsymbol{B} \cdot p_A(A) = \varnothing$.	Create the matrix $\boldsymbol{B}$ using the n independent row vectors $\underset{\sim}{\boldsymbol{P}}_k'$. *Then the matrix* $\boldsymbol{B}$ *is nonsingular and* $\boldsymbol{B} \cdot p_A(A) = \varnothing$
5	**Hence we have the relationship** $p_A(A) = \varnothing$.	Since $\boldsymbol{B}$ is nonsingular and nonzer we must have the relationship $p_A(A) = \varnothing$.

below.

If we express the coefficients of each power of matrix $\boldsymbol{A}$ as a_k, then the polynomial matrix can be alternately expressed as

$$p_A(A) = \sum_{k=0}^{n} a_k \cdot \boldsymbol{A}^k = \varnothing \quad .$$

(Exercises)

These exercises provide an opportunity for the reader to test his/her skill at using the characteristic equations of the associated matrix to express the $\mathbf{n}^{th}$ or $(\mathbf{n+1})^{th}$ power of an **n-by-n** matrix.

1. Given the matrix $\begin{bmatrix} -148 & 202 & -58 \\ 70 & -100 & 28 \\ 645 & -897 & 255 \end{bmatrix}$ and its associated characteristic equation

$\rho_A(x) = -x^3 + 7 \cdot x^2 + 54 \cdot x + 72 = 0$, derive an equation for A^3 as a linear combination of I, A and A^2.

2. Given the matrix $\begin{bmatrix} 56 & -27 & -6 \\ 102 & -45 & -4 \\ -39 & 11 & -9 \end{bmatrix}$ and its associated characteristic equation

$\rho_A(x) = -x^3 + 2 \cdot x^2 + 55 \cdot x - 56 = 0$, derive an equation for A^3 as a linear combination of I, A and A^2.

3. Using the matrix $\begin{bmatrix} -148 & 202 & -58 \\ 70 & -100 & 28 \\ 645 & -897 & 255 \end{bmatrix}$ from exercise 1 above and its associated characteristic equation $\rho_A(x) = -x^3 + 7 \cdot x^2 + 54 \cdot x + 72 = 0$, derive an equation for $\boldsymbol{A}^4$ as a linear combination of $\boldsymbol{I}$, $\boldsymbol{A}$ and $\boldsymbol{A}^2$.

4. Given the matrix $\begin{bmatrix} 56 & -27 & -6 \\ 102 & -45 & -4 \\ -39 & 11 & -9 \end{bmatrix}$ from exercise 2 above and its associated characteristic equation $\rho_A(x) = -x^3 + 2 \cdot x^2 + 55 \cdot x - 56 = 0$, derive an equation for $\boldsymbol{A}^4$ as a linear combination of $\boldsymbol{I}$, $\boldsymbol{A}$ and $\boldsymbol{A}^2$.

5. Using the results of exercise 4 above, derive an equation for $\boldsymbol{A}^5$ as a linear combination of $\boldsymbol{I}$, $\boldsymbol{A}$ and $\boldsymbol{A}^2$.

5.5 Powers Of A Via Characteristic Equation

In the previous section, we derived the Cayley-Hamilton theorem, the matrix **A** satisfies the characteristic equation. Expressed in matrix notation we have

$\rho_A(A) = \sum_{k=0}^{n} a_k \cdot A^k = \varnothing$. Recall the coefficient of the highest term of the polynomial equation is non-zero; in fact it has a value of $(-1)^n$. By separating the highest term from the lower terms of the equation, we get $\sum_{k=0}^{n-1} a_k \cdot A^k = (-1)^{n+1} \cdot A^n$. Making the substitution $c_k(0) = (-1)^{n+1} \cdot a_k$, we can express the **n**th power of the matrix A *as polynomial expression of the lower powers of the matrix;* $A^n = \sum_{k=0}^{n-1} c_k(0) \cdot A^k$.

Notice that we assigned a value of zero to the coefficients of the matrix polynomial equation that expresses the relationship for A^n. *As we derive an expression for the next power,* A^{n+1} *, we assign a value of one to the coefficients. Hence the expression for the (**n**+1)th power of the matrix is given as the following:* $A^{n+1} = \sum_{k=0}^{n-1} c_k(1) \cdot A^k$. To compute the coefficients for this powers of A *, we will start with the previous equation and multiply by* A. *This multiplication yields the following relationship:*

$A^{n+1} = \sum_{k=0}^{n-1} c_k(0) \cdot A^{k+1} = c_{n-1}(0) \cdot A^n + \sum_{k=1}^{n-1} c_{k-1}(0) \cdot A^k$. By expanding the A^n. *term in this equation and combining like powers we can further refine this relationship as*

$$A^{n+1} = c_{n-1}(0) \cdot \sum_{k=0}^{n-1} c_k(0) \cdot A^k + \sum_{k=1}^{n-1} c_{k-1}(0) \cdot A^k$$ and hence we have

$$A^{n+1} = \sum_{k=0}^{n-1} (c_{n-1}(0) \cdot c_k(0) + c_{k-1}(0)) \cdot A^k + c_{n-1}(0) \cdot c_0(0) \cdot I.$$

With this relationship established, we now have an expression of coefficients $c_k(1)$. *These coefficients are calculated as follows:*

$$c_k(1) = \left\{ \begin{array}{ll} c_{n-1}(0) \cdot c_k(0) + c_{k-1}(0), & k > 0 \\ c_{n-1}(0) \cdot c_0(0), & k = 0 \end{array} \right\}.$$

The approach used for computing the coefficients $c_k(1)$, *we also use for relating the coefficients* $c_k(m)$ to $c_k(m+1)$. *Using the above approach, we get the following relationship:*

$$c_k(m+1) = \left\{ \begin{array}{ll} c_{n-1}(m) \cdot c_k(0) + c_{k-1}(m), & k > 0 \\ c_{n-1}(0) \cdot c_0(m), & k = 0 \end{array} \right\}.$$ With these coefficients we get the general expression for a matrix raised to a power in the following expression: $A^{n+m} = \sum_{k=0}^{n-1} c_k(m) \cdot A^k$. Hence knowing the powers of the matrix below the value of "**n**" and knowing the coefficients of the characteristic equation, we can compute higher powers of the matrix. without physically multiplying the matrix that number of times.

(Example)

Consider the "4-by-4" matrix with the polynomial characteristic equation $\rho_A(x) = x^4 - x^3 + 4 \cdot x^2 - 2 \cdot x + 2 = 0$. Placing the matrix in this equation and solving for $\boldsymbol{A}^4$ *yields the matrix equation* $\boldsymbol{A}^4 = \boldsymbol{A}^3 - 4 \cdot \boldsymbol{A}^2 + 2 \cdot \boldsymbol{A} - 2 \cdot \boldsymbol{I}$. Derive the coefficients that will allow us to compute $\boldsymbol{A}^{14}$. *Using the sequential relationships shown above, the coefficients are computed as shown in this table.*

TABLE 11. **Computation of** $c_k(m)$

m	0	1	2	3
0	-2	2	-4	1
1	-2	0	-4	-3
2	-2	-8	-4	1
3	-2	0	-4	-3
4	-2	-8	-4	1
5	-2	0	-4	-3
6	-2	-8	-4	1
7	-2	0	-4	-3
8	-2	-8	-4	1
9	-2	0	-4	-3
10	-2	-8	-4	1

Using the bottom row of this table, we can compute the 14th power of the matrix as the following summation: $$A^{14} = A^3 - 4 \cdot A^2 - 8 \cdot A - 2 \cdot I$$.

Exercises

For these exercises, an Excel™ worksheet is described in the Appendix. It will be useful for construction of the table of coefficients $C_k(\mathbf{m})$, described in this section. In exercises one through ten below for each specified characteristic equation, construct a table of coefficients for $C_k(\mathbf{m})$ for **m** equal zero to ten.

1. $\rho_A(x) = x^2 - 25 = 0$

2. $\rho_A(x) = x^2 - 4 \cdot x + 3 = 0$

3. $\rho_A(x) = x^2 + 3 \cdot x + 2 = 0$

4. $\rho_A(x) = x^2 + 4 \cdot x - 5 = 0$

5. $\rho_A(x) = -x^3 - 4 \cdot x^2 - 5 \cdot x - 2 = 0$

6. $\rho_A(x) = -x^3 + 9 \cdot x^2 - 24 \cdot x + 20 = 0$

7. $\rho_A(x) = -x^3 - 3 \cdot x^2 + 10 \cdot x + 24 = 0$

8. $\rho_A(x) = -x^3 - 4 \cdot x^2 + 7 \cdot x + 10 = 0$

9. $\rho_A(x) = x^4 - 6 \cdot x^2 + 8 \cdot x - 3 = 0$

10. $\rho_A(x) = x^4 - x^3 - 3 \cdot x^2 + x + 2 = 0$

5.6 Limits Of A^N as N Approaches Infinity

In the previous sections, we showed that any matrix A could be represented as a linear combination of pair-wise orthogonal idempotent matrices. Recall that a matrix J is idempotent if the square of the matrix is equivalent to a scalar multiple of the original matrix; $J^2 = c \cdot J$. Matrices $J_1, \ldots, J_n$ are pair-wise orthogonal, if for any two distinct matrices we have the relationship $J_i \cdot J_k = \varnothing$. With this notation and the results of the previous sections, we have the following spectral decomposition for A :

$A = \sum_{k=1}^{n} x_k \cdot J_k$. In this notation, the x_k term are the characteristic roots and the idempotent matrices in addition to being pair-wise orthogonal, we have $J_k^2 = J_k$.

Using the above representation for computing the powers of the matrix A , gives us the following: $A^m = \sum_{k=1}^{n} (x_k)^m \cdot J_k$. Using the distributive properties of real numbers, if we have a finite ranked polynomial $f_m(x)$, then by constructing a corresponding polynomial function of the matrix as an argument we get $f_m(A) = \sum_{k=1}^{n} f_m(x_k) \cdot J_k$. This relationship will prove to be very important as we, more closely, examine the limiting values of the power of A^N as N approaches Infinity (∞). But for now let's concentrate on determining an expression for $\lim_{N \to \infty} A^N$.

By considering the spectral decomposition shown above, the above limit is expressed as $\lim_{N \to \infty} A^N = \lim_{N \to \infty} \sum_{k=1}^{n} (x_k)^N \cdot J_k$. Since the right side of the equation is a finite sum, the limit function is moved under the summation sign and this limit becomes the following: $\lim_{N \to \infty} A^N = \sum_{k=1}^{n} \lim_{N \to \infty} (x_k)^N \cdot J_k$. Notice that the limit depends solely on the values of the characteristic roots of the matrix. Since any value **x** raised to the power of **N** converges to the value of zero when $|x| < 1$, we can state that $\lim_{N \to \infty} A^N = \varnothing$ when all characteristic roots have an absolute value that is less than one. When $x = 1$, then **x** raised to the power of **N** converges to the value of one. Hence if at least one of the characteristic roots has a value of one and the remaining roots satisfy the inequality $|x| < 1$, then $\lim_{N \to \infty} A^N = J$ where the matrix J is an nonzero idempotent matrix. If the multiplicity of the characteristic root of value one is **m**, the idempotent matrix J is the sum of the **m** characteristic matrices associated with the root of value one. Note that if a characteristic root is an imaginary number y such that $|y| < 1$ then **y** raised to the power of **N** converges to the value of zero as **N** approaches infinity. Hence the above conditions for convergence of the limit includes both real and imaginary numbers. For all other conditions $\lim_{N \to \infty} A^N$ diverges.

With the above observations concerning the convergence and divergence of the function $f(x) = x^N$,corollary 5.6.1 below gives conditions for the convergence of $\lim_{N \to \infty} A^N$ and the meaning of the limit.

Corollary 5.6.1 If the '**n-by-n**' square matrix A has only characteristic roots such that $|x| < 1$, then we have $\lim_{N \to \infty} A^N = \varnothing$. Let the matrix A be represented as the linear sum of idempotent matrices such as $A = \sum_{k \in \chi} x_k \cdot J_k + \sum_{k \in \Upsilon} x_k \cdot J_k$.

The set χ be the set of integers '**k**' such that $|x_k| < 1$ and let the set Υ be the set of integers '**k**' such that $x_k = 1$. Then the limit of the matrix A raised to the power of '**N**' as 'N' approaches infinity is $\lim_{N \to \infty} A^N = \sum_{k \in \Upsilon} J_k$.

Proof: The proof of this corollary follows from the discussion of the function $f(x) = x^N$ and its limit as **N** approaches infinity. The details of this proof is left as an exercise for the reader.

We illustrated in the previous section that any power of the **n-by-n** matrix A raised to the power of $n + m$ is expressed as a polynomial expression of powers of A raised to integer powers of zero through **n-1**. Using the sequence of variables defined in section 5.5, the power of A^{n+m} is expressed using the polynomial expression $A^{n+m} = \sum_{k=0}^{n-1} c_k(m) \cdot A^k$. The coefficients in this equation are derived sequentially by multiplying each side of the equation by the original matrix A and expanding A^n using the original polynomial expression. Using a similar polynomial expansion for the 'n+m+1' power of the matrix, we get

$A^{n+m+1} = \sum_{k=0}^{n-1} c_k(m+1) \cdot A^k$. By multiplying the equation above by the original matrix A we get $A^{n+m+1} = \sum_{k=0}^{n-1} c_k(m) \cdot A^{k+1}$.

Expanding the right side of the second equation, we can define the '**m+1**' coefficients as the following (As we have seen in section 5.5.):

$$c_k(m+1) = \begin{cases} c_{n-1}(m) \cdot c_k(0) + c_{k-1}(m), & k > 0 \\ c_{n-1}(0) \cdot c_0(m), & k = 0 \end{cases}$$

. The coefficients

$c_k(0)$ are the coefficients of a polynomial derived from the characteristic polynomial

$\rho_A(x) = \sum_{k=0}^{n} a_k \cdot x^k = 0$ as follows: $\rho_{n-1}(x) = x^n - \rho_A(x) / a_n$.

Corollary 5.6.2 Let the **n-by-n** matrix A have at least one characteristic root equal to one and the remaining roots satisfy the inequality $|x_k| < 1$, then $\lim_{N \to \infty} A^N$ converges to a nonzero idempotent matrix $J = \alpha^* \cdot \left(\sum_{k=0}^{n-1} c^*_k \cdot A^k \right)$.

The coefficients of this polynomial expression is defined below.

$$c^*_0 = c_0(0)$$
$$c^*_1 = c_1(0) + c_0(0)$$
$$\ldots$$
$$c^*_{n-2} = c_{n-2}(0) + \ldots + c_1(0) + c_0(0)$$
$$c^*_{n-1} = c_{n-1}(0) + \ldots + c_1(0) + c_0(0) = 1$$

Proof: Using the sequential relationship of the coefficients relating the sequential powers of the matrix $\boldsymbol{A}$, we obtain the following equations:

$$c_k(N-n+1) = \left\{ \begin{matrix} c_{n-1}(N-n) \cdot c_k(0) + c_{k-1}(N-n), \ k > 0 \\ c_{n-1}(0) \cdot c_0(N-n), \qquad k = 0 \end{matrix} \right\}.$$

Using these coefficients, we can compute the $\mathbf{N}^{\text{th}}$ power of $\boldsymbol{A}$ as follows: $\boldsymbol{A}^N = \sum_{k=0}^{n-1} c_k(N-n) \cdot \boldsymbol{A}^k$. In the statement of this corollary for matrix $\boldsymbol{A}$, have at least one characteristic root equal to one and the remaining roots satisfy the inequality $|x_k| < 1$. From corollary 5.6.1 we get $\lim_{N \to \infty} \boldsymbol{A}^N$ converges to a nonzero idempotent matrix $\boldsymbol{J} = \sum_{k \in \Upsilon} \boldsymbol{J}_k$. Having established the convergence of $\boldsymbol{A}^N$ as **N** approaches infinity, we can assume that as N becomes sufficiently large the respective coefficients are independent of the value of **N**. This condition implies that $c_k(N-n+1) = c_k(N-n)$ as **N** becomes sufficiently large. This phenomenon is expressed in the following set of equations: $c_k = \left\{ \begin{matrix} c_{n-1} \cdot c_k(0) + c_{k-1}, & k > 0 \\ c_{n-1}(0) \cdot c_0, & k = 0 \end{matrix} \right\}$.

Using these relationships and the knowledge that A^N converges as **N** approaches infinity, we have $\lim_{N \to \infty} A^N = \sum_{k=0}^{n-1} c_k \cdot A^k$. Solving for the coefficients of this polynomial expression, we expand the relationships above to get the following:

$$\begin{aligned} c_0 &= c_{n-1} \cdot c_0(0) \\ c_1 &= c_{n-1} \cdot c_1(0) + c_0 \\ &\dots \\ c_{n-2} &= c_{n-1} \cdot c_{n-2}(0) + c_{n-3} \\ c_{n-1} &= c_{n-1} \cdot c_{n-1}(0) + c_{n-2} \end{aligned}$$

. By starting with the zero coefficient and making substitutions, we get

$$\begin{aligned} c_0 &= c_{n-1} \cdot c_0(0) \\ c_1 &= c_{n-1} \cdot (c_1(0) + c_0(0)) \\ &\dots \\ c_{n-1} &= c_{n-1} \cdot (c_{n-1}(0) + \dots + c_0(0)) \end{aligned}$$

. Since this sequence converges to a nonzero matrix, we have $c_{n-1} \neq 0$ which implies the following equality: $c_0(0) + \dots + c_{n-1}(0) = 1$. This condition is true since these are the coefficients of the polynomial $\rho_{n-1}(x) = x^n - \rho_A(x) / a_n$

and $x = 1$ is a solution of $\rho_A(x)$ which implies

$$\rho_{n-1}(1) = 1^n - \rho_A(1)/a_n = 1 .$$

By designating $\alpha^* = c_{n-1}$ and $c^*_k = \sum_{i=0}^{\kappa} c_i(0)$ we define

$$J = \alpha^* \cdot \left(\sum_{k=0}^{n-1} c^*_k \cdot A^k \right)$$ which gives the results $\lim_{N \to \infty} A^N = J$.

Hence we have proven the corollary to be true.

The conditions stated in corollary 5.6.2 are very restrictive and apply only to a small subset of square matrices. At this point you are probably questioning the usefulness of such a corollary. If we examine the idempotent matrix J we note that it is a characteristic matrix of the matrix A corresponding to the characteristic root having a value of one. Recall that $A \cdot J = J$ which is equivalent to $(A - I) \cdot J = \varnothing$. This relationship holds when at least one characteristic root of the matrix is equivalent to a value of one.

(Computing α^* of Matrix J)

We stated earlier that the limit of the Nth power of the matrix A is computed to be the sum of the idempotent matrices corresponding to the characteristic root of value 1. Since the matrices are pair-wise orthogonal and the square of each matrix is equal to the original matrix, we must have $J^2 = J$. For this condition to be true we must have $J^2 = \alpha^* \cdot \left(\sum_{k=0}^{n-1} c^*_k \cdot A^k \right) \cdot J$. Using the distributive properties of matrix multiplication and the fact that $A \cdot J = J$, we rewrite this equation to give the following: $J^2 = \alpha^* \cdot \left(\sum_{k=0}^{n-1} c^*_k \cdot J \right)$.

This leads to the equality $J = \alpha^* \cdot \left(\sum_{k=0}^{n-1} c^*_k \right) \cdot J$. The scalar term on the right side of the equation must be equal to one for this equality to be true. Hence we have the following derivation $\alpha^* = 1 / \left(\sum_{k=0}^{n-1} c^*_k \right)$. In terms of the coefficients of the polynomial $\rho_{n-1}(x)$ we get $\alpha^* = 1 / \left(\sum_{k=0}^{n-1} \sum_{i=0}^{k} c_i(0) \right)$. Interchanging the summation signs in this equation, we obtain

$$\alpha^* = 1 / \left(\sum_{i=0}^{n-1} \sum_{k=i}^{n-1} c_i(0) \right) \text{ or } \alpha^* = 1 / \left(n - \sum_{i=0}^{n-1} i \cdot c_i(0) \right) .$$

After examining the limiting idempotent matrix denoted as $J = \alpha^* \cdot \left(\sum_{k=0}^{n-1} c^*_k \cdot A^k \right)$, we now use the properties of this matrix to derive corollary 5.6.3 below.

Corollary 5.6.3 Let A be an **n-by-n** matrix with a nonzero real characteristic root λ , then the matrix formed by dividing the original matrix by λ has a characteristic root of one (1). Furthermore, the matrix given by the expression

$$J = \alpha^* \cdot \left(\sum_{k=0}^{n-1} c^*_k \cdot \left(\frac{A}{\lambda} \right)^k \right) \text{ is idempotent and}$$

is a characteristic matrix associated with the root of value one of the matrix A / λ . The coefficients beneath the summation sign are defined as follows: $c^*_k = \sum_{i=0}^{k} \frac{c_i(0) \cdot \lambda^i}{\lambda^n}$. The coefficient α^* for this matrix is defined to be

$$\alpha^* = 1 / \left(\sum_{k=0}^{n-1} c^*_k \right) \text{ or equivalently}$$

$$\alpha^* = 1 / \left(n - \sum_{i=0}^{n-1} i \cdot \frac{c_i(0) \cdot \lambda^i}{\lambda^n} \right) .$$

Proof: Using the logic discussed above, the particulars of this proof is left for the reader as an exercise.

(Alternate Approach to Computing Matrix J)

Above we used the formula derived while computing the limit $\lim_{N \to \infty} A^N$ to compute the matrix J . After observing that the same formula that is applicable for computing the limit under the restrictive conditions of convergence, we noted that the computed matrix could be applied in the computation of characteristic matrices associated with real characteristic roots of the original matrix.

We now introduce an alternate approach for computing the characteristic matrices associated with real roots. This alternate approach begins with the characteristic polynomial $\rho_A(x)$. From previous discussions, if we evaluate this polynomial with the matrix A, we get $\rho_A(A) = \varnothing$. At the beginning of this section, we illustrated that the evaluation of matrix polynomial is equivalent to evaluating all the roots of the matrix and multiplying the results times the corresponding characteristic matrix of the root. Hence the mathematically representation of the above equality is $\rho_A(A) = \sum_{k=1}^{n} \rho_A(x_k) \cdot J_k = \varnothing$. This relationship leads to corollary 5.6.5 shown below.

Corollary 5.6.4 If λ is a non-zero characteristic root of the matrix A and it is of multiplicity **m**, then the polynomial defined as $\rho_1(x) = \rho_A(x)/(x-\lambda)^m$ gives us a new method for computing J, namely $J = \rho_1(A)/\rho_1(\lambda)$.

Proof: Using the fact that $\rho_1(A) = \sum_{k=1}^{n} \rho_1(x_k) \cdot J_k$ we can partition the roots of the matrix into the set of integers such that $k \in \Upsilon \Rightarrow (x_k = \lambda)$ and $k \in \chi \Rightarrow (x_k \neq \lambda)$. With this partition, we have the following notation $\rho_1(A) = \sum_{k \in \chi} \rho_1(x_k) \cdot J_k + \sum_{k \in \Upsilon} \rho_1(\lambda) \cdot J_k$. By definition of each partition this equation is rewritten as

$\rho_1(A) = \sum_{k \in X} \rho_1(x_k) \cdot J_k + \rho_1(\lambda) \cdot \sum_{k \in \Upsilon} J_k$. For the value of λ the evaluation of the polynomial is nonzero; $\rho_1(\lambda) \neq 0$. For all other values of x_k we have $\rho_1(x_k) = 0$. With these equalities and inequalities, the evaluation of the polynomial with the matrix yields $\rho_1(A) = \rho_1(\lambda) \cdot \sum_{k \in \Upsilon} J_k$. Hence the statement of the corollary is proved true and we have an alternate approach for computing the idempotent characteristic matrices associated with the real characteristic roots of the original matrix.

Another polynomial function of interest is polynomial defined as the minimum ranked polynomial that factors into $\rho_A(x)$ evenly and all real roots of the matrix are zero roots of the polynomial. We shall specify this polynomial as $\rho_2(x)$. Then we are able to compute a linear combination of the characteristic idempotent matrices associated with complex roots using $\rho_2(A)$.

Exercises: Convergence/Divergence Of $\mathbf{A}^N$

In exercises 1 through 5 below, determine whether the matrix **A** raised to the power of **N** ($\mathbf{A}^N$) diverges or converges as N approaches positive infinity. (You will need to determine the characteristic equation of the matrix and solve for each characteristic root.)

1. $\begin{bmatrix} -\frac{7}{6} & -\frac{3}{2} & \frac{2}{3} \\ 5 & 5 & -2 \\ \frac{17}{3} & 5 & \frac{15}{3} \end{bmatrix}$

2. $\begin{bmatrix} -8 & -9 & -3 \\ 26 & \frac{55}{2} & 9 \\ -\frac{111}{2} & -57 & -\frac{37}{2} \end{bmatrix}$

3. $\begin{bmatrix} -2 & -\frac{3}{2} & 0 \\ 3 & \frac{5}{2} & 0 \\ 9 & 9 & -\frac{1}{2} \end{bmatrix}$

4. $\begin{bmatrix} -\frac{71}{6} & \frac{49}{6} & -\frac{20}{3} \\ -4 & 2 & -2 \\ \frac{55}{3} & -\frac{41}{3} & \frac{32}{3} \end{bmatrix}$

5. $\begin{bmatrix} \frac{25}{3} & -4 & \frac{52}{3} & -6 \\ -37 & 16 & -63 & 21 \\ -56 & 24 & -105 & 36 \\ -\frac{374}{3} & 54 & -\frac{722}{3} & 83 \end{bmatrix}$

Exercises: Computing Limits of A^N as N Approaches Infinity

All matrices, when raised to a power of N, in exercises 6 through 10 below, converges as N approaches infinity. Use the computation of the coefficients described in this section to compute the limiting matrices.

6. $$\begin{bmatrix} -1 & -2 & 2 \\ -5 & -4 & 5 \\ -6 & -6 & 7 \end{bmatrix}$$

7. $$\begin{bmatrix} -8 & -\frac{17}{2} & -\frac{9}{2} \\ \frac{129}{4} & \frac{141}{4} & \frac{75}{4} \\ -\frac{93}{2} & -\frac{103}{2} & -\frac{55}{2} \end{bmatrix}$$

8. $$\begin{bmatrix} 1 & 0 & 0 \\ 0 & -9 & -5 \\ 0 & \frac{50}{3} & \frac{28}{3} \end{bmatrix}$$

9. $$\begin{bmatrix} \frac{4}{5} & 0 & 0 \\ 0 & -9 & -5 \\ 0 & \frac{50}{3} & \frac{28}{3} \end{bmatrix}$$

10. $$\begin{bmatrix} 11.183 & -14.65 & -9.1 & -4.383 \\ 19.03 & -26.3 & -17.2 & -8.433 \\ -38.60 & 55.4 & 36.6 & 17.8 \\ 43.67 & -63 & -41 & -19667 \end{bmatrix}$$

CHAPTER 6

Applications/ Special Matrices

Through the first five chapters of this book, the matrix theory concepts such as addition, multiplication, inverses and characteristic roots were formalized. For the concept of matrix inverse, two approaches were discussed:

(1). Inverse by Row/Column Echelon Reductions (The use of elementary matrices) and

(2). Inverse using partitions.

Chapter six will reinvestigate the computation of matrix inverses by introducing two additional approaches to computing inverses. One approach uses characteristic equations/roots and the second approach uses idempotent matrices. The two approaches gives us additional approaches for computing inverses; which eases the burden of computing solutions for the myriad of square matrices that will be encountered.

The special matrices that are examined in this chapter are taken from the theory of statistics and probability. A concept in statistics that eases the work of the statistician or data analyst is that of independence. Working with the normal probability distribution, independence is synonymous with the covariant matrix of the variates being a diagonal matrix. The variates of the normal distribution are assumed to be discrete time series estimates. The work in the associated sections will be the reduction of the covariate matrix to that of a diagonal matrix by pre-multiplication and post-multiplication of elementary matrices. The techniques used in these sections provide a general framework in matrix theory for the handling of such matrices.

6.1 Matrix Inverse By Characteristic Equation

In chapter five, we discovered how to determine the real characteristic roots associated with a square matrix. We also discovered how to determine the characteristic matrices associated with the matrix when the roots are known. When the characteristic roots and the characteristic matrices are known, we are able to create a basis for the space of all positive powers of the original matrix. So, for the **n-by-n** matrix $\boldsymbol{A}$, the set of values $(x_1, x_2, \ldots, x_n)$ represent the "**n**" characteristic roots of the matrix and the matrices $\boldsymbol{J}_1, \boldsymbol{J}_2, \ldots, \boldsymbol{J}_n$ represent the respective characteristic matrices associated with the roots. Then we can represent the positive powers of the matrix as the following:

$\boldsymbol{A}^r = x_1^r \cdot \boldsymbol{J}_1 + x_2^r \cdot \boldsymbol{J}_2 + \ldots + x_n^r \cdot \boldsymbol{J}_n$. To restrict the space to that of real values, we impose the conditions that the characteristic roots of the matrix are real non-negative values. This assures us that the characteristic roots raised to a non-negative value is a real value.

A natural extension of representing positive powers of the matrix is the representation of negative powers of the matrix. If we replace the positive value r with the negative value $-r$, we can express negative powers of the matrix using $\boldsymbol{A}^{-r} = x_1^{-r} \cdot \boldsymbol{J}_1 + x_2^{-r} \cdot \boldsymbol{J}_2 + \ldots + x_n^{-r} \cdot \boldsymbol{J}_n$. When we have the special case, where the value of r is one, we get the expression

$\boldsymbol{A}^{-1} = x_1^{-1} \cdot \boldsymbol{J}_1 + x_2^{-1} \cdot \boldsymbol{J}_2 + \ldots + x_n^{-1} \cdot \boldsymbol{J}_n$. Hence, we have another expression for the inverse of a matrix. To verify that these matrices are indeed inverses, we multiply these matrices together to get the following sequence of operations: $(A) \cdot (A^{-1}) = \left(\sum_{k=1}^{n} x_k \cdot J_k\right) \cdot \left(\sum_{k=1}^{n} x_k^{-1} \cdot J_k\right)$,

$(A) \cdot (A^{-1}) = \left(\sum_{k=1}^{n} x_k \cdot x_k^{-1} \cdot J_k\right)$ (since characteristic matrices are orthogonal),

$(A) \cdot (A^{-1}) = \left(\sum_{k=1}^{n} J_k\right)$ (using multiplicative inverse property of real numbers) and

$(A) \cdot (A^{-1}) = I$ (all characteristic roots equal to one implies identity matrix).

To apply this representation of an inverse, we shall consider the following examples.

(Example 1)

Given the following matrix $\begin{bmatrix} 2.88 & 1.85 & 0.5 \\ 12.21 & 2.61 & 1.25 \\ -19.27 & 8.31 & 8.5 \end{bmatrix}$, compute the characteristic roots and matrices, then compute the inverse of this matrix.

Using the concepts from chapter 5, the characteristic equation for this matrix is determined to be $\rho_A(x) = -x^3 + 14 \cdot x^2 - 31 \cdot x - 126 = 0$. Solving this equations for the real roots, we determine the value of the roots to be -2, 7 and 9. The characteristic matrices corresponding to these roots are

$\begin{bmatrix} 0.38 & -0.15 & 0 \\ -1.54 & 0.62 & 0 \\ -19.27 & -0.77 & 0 \end{bmatrix}$, $\begin{bmatrix} 0.94 & -0.08 & -0.25 \\ 2.36 & -0.19 & -0.63 \\ -0.94 & 0.08 & 0.25 \end{bmatrix}$ and $\begin{bmatrix} -0.33 & 0.23 & 0.25 \\ -0.82 & 0.58 & 0.62 \\ -0.98 & 0.69 & 0.75 \end{bmatrix}$. The inverse of this matrix is determined as follows:

$$-\begin{bmatrix} 0.38 & -0.15 & 0 \\ -1.54 & 0.62 & 0 \\ -19.27 & -0.77 & 0 \end{bmatrix} \div 2 + \begin{bmatrix} 0.94 & -0.08 & -0.25 \\ 2.36 & -0.19 & -0.63 \\ -0.94 & 0.08 & 0.25 \end{bmatrix} \div 7 + \begin{bmatrix} -0.33 & 0.23 & 0.25 \\ -0.82 & 0.58 & 0.62 \\ -0.98 & 0.69 & 0.75 \end{bmatrix} \div 9 \ .$$

After performing these operations the results or inverse of the original matrix is $\begin{bmatrix} -0.094 & 0.092 & -0.008 \\ 1.015 & -0.271 & -0.02 \\ -1.205 & 0.473 & 0.119 \end{bmatrix}$.

(Example 2)

Given the following matrix $\begin{bmatrix} 26 & 9.5 \\ -76 & -3 \end{bmatrix}$, compute the characteristic roots and matrices, then compute the inverse of this matrix.

The characteristic equation for this matrix is

$\rho_A(x) = x^2 + 5 \cdot x - 84 = 0$. Solving this equations for the real roots, we determine the value of the roots to be -12 and 7. The characteristic matrices corresponding to these roots are $\begin{bmatrix} -1 & -\frac{1}{2} \\ 4 & 2 \end{bmatrix}$ and $\begin{bmatrix} 2 & \frac{1}{2} \\ -4 & -1 \end{bmatrix}$. The inverse of this matrix is determined as follows: $-\begin{bmatrix} -1 & -\frac{1}{2} \\ 4 & 2 \end{bmatrix} \div 12 + \begin{bmatrix} 2 & \frac{1}{2} \\ -4 & -1 \end{bmatrix} \div 7$. After performing these operations the results or inverse of the original matrix is $\begin{bmatrix} 0.369 & 0.113 \\ -0.905 & -0.310 \end{bmatrix}$.

The approach of determining the characteristic equation, determining the characteristic roots and determining the characteristic matrices to compute the inverse of a matrix is a lengthy process. Although, the use of high speed desktop computers has simplified such computations, it is still in our nature to search for easier ways to accomplish the work. Using the characteristic equation, when the matrix is non-singular (determinant not equal to zero), we can apply the Cayley-Hamilton theorem to derive the inverse of **A**. Mathematically, we write the equation:

$\rho_A(A) = a_n \cdot A^n + a_{n-1} \cdot A^{n-1} + \ldots + a_0 I = 0$. Since the matrix A is non-singular, its inverse exist and if we multiply the characteristic equation by the inverse, we have the following relationship:

$A^{-1} \cdot \rho_A(A) = a_n \cdot A^{n-1} + a_{n-1} \cdot A^{n-2} + \ldots + a_0 A^{-1} = \varnothing$. Using this equation to solve for A^{-1}, we get

$A^{-1} = -\left(\frac{a_n}{a_0}\right) \cdot A^{n-1} + -\left(\frac{a_{n-1}}{a_0}\right) \cdot A^{n-2} - \ldots - \left(\frac{a_1}{a_0}\right) \cdot I$. Now we have an easier approach for computing inverses of matrices. After determining the characteristic equation of the matrix, we define the function

$P_A(x) = \left(\frac{a_0 - \rho_A(x)}{a_0 \cdot x}\right)$. Evaluating this function using the matrix A as an argument, we get $A^{-1} = P_A(A)$.

(Example 3)

Using the matrix shown in example 1, compute the inverse of the matrix.

The matrix from example 1 is $\begin{bmatrix} 2.88 & 1.85 & 0.5 \\ 12.21 & 2.61 & 1.25 \\ -19.27 & 8.31 & 8.5 \end{bmatrix}$ and the characteristic equation of matrix is $\rho_A(x) = -x^3 + 14 \cdot x^2 - 31 \cdot x - 126 = 0$. Applying the formula to compute the function $\boldsymbol{P}_A(x)$, we have $\boldsymbol{P}_A(x) = (-x^2 + 14 \cdot x - 31 \cdot x) / 126$. To compute the inverse of the matrix $\boldsymbol{A}$, we evaluate this function with the matrix as its argument. Hence we get $\boldsymbol{A}^{-1} = (-\boldsymbol{A}^2 + 14 \cdot \boldsymbol{A} - 31 \cdot \boldsymbol{I}) / 126$. To evaluate this expression we begin by computing $\boldsymbol{A}^2 = \begin{bmatrix} 21.2479 & 14.33 & 8.0025 \\ 43.0675 & 39.8404 & 20.005 \\ -117.828 & 56.7577 & 73.0025 \end{bmatrix}$. By including these matrices in the above equation, we get

$$\boldsymbol{A}^{-1} = \left(- \begin{bmatrix} 21.2479 & 14.33 & 8.0025 \\ 43.0675 & 39.8404 & 20.005 \\ -117.828 & 56.7577 & 73.0025 \end{bmatrix} + 14 \cdot \begin{bmatrix} 2.88 & 1.85 & 0.5 \\ 12.21 & 2.61 & 1.25 \\ -19.27 & 8.31 & 8.5 \end{bmatrix} - 31 \cdot \boldsymbol{I} \right) / 126$$

. As we evaluate the matrix term within the parenthesis, we reduce this expression to $\boldsymbol{A}^{-1} = \begin{bmatrix} -11.9279 & 11.57 & -1.0025 \\ 127.8725 & -34.1604 & -2.505 \\ -151.953 & 59.5823 & 14.9975 \end{bmatrix} / 126$ or

$$A^{-1} = \begin{bmatrix} -0.095 & 0.092 & -0.008 \\ 1.015 & -0.271 & -0.02 \\ -1.206 & 0.473 & 0.119 \end{bmatrix}$$. Note that this result corresponds to the inverse computed in example 1 of this section.

Exercises: Matrix Inverse Using Spectral Decomposition

In exercises 1 through 5 below, use the associated idempotent matrices J_k of the matrix A to compute the inverse A^{-1} using characteristic roots.

1. Compute the inverse of the matrix $A = \begin{bmatrix} -16 & 14 \\ -21 & 19 \end{bmatrix}$, where $J_1 = \begin{bmatrix} 3 & -2 \\ 3 & -2 \end{bmatrix}$ and $J_2 = \begin{bmatrix} -2 & 2 \\ -3 & 3 \end{bmatrix}$.

2. Compute the inverse of the matrix $A = \begin{bmatrix} -6 & -2 \\ 1 & -3 \end{bmatrix}$, where $J_1 = \begin{bmatrix} 2 & 2 \\ -1 & -1 \end{bmatrix}$ and $J_2 = \begin{bmatrix} -1 & -2 \\ 1 & 2 \end{bmatrix}$.

3. Compute the inverse of the matrix $A = \begin{bmatrix} 0 & 6 & -3 \\ 1 & 19 & -9 \\ 4 & 48 & -23 \end{bmatrix}$, where

$$J_1 = \begin{bmatrix} -2 & -6 & 3 \\ -4 & -12 & 6 \\ -10 & -30 & 15 \end{bmatrix}, \ J_2 = \begin{bmatrix} 3 & 6 & -3 \\ 5 & 10 & -5 \\ 12 & 24 & -12 \end{bmatrix} \text{ and } J_3 = \begin{bmatrix} 0 & 0 & 0 \\ -1 & 3 & -1 \\ -2 & 6 & -2 \end{bmatrix}.$$

4. Compute the inverse of the matrix $A = \begin{bmatrix} 5 & 12 & 3 \\ -6 & -17 & -7 \\ 6 & 18 & 8 \end{bmatrix}$, where

$$J_1 = \begin{bmatrix} -2 & -6 & -3 \\ 2 & 6 & 3 \\ -2 & -6 & -3 \end{bmatrix}, \ J_2 = \begin{bmatrix} 3 & 9 & 6 \\ -2 & -6 & -4 \\ 2 & 6 & 4 \end{bmatrix} \text{ and } J_3 = \begin{bmatrix} 0 & -3 & -3 \\ 0 & 1 & 1 \\ 0 & 0 & 0 \end{bmatrix}.$$

5. Compute the inverse of the matrix $A = \begin{bmatrix} -46 & 71 & -43 & 6 \\ -10 & -9 & 25 & -10 \\ 64 & -152 & 132 & -32 \\ 212 & -446 & 358 & -80 \end{bmatrix}$, where

$$J_1 = \begin{bmatrix} 4 & -10 & 11 & -3 \\ 2 & 2 & -5 & 2 \\ -2 & 22 & -31 & 10 \\ -10 & 64 & -86 & 27 \end{bmatrix} \text{ (associated root of multiplicity 2),}$$

$$J_2 = \begin{bmatrix} 6 & 3 & -15 & 6 \\ -2 & -1 & 5 & -2 \\ -16 & -8 & 40 & -16 \\ -44 & -22 & 110 & -44 \end{bmatrix} \text{ and } J_3 = \begin{bmatrix} -9 & 7 & 4 & -3 \\ 0 & 0 & 0 & 0 \\ 18 & -14 & -8 & 6 \\ 54 & -42 & -24 & 18 \end{bmatrix}.$$

Exercises: Matrix Inverse Via Cayley-Hamilton Theorem

For exercises 6 through 10, use the associated characteristic equation of the matrix $\boldsymbol{A}$ to compute the inverse $\boldsymbol{A}^{-1}$.

6. Given the matrix $\boldsymbol{A} = \begin{bmatrix} 4 & 0 \\ 16 & -4 \end{bmatrix}$ and its associated characteristic equation

$\rho_A(x) = x^2 - 16 = 0$, compute $\boldsymbol{A}^{-1}$.

7. Given the matrix $\boldsymbol{A} = \begin{bmatrix} 12 & -3 & 5 \\ 0 & -4 & 0 \\ -48 & 9 & -19 \end{bmatrix}$ and its associated characteristic

equation $\rho_A(x) = -x^3 - 11 \cdot x^2 - 40 \cdot x - 16 = 0$, compute $\boldsymbol{A}^{-1}$.

8. Given the matrix $\boldsymbol{A} = \begin{bmatrix} -4 & 0 & -1 \\ 6 & -6 & 3 \\ 6 & -2 & 1 \end{bmatrix}$ and its associated characteristic

equation $\rho_A(x) = -x^3 - 9 \cdot x^2 - 26 \cdot x - 24 = 0$, compute $\boldsymbol{A}^{-1}$.

9. Given the matrix $A = \begin{bmatrix} -8 & -20 & 0 & -10 \\ -12 & -15 & -10 & -12 \\ -5 & -5 & -6 & -5 \\ 29 & 44 & 20 & 31 \end{bmatrix}$ and its associated characteristic equation $\rho_A(x) = x^4 - 2 \cdot x^3 - 13 \cdot x^2 + 14 \cdot x + 16 = 0$, compute A^{-1}.

10. Given the matrix $A = \begin{bmatrix} 8 & -30 & 9 & -3 \\ 6 & -16 & 5 & -1 \\ 9 & -15 & 3 & -2 \\ -9 & 15 & -6 & -1 \end{bmatrix}$ and its associated characteristic equation $\rho_A(x) = x^4 + 6 \cdot x^3 + 3 \cdot x^2 - 26 \cdot x - 24 = 0$, compute A^{-1}.

6.2 *Inverse By Idempotent Matrices*

It seems that there is no end to the number of different ways to compute the inverse of a matrix. This approach, the use of idempotent matrices was developed when the computer's random access memory was a scarce commodity and the computation of the inverse of a **20-by-20** matrix was considered a memory intense application. This approach provides a sequential method of computing the inverse; one row at a time and when programmed in a computer or a programmable calculator, it requires a minimal amount of memory. Before this approach is revealed, a discussion about idempotent matrices is required.

Recall that an idempotent matrix is an **n-by-n** square matrix $\boldsymbol{J}$ such that $\boldsymbol{J}^2 = \alpha \cdot \boldsymbol{J}$ and the number α is nonzero. From corollary 4.3.7, the determinant of an idempotent matrix is either zero or it is equal to α^n. Having gained additional insight into the properties of idempotent matrices, we provide the following corollary as an extension of corollary 4.3.7.

Corollary 6.2.1 If the **n-by-n** square matrix $\boldsymbol{J}$ is idempotent such that $\boldsymbol{J}^2 = \alpha \cdot \boldsymbol{J}$, then either $|\boldsymbol{J}| = 0$ or $\boldsymbol{J}$ is equal to α times the identity matrix; $\boldsymbol{J} = \alpha \cdot \boldsymbol{I}$,

Proof: To prove this corollary, we begin with the mathematical definition and collect all matrix's terms to get the equation:

$\boldsymbol{J}^2 - \alpha \cdot \boldsymbol{J} = 0$ or $\boldsymbol{J} \cdot (\boldsymbol{J} - \alpha \cdot \boldsymbol{I}) = 0$. From corollary 4.3.7 either $|\boldsymbol{J}| = 0$ or $|\boldsymbol{J}| = \alpha^n$. If the value of the determinant is nonzero, then the matrix is non-singular and for the equation $\boldsymbol{J} \cdot (\boldsymbol{J} - \alpha \cdot \boldsymbol{I}) = 0$ to be true, we must have

$J - \alpha \cdot I = 0$ or $J = \alpha \cdot I$. Hence we can conclude that for an idempotent matrix that either $|J| = 0$ or $J = \alpha \cdot I$.

Another matrix that is very useful is the matrix formed as a linear combination of the identity matrix and the idempotent matrix J. Mathematically we write such a matrix as $I + \beta \cdot J$. The coefficient for the identity matrix was purposely omitted, since by division by the nonzero value will convert it to this form for convenience. The first property of interest for such a matrix, is the determination of non-singularity. This is determined by evaluating the determinant of such a matrix. Using the properties of characteristic roots and matrices, discussed in chapter 5, corollary 6.2.2 gives an expression for the determinant.

Corollary 6.2.2 Let J be an idempotent matrix such that $J^2 = \alpha \cdot J$. If the multiplicity of the root α is **m**, then the determinant of the matrix $A = I + \beta \cdot J$ is $(1 + \alpha \cdot \beta)^m$.

Proof: We illustrated in chapter 5 that any matrix **n-by-n** square matrix can be represented as the linear combination of **n** pair-wise orthogonal idempotent matrix. For example, we can express the idempotent matrix as the sum $J = \sum_{k=1}^{n} x_k \cdot J_k$. Then by squaring the idempotent matrix J, we have the relationship $J^2 = \alpha \cdot J$ by the statement of the corollary. Expanding this equation in terms of the pair-wise orthogonal idempotent matrices we get

$\sum_{k=1}^{n} (x_k)^2 \cdot \boldsymbol{J}_k = \sum_{k=1}^{n} \alpha \cdot x_k \cdot \boldsymbol{J}_k$. This implies that for each value of **k**, either $x_k = 0$ or $x_k = \alpha$. Since we have, by statement of the corollary, the multiplicity of root α is **m** we rewrite the expression for $\boldsymbol{J}$ as $\boldsymbol{J} = \sum_{k=1}^{m} \alpha \cdot \boldsymbol{J}_k$. If we construct the matrix by adding **n** pair-wise orthogonal idempotent matrices with all coefficients equal to one, we get $\boldsymbol{I} = \sum_{k=1}^{n} \boldsymbol{J}_k$. If we construct the matrix $\boldsymbol{A} = \boldsymbol{I} + \beta \cdot \boldsymbol{J}$ in terms of pair-wise orthogonal idempotent matrices, we get the following:

$\boldsymbol{A} = \sum_{k=m+1}^{n} \boldsymbol{J}_k + \sum_{k=1}^{m} (1 + \beta \cdot \alpha) \cdot \boldsymbol{J}_k$. Since the determinant of this matrix is the product of the characteristic roots, we get $|\boldsymbol{A}| = (1 + \alpha \cdot \beta)^m$.

With corollary 6.2.2, we can state that the matrix $\boldsymbol{I} + \beta \cdot \boldsymbol{J}$ is singular when $\beta = -1/\alpha$ and non-singular otherwise. Knowing when the matrix is non-singular gives us the conditions under which we can derive the inverse of the matrix. In the past my approach to deriving the inverse would be to assume that the inverse was of the form $\boldsymbol{I} + \kappa \cdot \boldsymbol{J}$ and multiply it to the original matrix to solve for the value of κ . I will leave this approach as an exercise to the reader and use another approach that I consider a finesse approach. The results of either approach leads to the results expressed in corollary 6.2.3 below.

Corollary 6.2.3 Let $\boldsymbol{J}$ be an idempotent matrix such that $\boldsymbol{J}^2 = \alpha \cdot \boldsymbol{J}$. If β is a number such that $\beta \neq (-1/\alpha)$ then the inverse of the matrix $\boldsymbol{I} + \beta \cdot \boldsymbol{J}$ is the matrix specified as

$$\boldsymbol{I} - \left(\frac{\beta}{(1 + \alpha \cdot \beta)}\right) \cdot \boldsymbol{J} .$$

Proof: Since the characteristic roots of the matrix $\boldsymbol{I} + \beta \cdot \boldsymbol{J}$ is either one 1 or $(1 + \beta \cdot \alpha)$, a minimum order polynomial such that the matrix also satisfy the equality of the polynomial is $p(x) = (x-1) \cdot (x - 1 - \beta \cdot \alpha)$ or equivalently $p(x) = x^2 - (2 + \beta \cdot \alpha) \cdot x + (1 + \beta \cdot \alpha)$. Evaluating the matrix in this equation and solving for the inverse we get $\boldsymbol{A}^{-1} = (-\boldsymbol{A} + (2 + \beta \cdot \alpha) \cdot \boldsymbol{I})/(1 + \beta \cdot \alpha)$. As we substitute the value $\boldsymbol{I} + \beta \cdot \boldsymbol{J}$ for the value $\boldsymbol{A}$ into the equation we get $\boldsymbol{A}^{-1} = \boldsymbol{I} - \left(\frac{\beta}{(1 + \alpha \cdot \beta)}\right) \cdot \boldsymbol{J}$.

The ease of using this corollary to produce inverses is seen in the examples below.

(Example 1)

Consider the row matrix $\underset{\sim}{\boldsymbol{R}}' = (1, -1, 1, 2)$ and the column matrix $\underset{\sim}{\boldsymbol{K}} = (\ 0, 2, 3, -2\ )^T$. If we define the matrix $\boldsymbol{A} = (\boldsymbol{I} + \underset{\sim}{\boldsymbol{K}} \cdot \underset{\sim}{\boldsymbol{R}}')$, use corollary 6.2.3 to compute its inverse.

The inverse is easily obtained by noting that $(\underset{\sim}{K} \cdot \underset{\sim}{R}')^2 = (-3) \cdot (\underset{\sim}{K} \cdot \underset{\sim}{R}')$. Applying corollary 6.2.3 yields $A^{-1} = (I + (\frac{1}{2}) \cdot \underset{\sim}{K} \cdot \underset{\sim}{R}')$, As we carry out the arithmetic of this formula, we get $A^{-1} = I + (\frac{1}{2}) \cdot \begin{bmatrix} 0 & 0 & 0 & 0 \\ 2 & -2 & 2 & 4 \\ 3 & -3 & 3 & 6 \\ -2 & 2 & -2 & -4 \end{bmatrix}$ or

$$A^{-1} = \begin{bmatrix} 1 & 0 & 0 & 0 \\ 1 & 0 & 1 & 2 \\ \frac{3}{2} & \frac{-3}{2} & \frac{5}{2} & 3 \\ -1 & 1 & -1 & -1 \end{bmatrix}.$$

Corollary 6.2.3 provides a quick and efficient way of computing the inverse of a matrix that is formed as a linear combination of an idempotent matrix and the identity matrix. By examining special cases of this corollary, it is possible to extend the inverse computations to the set of all non-singular matrices. This approach constructs the original matrix sequentially by beginning with the identity matrix and adding the **n** rows of original matrix. The mathematical expression for this representation of the matrix is

$A = I + \underset{\sim}{E}_1 \cdot (\underset{\sim}{A}^T{}_1 - \underset{\sim}{E}^T{}_1) + \ldots + \underset{\sim}{E}_n \cdot (\underset{\sim}{A}^T{}_n - \underset{\sim}{E}^T{}_n)$, The notation $\underset{\sim}{E}_k$ is a column matrix with all elements equal to zero but the kth element which is equivalent to one. The notation $\underset{\sim}{A}^T{}_k$ is the kth row of the matrix A . We introduce the matrix B_k, which is defined to be the sum

$B_k = I + \underset{\sim}{E}_1 \cdot (\underset{\sim}{A}^T{}_1 - \underset{\sim}{E}^T{}_1) + \ldots + \underset{\sim}{E}_k \cdot (\underset{\sim}{A}^T{}_k - \underset{\sim}{E}^T{}_k)$. With this notation we get $B_{k+1} = B_k + \underset{\sim}{E}_{k+1} \cdot (\underset{\sim}{A}^T{}_{k+1} - \underset{\sim}{E}^T{}_{k+1})$ and $A = B_n$. The labeling of the incremental part of this sequence of matrices as

$\boldsymbol{J}_k = \underset{\sim}{E}_k \cdot (\underset{\sim}{A}^T{}_k - \underset{\sim}{E}^T{}_k)$ simplifies the handling of the computations to follow using these terms. Upon squaring the kth incremental part we get the relationship $\boldsymbol{J}^2{}_k = \underset{\sim}{E}_k \cdot (\underset{\sim}{A}^T{}_k - \underset{\sim}{E}^T{}_k) \cdot \underset{\sim}{E}_k \cdot (\underset{\sim}{A}^T{}_k - \underset{\sim}{E}^T{}_k) = (\underset{\sim}{A}^T{}_k - \underset{\sim}{E}^T{}_k) \cdot \underset{\sim}{E}_k \cdot \boldsymbol{J}_k$; an idempotent matrix. This relationship holds for all matrices that is formed by pre-multiplying a row matrix by a column matrix. The resultant matrix is always an idempotent matrix.

By computing the inverse $\boldsymbol{B}^{-1}{}_k$, for **k** from 1 to **n**, we can construct a sequence of inverses such that $\boldsymbol{A}^{-1} = \boldsymbol{B}^{-1}{}_k$. On the surface, this approach seems to be more tedious than the previous approaches for computing inverses. But, we have shown that creating inverses using idempotent matrices is a relatively easy affair. Theorem 6.2.4 below gives a working formula for computing inverses of each sequence of matrices $\boldsymbol{B}_k$.

Theorem 6.2.4 Let A be a matrix such that the elements of the matrix are denoted as $[a_{i,j}]$ and the sequence of matrices $\boldsymbol{B}_k$ is defined such that $\boldsymbol{B}_{k+1} = \boldsymbol{B}_k + \boldsymbol{J}_{k+1}$, $\boldsymbol{B}_0 = \boldsymbol{I}$ and $\boldsymbol{J}_k = \underset{\sim}{E}_k \cdot (\underset{\sim}{A}^T{}_k - \underset{\sim}{E}^T{}_k)$. Then the inverse of $\boldsymbol{B}_{k+1}$ is

$$\boldsymbol{B}^{-1}{}_{k+1} = \boldsymbol{B}^{-1}{}_k - \left(\frac{1}{1+\alpha_{k+1}}\right) \cdot \boldsymbol{B}^{-1}{}_k \cdot \boldsymbol{J}_{k+1} \cdot \boldsymbol{B}^{-1}{}_k$$

where the scalar term α_{k+1} is given by the formula

$$\alpha_{k+1} = (\underset{\sim}{A}^T{}_{k+1} - \underset{\sim}{E}^T{}_{k+1}) \cdot \boldsymbol{B}^{-1}{}_k \cdot \underset{\sim}{E}_{k+1}.$$

Proof: By writing $\boldsymbol{B}_{k+1}$ as the product $\boldsymbol{B}_k \cdot (\boldsymbol{I} + \boldsymbol{B}^{-1}{}_k \cdot \underset{\sim}{E}_{k+1} \cdot (\underset{\sim}{A}^T{}_{k+1} - \underset{\sim}{E}^T{}_{k+1}))$, we compute the inverse of $\boldsymbol{B}_{k+1}$ to be $(\boldsymbol{I} + \boldsymbol{B}^{-1}{}_k \cdot \underset{\sim}{E}_{k+1} \cdot (\underset{\sim}{A}^T{}_{k+1} - \underset{\sim}{E}^T{}_{k+1}))^{-1} \cdot \boldsymbol{B}^{-1}{}_k$. Since the matrix $\boldsymbol{B}^{-1}{}_k \cdot \underset{\sim}{E}_k \cdot (\underset{\sim}{A}^T{}_k - \underset{\sim}{E}^T{}_k)$ is an idempotent matrix, we can apply corollary 6.2.3 to get

$$\boldsymbol{B}^{-1}{}_{k+1} = \boldsymbol{B}^{-1}{}_k - \left(\frac{1}{1+\alpha_{k+1}}\right) \cdot \boldsymbol{B}^{-1}{}_k \cdot \boldsymbol{J}_{k+1} \cdot \boldsymbol{B}^{-1}{}_k$$

where the scalar term α_{k+1} is given by the formula $(\underset{\sim}{A}^T{}_{k+1} - \underset{\sim}{E}^T{}_{k+1}) \cdot \boldsymbol{B}^{-1}{}_k \cdot \underset{\sim}{E}_{k+1}$. Hence we have verified the statement of the theorem.

A simpler way of looking at the constant term α_{k+1} is that it is the product of the $(k+1)^{th}$ row of matrix $\boldsymbol{A}$ and the $(k+1)^{th}$ column of $\boldsymbol{B}^{-1}{}_k$ minus the $(k+1)^{th}$ diagonal element of the matrix $\underset{\sim}{\boldsymbol{B}}^{-1}{}_k$;

$$\underset{\sim}{A}^T{}_{k+1} \cdot (\boldsymbol{B}^{-1}{}_k \cdot \underset{\sim}{E}_{k+1}) - [\boldsymbol{B}^{-1}{}_{k+1,k+1}].$$

(Example 2)

Applying theorem 6.2.4, compute the inverse of the following matrix using a sequential approach: $A = \begin{bmatrix} 2.88 & 1.85 & 0.5 \\ 12.21 & 2.61 & 1.25 \\ -19.27 & 8.31 & 8.5 \end{bmatrix}$.

Using the sequential approach, we define the first matrix of the sequence as $B_1 = \begin{bmatrix} 2.88 & 1.85 & 0.5 \\ 0 & 1 & 0 \\ 0 & 0 & 1 \end{bmatrix}$. Applying theorem 6.2.4, gives the following inverse

$$B^{-1}_1 = \begin{bmatrix} 1 & 0 & 0 \\ 0 & 1 & 0 \\ 0 & 0 & 1 \end{bmatrix} - \left(\frac{1}{1+1.88}\right) \cdot \begin{bmatrix} 1.88 & 1.85 & 0.5 \\ 0 & 0 & 0 \\ 0 & 0 & 0 \end{bmatrix} \text{ or}$$

$$B^{-1}_1 = \begin{bmatrix} 1 & 0 & 0 \\ 0 & 1 & 0 \\ 0 & 0 & 1 \end{bmatrix} - \left(\frac{1}{1+1.88}\right) \cdot \begin{bmatrix} 1.88 & 1.85 & 0.5 \\ 0 & 0 & 0 \\ 0 & 0 & 0 \end{bmatrix} = \begin{bmatrix} 0.3472 & -0.6424 & -0.1736 \\ 0 & 1 & 0 \\ 0 & 0 & 1 \end{bmatrix}.$$

The second matrix of the sequence is $B_2 = \begin{bmatrix} 2.88 & 1.85 & 0.5 \\ 12.21 & 2.61 & 1.25 \\ 0 & 0 & 1 \end{bmatrix}$. Applying theorem 6.2.4 gives the inverse

$$B^{-1}_2 = \begin{bmatrix} 0.3472 & -0.6424 & -0.1736 \\ 0 & 1 & 0 \\ 0 & 0 & 1 \end{bmatrix} - \left(\frac{1}{1-6.2288}\right) \cdot \begin{bmatrix} -0.6424 \\ 1 \\ 0 \end{bmatrix} \cdot \begin{bmatrix} 4.2393 & -6.2288 & -0.8697 \end{bmatrix}.$$

This inverse is reduced to

$$B^{-1}{}_2 = \begin{bmatrix} 0.3472 & -0.6424 & -0.1736 \\ 0 & 1 & 0 \\ 0 & 0 & 1 \end{bmatrix} + \begin{bmatrix} -0.5208 & 0.7653 & 0.1068 \\ 0.8108 & -1.1912 & -0.1663 \\ 0 & 0 & 0 \end{bmatrix} \text{ or}$$

$$B^{-1}{}_2 = \begin{bmatrix} -0.1736 & 0.1229 & -0.0668 \\ 0.8108 & -0.1912 & -0.1663 \\ 0 & 0 & 1 \end{bmatrix}.$$

The third matrix in the sequence is $B_3 = \begin{bmatrix} 2.88 & 1.85 & 0.5 \\ 12.21 & 2.61 & 1.25 \\ -19.27 & 8.31 & 8.5 \end{bmatrix}$. Applying theorem 6.2.4 gives the inverse

$$B^{-1}{}_3 = \begin{bmatrix} -0.1736 & 0.1229 & -0.0668 \\ 0.8108 & -0.1912 & -0.1663 \\ 0 & 0 & 1 \end{bmatrix} - \left(\frac{1}{1 + 7.4053}\right) \cdot \begin{bmatrix} -0.0668 \\ -0.1663 \\ 1 \end{bmatrix} \cdot \begin{bmatrix} 10.0830 & -3.9572 & 7.4053 \end{bmatrix}$$

.

This inverse is reduced to

$$B^{-1}{}_3 = \begin{bmatrix} -0.1736 & 0.1229 & -0.0668 \\ 0.8108 & -0.1912 & -0.1663 \\ 0 & 0 & 1 \end{bmatrix} + \begin{bmatrix} 0.0801 & -0.0315 & 0.0589 \\ 0.1995 & -0.0783 & 0.1465 \\ -1.1996 & 0.4708 & -0.8810 \end{bmatrix} = \begin{bmatrix} -0.09347 & 0.0915 & -0.008 \\ 1.0103 & -0.2695 & -0.0198 \\ -1.1996 & 0.4708 & 0.119 \end{bmatrix}$$

. Hence the inverse of the matrix A is computed as the inverse of the third matrix of the sequence. Also note that this inverse compares favorably to the results of example 3 of section 6.1.

As a final note in this section, we shall review the special case of the inverse by partitions in the special case where the k=1. In this instance, the minor matrix of the inverse matrix, namely $\boldsymbol{B}_{2,2}$, is defined to be

$B_{2,2} = (A_{2,2} - \underset{\sim}{A}_{2,1} \cdot \underset{\sim}{A}'_{1,2} / a_{1,1})^{-1}$. Using the approach shown in theorem 6.2.4, we get the following modification:

$$B_{2,2} = \left(A^{-1}_{2,2} + \frac{A^{-1}_{2,2} \cdot \underset{\sim}{A}_{2,1} \cdot \underset{\sim}{A}'_{1,2} \cdot A^{-1}_{2,2}}{(a_{1,1} - \underset{\sim}{A}'_{1,2} \cdot \underset{\sim}{A}_{2,1})} \right)$$

. This modification will prove to be very useful as we examine special matrices and various applications of matrices.

Exercises: Inverse of (I + α J)

For exercises 1 through 5, use corollary 6.2.3 to compute the inverse of matrix expressed as **(I + α J)** where the matrix **J** is an idempotent matrix.

1. Compute $(\boldsymbol{I} - 5 \cdot \boldsymbol{J})^{-1}$ where **J** is the idempotent matrix $\boldsymbol{J} = \begin{bmatrix} 4 & 0 & 5 \\ -8 & 0 & -10 \\ 4 & 0 & 5 \end{bmatrix}$.

2. Compute $(\boldsymbol{I} - 0.25 \cdot \boldsymbol{J})^{-1}$ where **J** is the idempotent matrix

$$\boldsymbol{J} = \begin{bmatrix} 2 & 2 & 10 & 16 \\ -3 & -3 & -15 & -24 \\ 4 & 4 & 20 & 32 \\ 3 & 3 & 15 & 24 \end{bmatrix}.$$

3. Compute $(\boldsymbol{I}-3\cdot\boldsymbol{J})^{-1}$ where **J** is the idempotent matrix $\boldsymbol{J} = \begin{bmatrix} 1 & 1 & 1 & 1 & 1 \\ 1 & 1 & 1 & 1 & 1 \\ 1 & 1 & 1 & 1 & 1 \\ 1 & 1 & 1 & 1 & 1 \\ 1 & 1 & 1 & 1 & 1 \end{bmatrix}$.

4. Let **J** be an **n-by-n** matrix having the form $\boldsymbol{J} = \begin{bmatrix} 1 & 1 & 1 & \dots & 1 \\ 1 & 1 & 1 & \dots & 1 \\ 1 & 1 & 1 & \dots & 1 \\ \dots & \dots & \dots & \dots & \dots \\ 1 & 1 & 1 & \dots & 1 \end{bmatrix}$, derive a formula

for the inverse $(\boldsymbol{I}+c\cdot\boldsymbol{J})^{-1}$.

5. Let **J** be an **n-by-n** matrix having the form $\boldsymbol{J} = \begin{bmatrix} 1 & p & p^2 & \dots & p^{n-1} \\ p & p^2 & p^3 & \dots & p^n \\ p^2 & p^3 & p^4 & \dots & p^{n+1} \\ \dots & \dots & \dots & \dots & \dots \\ p^{n-1} & p^n & p^{n+1} & \dots & p^{2\cdot(n-1)} \end{bmatrix}$,

show that a formula for the inverse is

$$(\boldsymbol{I}+c\cdot\boldsymbol{J})^{-1} = \boldsymbol{I} - \frac{c\cdot(1-p^2)}{(1-p^2)+c\cdot(1-p^{2\cdot(n-1)})}\cdot\boldsymbol{J} \ .$$

Exercises: Inverse of A Using Idempotent Matrices

Applying theorem 6.2.4 for each matrix **A** in exercises 6 through 10, compute the sequence of matrices $\mathbf{B_k}$ and $\mathbf{B_k}^{-1}$.

6. $$A = \begin{bmatrix} -65 & 10 & -34 \\ -210 & 34 & -105 \\ 68 & -10 & 37 \end{bmatrix}$$

7. $$A = \begin{bmatrix} -16 & -13 & -9 \\ 16 & 13 & 8 \\ 2 & 2 & 3 \end{bmatrix}$$

8. $$A = \begin{bmatrix} -93 & 20 & -17 & -40 \\ 54 & -15 & 6 & 24 \\ 36 & -8 & 8 & 16 \\ 234 & -52 & 40 & 101 \end{bmatrix}$$

9. $$A = \begin{bmatrix} -31 & 9 & 0 & -16 \\ -108 & 32 & 0 & -50 \\ 274 & -86 & -2 & 122 \\ 0 & 0 & 0 & 3 \end{bmatrix}$$

10. $$A = \begin{bmatrix} -6 & -5 & -9 & 16 & -12 \\ 13 & 6 & 18 & -35 & 24 \\ -11 & -1 & -17 & 37 & -24 \\ -21 & -9 & -27 & 53 & -36 \\ -24 & -15 & -27 & 48 & -34 \end{bmatrix}$$

6.3 *Normal Distribution with Lag-1 Correlation*

In the study of statistics and probability, the most commonly used continuous distribution function is the normal distribution function. It is commonly called the distribution of physical phenomena. For example, if we wished to model the diameter of oranges produced from a particular orange grove over the lifetime of the orange grove, we would probably use the normal distribution with the random variate of the distribution being the diameter of an orange taken from the grove. In this instance, the diameter of an orange is called a single normal variate and we indicate the probability of the diameter of such an orange as being less than X as $F(X) = \int_{-\infty}^{X} \phi(x) \cdot dx$. In this equation, the expression $\phi(x)$ is called the probability density function (pdf) and has the form $\phi(x) = \frac{e^{-\left(\frac{x-\mu}{\sigma}\right)^2/2}}{\sqrt{2 \cdot \pi} \cdot \sigma}$. When multiple variates are involved (take "**n**" measurements of "**n**" distinct oranges in the grove), then the pdf's are combined to have the form $\phi(\underset{\sim}{X}) = \frac{e^{-(\underset{\sim}{X}-\underset{\sim}{\mu})^T \cdot \Sigma^{-1} \cdot (\underset{\sim}{X}-\underset{\sim}{\mu})/2}}{\left(\sqrt{2 \cdot \pi}\right)^n \cdot \sqrt{|\Sigma|}}$. Both $\underset{\sim}{X}$ and $\underset{\sim}{\mu}$ are **n-by-1** column matrices (or vectors) and $\underset{\sim}{\mu}$ is called the population mean of the variate vector $\underset{\sim}{X}$. The matrix Σ is called the covariant matrix of the variates. Most often this matrix is a multiple of the identity matrix and the computations involved are significantly simplified when we can assume that the covariant matrix is a multiple of the identity matrix.

As we examine the multiple variate normal distribution above, it is apparent that the computations of the probability density function or probability distribution is dependent upon the evaluation of Σ^{-1} and $|\Sigma|$. When the covariate matrix is not a multiple of the identity matrix, this is a non-trivial exercise and we must approach the evaluation of these quantities depending on the structure of the matrix. In this section, we will assume that each of the **n** variates correspond to measurements taken over **n** consecutive periods of time and that there is a fixed correlation associated with adjacent periods, but not with nonadjacent periods. This concept touches on the theory of time series analysis for distribution functions. The property of correlation, only between adjacent periods, is commonly referred to as a time series of l**ag-1**. The covariance matrix for this particular instance is $\Sigma = \sigma^2 \cdot \begin{bmatrix} 1 & p & 0 & 0 & \dots & 0 \\ p & 1 & p & 0 & \dots & 0 \\ 0 & p & 1 & p & \dots & 0 \\ 0 & 0 & p & 1 & \dots & 0 \\ \dots & \dots & \dots & \dots & \dots & \dots \\ 0 & 0 & 0 & 0 & \dots & 1 \end{bmatrix}$. By letting $\boldsymbol{A}$ represent this matrix, we can express the determinant of Σ as $|\Sigma| = \sigma^{2n} \cdot |\boldsymbol{A}|$. The inverse of this matrix is expressed as $\Sigma^{-1} = \boldsymbol{A}^{-1} \div \sigma^2$. Hence our discussions concerning the lag-1 time series will principally be concerned with the evaluation of the determinant and the inverse of the matrix $\boldsymbol{A}$.

6.3.1 Determinant

To facilitate the computation of the determinant of the matrix $\boldsymbol{A}$, we introduce the notation $\varphi_n(p) = |\boldsymbol{A}|$. Using minor matrices to compute the determinant we have, for this matrix, $\varphi_n(p) = |\boldsymbol{A}| = |\boldsymbol{A}_{1,1}| - p \cdot |\boldsymbol{A}_{1,2}|$. The minor matrices are the **(n-1) - by - (n-1)** matrices

$$A_{1,1} = \begin{bmatrix} 1 & p & 0 & 0 & \dots & 0 \\ p & 1 & p & 0 & \dots & 0 \\ 0 & p & 1 & p & \dots & 0 \\ 0 & 0 & p & 1 & \dots & 0 \\ \dots & \dots & \dots & \dots & \dots & \dots \\ 0 & 0 & 0 & 0 & \dots & 1 \end{bmatrix} \quad \text{and} \quad A_{1,2} = \begin{bmatrix} p & p & 0 & 0 & \dots & 0 \\ 0 & 1 & p & 0 & \dots & 0 \\ 0 & p & 1 & p & \dots & 0 \\ 0 & 0 & p & 1 & \dots & 0 \\ \dots & \dots & \dots & \dots & \dots & \dots \\ 0 & 0 & 0 & 0 & \dots & 1 \end{bmatrix}.$$

The minor matrix $A_{1,1}$ is a time series lag-1 **(n-1)-by-(n-1)** matrix. Hence we can represent its determinant as $\varphi_{n-1}(p)$. If we post multiply the minor matrix $A_{1,2}$ by the elementary matrix E_{2-1}, which subtracts column one from column two yields the matrix $\begin{bmatrix} p & 0 & 0 & 0 & \dots & 0 \\ 0 & 1 & p & 0 & \dots & 0 \\ 0 & p & 1 & p & \dots & 0 \\ 0 & 0 & p & 1 & \dots & 0 \\ \dots & \dots & \dots & \dots & \dots & \dots \\ 0 & 0 & 0 & 0 & \dots & 1 \end{bmatrix}$. Upon closer observation of this matrix we compute its determinant to be $p \cdot \varphi_{n-2}(p)$. Since the determinant of the type III elementary matrix is one we get $|A_{1,2}| = p \cdot \varphi_{n-2}(p)$. Combining the results of the determinants for the minor matrices we now have the relationship $\varphi_n(p) = \varphi_{n-1}(p) - p^2 \cdot \varphi_{n-2}(p)$. In the list below, you will find the evaluation of the first eight of these functions. Eight functions are evaluated to give a sense of the pattern of the coefficients for each function.

The initial eight functions are as follows:

$$\varphi_1(p) = 1$$

$$\varphi_2(p) = 1 - p^2$$

$$\varphi_3(p) = 1 - 2 \cdot p^2$$

$$\varphi_4(p) = 1 - 3 \cdot p^2 + p^4$$

$$\varphi_5(p) = 1 - 4 \cdot p^2 + 3 \cdot p^4$$

$$\varphi_6(p) = 1 - 5 \cdot p^2 + 6 \cdot p^4 - p^6$$

$$\varphi_7(p) = 1 - 6 \cdot p^2 + 3 \cdot p^4 - 4 \cdot p^6$$

$$\varphi_8(p) = 1 - 7 \cdot p^2 + 8 \cdot p^4 - 10 \cdot p^6 + p^8$$

. The coefficients shown in the above equations look very familiar. If you were to construct Pascal's triangle, you would see these coefficients there. Using mathematical induction and the empirical observation, the resultant formula for the **n**th function is

$$\varphi_n(p) = \sum_{k=0}^{[n/2]} (-1)^k \cdot \binom{n-k}{k} \cdot p^{2 \cdot k}$$

. The expression denoted as $[n/2]$ is the largest integer less than **n** divided by two. Hence with this formula we can express the determinant of the matrix $\boldsymbol{A}$ as

$$|\boldsymbol{A}| = \sum_{k=0}^{[n/2]} (-1)^k \cdot \binom{n-k}{k} \cdot p^{2 \cdot k}.$$

6.3.2 Inverse of Lag-1 Matrix

To compute the inverse of this matrix, the most appropriate approach is that of row and column reduction through the use of elementary matrices. We begin the computation by pre-multiplying and post-multiplying the matrix by elementary matrices to place zeros in positions first row-second column and second row-first column. The first step in this approach has the following appearance:

$$\boldsymbol{E}_{2-(p)\cdot 1}\cdot\begin{bmatrix} 1 & p & 0 & 0 & \dots & 0 \\ p & 1 & p & 0 & \dots & 0 \\ 0 & p & 1 & p & \dots & 0 \\ 0 & 0 & p & 1 & \dots & 0 \\ \dots & \dots & \dots & \dots & \dots & \dots \\ 0 & 0 & 0 & 0 & \dots & 1 \end{bmatrix} = \begin{bmatrix} 1 & p & 0 & 0 & \dots & 0 \\ 0 & 1-p^2 & p & 0 & \dots & 0 \\ 0 & p & 1 & p & \dots & 0 \\ 0 & 0 & p & 1 & \dots & 0 \\ \dots & \dots & \dots & \dots & \dots & \dots \\ 0 & 0 & 0 & 0 & \dots & 1 \end{bmatrix}$$

and

$$\begin{bmatrix} 1 & p & 0 & 0 & \dots & 0 \\ 0 & 1-p^2 & p & 0 & \dots & 0 \\ 0 & p & 1 & p & \dots & 0 \\ 0 & 0 & p & 1 & \dots & 0 \\ \dots & \dots & \dots & \dots & \dots & \dots \\ 0 & 0 & 0 & 0 & \dots & 1 \end{bmatrix}\cdot\boldsymbol{E}^T{}_{2-(p)\cdot 1} = \begin{bmatrix} 1 & 0 & 0 & 0 & \dots & 0 \\ 0 & 1-p^2 & p & 0 & \dots & 0 \\ 0 & p & 1 & p & \dots & 0 \\ 0 & 0 & p & 1 & \dots & 0 \\ \dots & \dots & \dots & \dots & \dots & \dots \\ 0 & 0 & 0 & 0 & \dots & 1 \end{bmatrix}.$$

By continuing this approach until the resultant matrix is a diagonal matrix, we can easily represent the inverse of the matrix $\boldsymbol{A}$ as the product of elementary matrices and the resultant diagonal matrix. This representation will be listed below. Rather than attempting to display a rigorous proof for the representation of all the elementary matrices, we assume a given form for the kth elementary matrix pair and use mathematical induction to verify that the form holds for the (k+1)th elementary matrix pair. Before we jump into the proof by mathematical induction, let's introduce some new notation and derive the 2nd elementary matrix pair.

Then next elementary matrix pair, when pre-multiplying and post-multiplying the resultant matrix above, places all zeros in column 2 and row 2 except for the diagonal element. This is accomplished as follows:

$$\boldsymbol{E}_{3-\left(\frac{p}{1-p^2}\right)\cdot 2} \cdot \begin{bmatrix} 1 & 0 & 0 & 0 & \dots & 0 \\ 0 & 1-p^2 & p & 0 & \dots & 0 \\ 0 & p & 1 & p & \dots & 0 \\ 0 & 0 & p & 1 & \dots & 0 \\ \dots & \dots & \dots & \dots & \dots & \dots \\ 0 & 0 & 0 & 0 & \dots & 1 \end{bmatrix} = \begin{bmatrix} 1 & 0 & 0 & 0 & \dots & 0 \\ 0 & 1-p^2 & p & 0 & \dots & 0 \\ 0 & 0 & \frac{1-2\cdot p^2}{1-p^2} & p & \dots & 0 \\ 0 & 0 & p & 1 & \dots & 0 \\ \dots & \dots & \dots & \dots & \dots & \dots \\ 0 & 0 & 0 & 0 & \dots & 1 \end{bmatrix}$$

and

$$\begin{bmatrix} 1 & 0 & 0 & 0 & \dots & 0 \\ 0 & 1-p^2 & p & 0 & \dots & 0 \\ 0 & 0 & \frac{1-2\cdot p^2}{1-p^2} & p & \dots & 0 \\ 0 & 0 & p & 1 & \dots & 0 \\ \dots & \dots & \dots & \dots & \dots & \dots \\ 0 & 0 & 0 & 0 & \dots & 1 \end{bmatrix} \cdot \boldsymbol{E}^T_{3-\left(\frac{p}{1-p^2}\right)\cdot 2} = \begin{bmatrix} 1 & 0 & 0 & 0 & \dots & 0 \\ 0 & 1-p^2 & 0 & 0 & \dots & 0 \\ 0 & 0 & \frac{1-2\cdot p^2}{1-p^2} & p & \dots & 0 \\ 0 & 0 & p & 1 & \dots & 0 \\ \dots & \dots & \dots & \dots & \dots & \dots \\ 0 & 0 & 0 & 0 & \dots & 1 \end{bmatrix}$$

.

From the discussion of determinants we introduced the notation for the determinants of a lag-1 matrix of k-dimension to be $\varphi_k(p)$ and the terms in the above representation are

$$\begin{aligned} \varphi_1(p) &= 1 \\ \varphi_2(p) &= 1-p^2 \\ \varphi_3(p) &= 1-2\cdot p^2 \end{aligned}$$

. Next we introduce the notation $\psi_k(p) = \dfrac{\varphi_k(p)}{\varphi_{k-1}(p)}$. Now we hypothesize on the general form of the elementary matrices and the elements along the diagonal of the resulting diagonal matrix. For the **k**th elementary matrix, we hypothesize that the **k**th element along the diagonal matrix is $\psi_k(p)$ and hence the pre-multiplication elementary matrix is $E_{(k+1)-\left(\frac{p}{\psi_k(p)}\right)\cdot k}$ and the post-multiplication elementary matrix is the transpose of the pre-multiplication elementary matrix. To verify this hypothesis, assume that it is true for **k** and illustrate that it holds for **k+1**. Assuming that the resultant matrix has zero values for its non-diagonal elements in columns 1 through **k-1** and rows 1 through **k-1** In row **k**, the **k+1** element of the row is p (similarly for the **k+1** element of the **k** column). To convert the **k+1** element of the **k** row to zero, we pre-multiply the **k**th resultant matrix by the elementary matrix $E_{(k+1)-\left(\frac{p}{\psi_k(p)}\right)\cdot k}$. This operation will also convert the (**k+1**)th element along the diagonal of the resultant matrix to $1-\frac{p^2}{\psi_k(p)} = 1-\frac{p^2\cdot\varphi_{k-1}(p)}{\varphi_k(p)}$. In the derivation of the determinant of this matrix, we established the relationship $\varphi_{k+1}(p) = \varphi_k(p)-p^2\cdot\varphi_{k-1}(p)$. Hence we have that $1-\frac{p^2}{\psi_k(p)} = \frac{\varphi_{k+1}(p)}{\varphi_k(p)} = \psi_{k+1}(p)$. Hence the hypothesis is true.

Using this general form for the elementary matrices and the elements of the diagonal matrix we have the following relationship:

$$\prod_{k=1}^{n-1}\left(\boldsymbol{E}_{(k+1)-\left(\frac{p}{\psi_k(p)}\right)\cdot k}\right)\cdot \boldsymbol{A}\cdot \prod_{k=n-1}^{1}\left(\boldsymbol{E}^T_{(k+1)-\left(\frac{p}{\psi_k(p)}\right)\cdot k}\right) = \boldsymbol{D}$$. The diagonal elements are $d_{i,\,i} = \psi_i(p)$. Using the properties of inverses of products of matrices, we get

$$\boldsymbol{A}^{-1} = \prod_{k=n-1}^{1}\left(\boldsymbol{E}^T_{(k+1)-\left(\frac{p}{\psi_k(p)}\right)\cdot k}\right)\cdot \boldsymbol{D}^{-1}\cdot \prod_{k=1}^{n-1}\left(\boldsymbol{E}_{(k+1)-\left(\frac{p}{\psi_k(p)}\right)\cdot k}\right).$$

6.3.3 Change Of Variables

The actual computation of the inverse matrix is quite a formidable task; even with the explicit formula of the inverse given above. To make the work simpler, we use a common technique in statistics and probability called change of variables. As stated earlier, when the variables in the array are independent, the computations are much easier to perform. For the multi-variate normal distribution, the variates are independent when the covariate matrix is a diagonal matrix. This suggests that if we change from the original normal variates by multiplying the by the product of the elementary matrices above then the resulting variates are independent normal variates.

The suggested transformation for the vector $(\underset{\sim}{X} - \underset{\sim}{\mu})$ is the following

$$(\underset{\sim}{Y} - \underset{\sim}{U}) = \prod_{k=1}^{n-1}\left(\boldsymbol{E}_{(k+1)-\left(\frac{p}{\psi_k(p)}\right)\cdot k}\right)\cdot(\underset{\sim}{X} - \underset{\sim}{\mu})$$. Since the **k**th elementary matrix only affects the (k+1)th row of the vector, we have the following elements of the vectors $\underset{\sim}{Y}$ and $\underset{\sim}{U}$; $y_1 = x_1$, $u_1 = \mu_1$,

$y_{k+1} = x_{k+1} - \left(\frac{p}{\psi_k(p)}\right) \cdot x_k$ and $u_{k+1} = \mu_{k+1} - \left(\frac{p}{\psi_k(p)}\right) \cdot \mu_k$ where **k** is greater than 1 and less than **n**. The covariance matrix for this vector of variates is defined to be $\sigma^2 \cdot \boldsymbol{D}$.

With this transformation, we now have a vector of independent normal distributed variates with the following probability density function:

$$\phi(\underset{\sim}{Y}) = \frac{e^{-(\underset{\sim}{Y} - \underset{\sim}{U})^T \cdot (\sigma^2 \cdot \boldsymbol{D})^{-1} \cdot (\underset{\sim}{Y} - \underset{\sim}{U})/2}}{(\sqrt{\pi})^n \cdot \sigma^n \cdot \sqrt{\varphi_n(p)}}$$

. With this density function, we now have an expression that is easier to evaluate and properties of the original variates can be obtained by using the properties of the newly transformed variates.

Exercises: Determinants of Lag-1 Correlation Matrices

For the correlation matrices in exercises 1 through 5 below, compute the determinant of each lag-1 correlation matrix.

1. $\boldsymbol{A} = \begin{bmatrix} 1 & 0.5 & 0 \\ 0.5 & 1 & 0.5 \\ 0 & 0.5 & 1 \end{bmatrix}$

2. $\boldsymbol{A} = \begin{bmatrix} 1 & 0.25 & 0 \\ 0.25 & 1 & 0.25 \\ 0 & 0.25 & 1 \end{bmatrix}$

3. $$A = \begin{bmatrix} 1 & \frac{1}{3} & 0 & 0 \\ \frac{1}{3} & 1 & \frac{1}{3} & 0 \\ 0 & \frac{1}{3} & 1 & \frac{1}{3} \\ 0 & 0 & \frac{1}{3} & 1 \end{bmatrix}$$

4. $$A = \begin{bmatrix} 1 & 0.2 & 0 & 0 \\ 0.2 & 1 & 0.2 & 0 \\ 0 & 0.2 & 1 & 0.2 \\ 0 & 0 & 0.2 & 1 \end{bmatrix}$$

5. $$A = \begin{bmatrix} 1 & 0.1 & 0 & 0 & 0 \\ 0.1 & 1 & 0.1 & 0 & 0 \\ 0 & 0.1 & 1 & 0.1 & 0 \\ 0 & 0 & 0.1 & 1 & 0.1 \\ 0 & 0 & 0 & 0.1 & 1 \end{bmatrix}$$

Exercise: Associated Elementary Matrices of Lag-1 Matrix

For exercises 6 through 10 below, construct the elementary matrices and diagonal matrix using the matrices in exercises 1 through 5 above.

6. For the matrix $\boldsymbol{A} = \begin{bmatrix} 1 & 0.5 & 0 \\ 0.5 & 1 & 0.5 \\ 0 & 0.5 & 1 \end{bmatrix}$, construct the two elementary matrices that are used to convert the matrix to a diagonal matrix.

7. For the matrix $\boldsymbol{A} = \begin{bmatrix} 1 & 0.25 & 0 \\ 0.25 & 1 & 0.25 \\ 0 & 0.25 & 1 \end{bmatrix}$, construct the two elementary matrices that are used to convert the matrix to a diagonal matrix.

8. For the matrix $\boldsymbol{A} = \begin{bmatrix} 1 & \frac{1}{3} & 0 & 0 \\ \frac{1}{3} & 1 & \frac{1}{3} & 0 \\ 0 & \frac{1}{3} & 1 & \frac{1}{3} \\ 0 & 0 & \frac{1}{3} & 1 \end{bmatrix}$, construct the three elementary matrices that are used to convert the matrix to a diagonal matrix.

9. For the matrix $\boldsymbol{A} = \begin{bmatrix} 1 & 0.2 & 0 & 0 \\ 0.2 & 1 & 0.2 & 0 \\ 0 & 0.2 & 1 & 0.2 \\ 0 & 0 & 0.2 & 1 \end{bmatrix}$, construct the three elementary matrices that are used to convert the matrix to a diagonal matrix.

10. For the matrix $A = \begin{bmatrix} 1 & 0.1 & 0 & 0 & 0 \\ 0.1 & 1 & 0.1 & 0 & 0 \\ 0 & 0.1 & 1 & 0.1 & 0 \\ 0 & 0 & 0.1 & 1 & 0.1 \\ 0 & 0 & 0 & 0.1 & 1 \end{bmatrix}$, construct the four elementary matrices that are used to convert the matrix to a diagonal matrix.

Exercises: Inverse of Lag-1 Correlation Matrices

Compute the inverse of the following matrices shown in exercises 11 through 15.

11. $A = \begin{bmatrix} 1 & 0.5 & 0 \\ 0.5 & 1 & 0.5 \\ 0 & 0.5 & 1 \end{bmatrix}$

12. $A = \begin{bmatrix} 1 & 0.25 & 0 \\ 0.25 & 1 & 0.25 \\ 0 & 0.25 & 1 \end{bmatrix}$

13. $A = \begin{bmatrix} 1 & \frac{1}{3} & 0 & 0 \\ \frac{1}{3} & 1 & \frac{1}{3} & 0 \\ 0 & \frac{1}{3} & 1 & \frac{1}{3} \\ 0 & 0 & \frac{1}{3} & 1 \end{bmatrix}$

14. $A = \begin{bmatrix} 1 & 0.2 & 0 & 0 \\ 0.2 & 1 & 0.2 & 0 \\ 0 & 0.2 & 1 & 0.2 \\ 0 & 0 & 0.2 & 1 \end{bmatrix}$

15. $$A = \begin{bmatrix} 1 & 0.1 & 0 & 0 & 0 \\ 0.1 & 1 & 0.1 & 0 & 0 \\ 0 & 0.1 & 1 & 0.1 & 0 \\ 0 & 0 & 0.1 & 1 & 0.1 \\ 0 & 0 & 0 & 0.1 & 1 \end{bmatrix}$$

6.4 Normal Distribution: Exponential Correlation

The section is another example of a discrete time series of measured variates. As the number of periods increase between the variates the correlation is assumed to decrease exponentially. For example if the difference in the number of periods between variates is "**k**", the correlation between the two variates is proportional to $p^k;where(0 \le p < 1)$. As in the previous section, the distribution of concern is the normal distribution with mean vector $\underset{\sim}{\mu}$ and covariance matrix Σ. The covariance matrix for the exponentially decreasing, over time, correlations has the form

$$\Sigma = \sigma^2 \cdot \begin{bmatrix} 1 & p & p^2 & p^3 & \dots & p^{n-1} \\ p & 1 & p & p^2 & \dots & p^{n-2} \\ p^2 & p & 1 & p & \dots & p^{n-3} \\ p^3 & p^2 & p & 1 & \dots & p^{n-4} \\ \dots & \dots & \dots & \dots & \dots & \dots \\ p^{n-1} & p^{n-2} & p^{n-3} & p^{n-4} & \dots & 1 \end{bmatrix} = \sigma^2 \cdot A$$

. The computations of interest for this matrix is the computation of the determinant and the computation of the inverse.

6.4.1 Determinant

The determinant of the covariance matrix is $\Sigma = (\sigma^2)^n \cdot |A|$. To compute, the determinant of the matrix A, we will use elementary matrices to reduce the matrix to a diagonal matrix. If we subtract p times the second row of the matrix from the first row we get the following resultant matrix:

$$\boldsymbol{E}_{\mathbf{1}-(p)\cdot\mathbf{2}}\cdot\begin{bmatrix} 1 & p & p^2 & p^3 & \ldots & p^{n-1} \\ p & 1 & p & p^2 & \ldots & p^{n-2} \\ p^2 & p & 1 & p & \ldots & p^{n-3} \\ p^3 & p^2 & p & 1 & \ldots & p^{n-4} \\ \ldots & \ldots & \ldots & \ldots & \ldots & \ldots \\ p^{n-1} & p^{n-2} & p^{n-3} & p^{n-4} & \ldots & 1 \end{bmatrix} = \begin{bmatrix} 1-p^2 & 0 & 0 & 0 & \ldots & 0 \\ p & 1 & p & p^2 & \ldots & p^{n-2} \\ p^2 & p & 1 & p & \ldots & p^{n-3} \\ p^3 & p^2 & p & 1 & \ldots & p^{n-4} \\ \ldots & \ldots & \ldots & \ldots & \ldots & \ldots \\ p^{n-1} & p^{n-2} & p^{n-3} & p^{n-4} & \ldots & 1 \end{bmatrix}$$

. Performing the same operation on column 1 of the resultant matrix (subtract p times the second column from the first column). This operation is expressed as follows:

$$\begin{bmatrix} 1-p^2 & 0 & 0 & 0 & \ldots & 0 \\ p & 1 & p & p^2 & \ldots & p^{n-2} \\ p^2 & p & 1 & p & \ldots & p^{n-3} \\ p^3 & p^2 & p & 1 & \ldots & p^{n-4} \\ \ldots & \ldots & \ldots & \ldots & \ldots & \ldots \\ p^{n-1} & p^{n-2} & p^{n-3} & p^{n-4} & \ldots & 1 \end{bmatrix}\cdot\boldsymbol{E}^T_{\mathbf{1}-(p)\cdot\mathbf{2}} = \begin{bmatrix} 1-p^2 & 0 & 0 & 0 & \ldots & 0 \\ 0 & 1 & p & p^2 & \ldots & p^{n-2} \\ 0 & p & 1 & p & \ldots & p^{n-3} \\ 0 & p^2 & p & 1 & \ldots & p^{n-4} \\ \ldots & \ldots & \ldots & \ldots & \ldots & \ldots \\ 0 & p^{n-2} & p^{n-3} & p^{n-4} & \ldots & 1 \end{bmatrix}$$

. By continuing this approach we get the following product relationship:

$$\boldsymbol{P}\cdot\boldsymbol{A}\cdot\boldsymbol{P}^T = \begin{bmatrix} 1-p^2 & 0 & 0 & 0 & \ldots & 0 \\ 0 & 1-p^2 & 0 & 0 & \ldots & 0 \\ 0 & 0 & 1-p^2 & 0 & \ldots & 0 \\ 0 & 0 & 0 & 1-p^2 & \ldots & 0 \\ \ldots & \ldots & \ldots & \ldots & \ldots & \ldots \\ 0 & 0 & 0 & 0 & \ldots & 1 \end{bmatrix}$$ where the matrix $\boldsymbol{P}$ is product of

the pre-multiplication elementary matrices; $\boldsymbol{P} = \prod_{k=n-1}^{1}\left(\boldsymbol{E}_{\boldsymbol{k}-(p)\cdot(\boldsymbol{k}+1)}\right)$. If we compute the determinant of both sides of the above equation, we get $|\boldsymbol{P}| \cdot |\boldsymbol{A}| \cdot |\boldsymbol{P}^T| = (1 - p^2)^{n-1}$. Since the matrix $\boldsymbol{P}$ is the product of type III elementary matrices, we have $|\boldsymbol{P}| = 1$ and hence we get the determinant to be $|\boldsymbol{A}| = (1 - p^2)^{n-1}$.

6.4.2 Inverse and Change of Variables

Using the diagonal matrix equivalent representation for the matrix $\boldsymbol{A}$ and using the properties of the inverse of matrix products, we get the relationship $\boldsymbol{A}^{-1} = \boldsymbol{P}^T \cdot \boldsymbol{D}^{-1} \cdot \boldsymbol{P}$. The matrix $\boldsymbol{D}$ is the diagonal matrix derived in the computation of the determinant. The actual computation of the inverse is a little tedious and a bit involved, but fortunately a change of variates as in previous section reduces the work involved with applying the inverse of the matrix. For use in the evaluating of the normal probability density the dot products of vectors is $(\underset{\sim}{X} - \underset{\sim}{\mu})^T \boldsymbol{A}^{-1} (\underset{\sim}{X} - \underset{\sim}{\mu})$. Substituting the relationship

$$\boldsymbol{A}^{-1} = \prod_{k=n-1}^{1} \left(\boldsymbol{E}^T{}_{k-(p)\cdot(k+1)}\right) \cdot \boldsymbol{D}^{-1} \cdot \prod_{k=1}^{n-1} \left(\boldsymbol{E}_{k-(p)\cdot(k+1)}\right)$$ for the inverse and

then applying associative properties of matrices, we get $(\underset{\sim}{X} - \underset{\sim}{\mu})^T \boldsymbol{A}^{-1} (\underset{\sim}{X} - \underset{\sim}{\mu}) = (\underset{\sim}{Y} - \underset{\sim}{U})^T \boldsymbol{D}^{-1} (\underset{\sim}{Y} - \underset{\sim}{U})$. In this relationship we get $\underset{\sim}{Y} = \prod_{k=1}^{n-1} \left(\boldsymbol{E}_{k-(p)\cdot(k+1)}\right) \cdot \underset{\sim}{X}$ and $\underset{\sim}{U} = \prod_{k=1}^{n-1} \left(\boldsymbol{E}_{k-(p)\cdot(k+1)}\right) \cdot \underset{\sim}{\mu}$. As we evaluate the transformed variate vector, we get the relationship $y_k = x_k - p \cdot x_{k+1}$ for **k** less than **n** and $y_n = x_n$. Similarly for the mean vector, we have $U_k = \mu_k - p \cdot \mu_{k+1}$ for **k** less than **n** and $U_n = \mu_n$.

With this transformation, we now have a vector of independent normal distributed variates with the following probability density function:

$$\phi(\underset{\sim}{Y}) = \frac{e^{-(\underset{\sim}{Y} - \underset{\sim}{U})^T \cdot (\sigma^2 \cdot \boldsymbol{D})^{-1} \cdot (\underset{\sim}{Y} - \underset{\sim}{U})/2}}{(\sqrt{\pi})^n \cdot \sigma^n \cdot \sqrt{(1-p^2)^{n-1}}}$$

. With this density function, we now have an expression that is easier to evaluate and properties of the original variates can be obtained by using the properties of the newly transformed variates.

Exercises: Determinant of Exponential Correlation Matrix

For the correlation matrices in exercises 1 through 5 below, compute the determinant of each exponential correlation matrix.

1. $A = \begin{bmatrix} 1 & 0.5 & 0.25 \\ 0.5 & 1 & 0.5 \\ 0.25 & 0.5 & 1 \end{bmatrix}$

2. $A = \begin{bmatrix} 1 & 0.25 & 0.0625 \\ 0.25 & 1 & 0.25 \\ 0.0625 & 0.25 & 1 \end{bmatrix}$

3. $A = \begin{bmatrix} 1 & \frac{1}{3} & \frac{1}{9} & \frac{1}{27} \\ \frac{1}{3} & 1 & \frac{1}{3} & \frac{1}{9} \\ \frac{1}{9} & \frac{1}{3} & 1 & \frac{1}{3} \\ \frac{1}{27} & \frac{1}{9} & \frac{1}{3} & 1 \end{bmatrix}$

4. $A = \begin{bmatrix} 1 & 0.2 & 0.04 & 0.008 \\ 0.2 & 1 & 0.2 & 0.04 \\ 0.04 & 0.2 & 1 & 0.2 \\ 0.008 & 0.04 & 0.2 & 1 \end{bmatrix}$

5. $A = \begin{bmatrix} 1 & 0.1 & 0.01 & 0.001 & 0.0001 \\ 0.1 & 1 & 0.1 & 0.01 & 0.001 \\ 0.01 & 0.1 & 1 & 0.1 & 0.01 \\ 0.001 & 0.01 & 0.1 & 1 & 0.1 \\ 0.0001 & 0.001 & 0.01 & 0.1 & 1 \end{bmatrix}$

Exercise: Associated Elementary Matrices of Exponential Correlation Matrix

For exercises 6 through 10 below, construct the elementary matrices and the diagonal matrix using the matrices in exercises 1 through 5 above.

6. For the matrix $A = \begin{bmatrix} 1 & 0.5 & 0.25 \\ 0.5 & 1 & 0.5 \\ 0.25 & 0.5 & 1 \end{bmatrix}$, construct the two elementary matrices that are used to convert the matrix to a diagonal matrix.

7. For the matrix $A = \begin{bmatrix} 1 & 0.25 & 0.0625 \\ 0.25 & 1 & 0.25 \\ 0.0625 & 0.25 & 1 \end{bmatrix}$, construct the two elementary matrices that are used to convert the matrix to a diagonal matrix.

8. For the matrix $A = \begin{bmatrix} 1 & \frac{1}{3} & \frac{1}{9} & \frac{1}{27} \\ \frac{1}{3} & 1 & \frac{1}{3} & \frac{1}{9} \\ \frac{1}{9} & \frac{1}{3} & 1 & \frac{1}{3} \\ \frac{1}{27} & \frac{1}{9} & \frac{1}{3} & 1 \end{bmatrix}$, construct the three elementary matrices that are used to convert the matrix to a diagonal matrix.

9. For the matrix $A = \begin{bmatrix} 1 & 0.2 & 0.04 & 0.008 \\ 0.2 & 1 & 0.2 & 0.04 \\ 0.04 & 0.2 & 1 & 0.2 \\ 0.008 & 0.04 & 0.2 & 1 \end{bmatrix}$, construct the three elementary matrices that are used to convert the matrix to a diagonal matrix.

10. For the matrix $A = \begin{bmatrix} 1 & 0.1 & 0.01 & 0.001 & 0.0001 \\ 0.1 & 1 & 0.1 & 0.01 & 0.001 \\ 0.01 & 0.1 & 1 & 0.1 & 0.01 \\ 0.001 & 0.01 & 0.1 & 1 & 0.1 \\ 0.0001 & 0.001 & 0.01 & 0.1 & 1 \end{bmatrix}$, construct the four elementary matrices that are used to convert the matrix to a diagonal matrix.

Exercises: Inverse of Exponential Correlation Matrices

Compute the inverse of the following matrices shown in exercises 11 through 15.

11. $A = \begin{bmatrix} 1 & 0.5 & 0 \\ 0.5 & 1 & 0.5 \\ 0 & 0.5 & 1 \end{bmatrix}$

12. $A = \begin{bmatrix} 1 & 0.25 & 0 \\ 0.25 & 1 & 0.25 \\ 0 & 0.25 & 1 \end{bmatrix}$

13. $A = \begin{bmatrix} 1 & \frac{1}{3} & 0 & 0 \\ \frac{1}{3} & 1 & \frac{1}{3} & 0 \\ 0 & \frac{1}{3} & 1 & \frac{1}{3} \\ 0 & 0 & \frac{1}{3} & 1 \end{bmatrix}$

14. $A = \begin{bmatrix} 1 & 0.2 & 0 & 0 \\ 0.2 & 1 & 0.2 & 0 \\ 0 & 0.2 & 1 & 0.2 \\ 0 & 0 & 0.2 & 1 \end{bmatrix}$

15. $$A = \begin{bmatrix} 1 & 0.1 & 0 & 0 & 0 \\ 0.1 & 1 & 0.1 & 0 & 0 \\ 0 & 0.1 & 1 & 0.1 & 0 \\ 0 & 0 & 0.1 & 1 & 0.1 \\ 0 & 0 & 0 & 0.1 & 1 \end{bmatrix}$$

6.5 Least Squares Application

Chapter 2 of this book provides the solution of a system of linear equations through the use of matrix inverses. The format of the linear system took on the form of **k** unknowns and **k** equations. For example if we have **k-1** explanatory variables that are used to explain the behavior of a dependent variable, the linear model

$y = a_0 + a_1 \cdot x_1 + a_2 \cdot x_2 + a_3 \cdot x_3 + \ldots + a_{k-1} \cdot x_{k-1}$ expresses this phenomenon mathematically. As seen in the model, we have **k** coefficients in the model; provided that we are assuming a constant term in the explanatory model. When the constant coefficient is known, then the model has the form

$y^* = y - a_0 = a_1 \cdot x_1 + a_2 \cdot x_2 + a_3 \cdot x_3 + \ldots + a_{k-1} \cdot x_{k-1}$ and results in **k-1** coefficients. The first model is said to have a nonzero origin and the second model is said to have a zero origin.

When **n** observations of the dependent variable are available and each observed value is accompanied with a distinct set of **(k-1)** explanatory variables, then we can express the model in matrix notation, the model is written as

$\underset{\sim}{Y} = X \cdot \underset{\sim}{\beta}$. For this equation, we have the following notation $\underset{\sim}{Y} = \begin{bmatrix} y_1 \\ y_2 \\ \ldots \\ y_n \end{bmatrix}$,

$\underset{\sim}{\beta} = \begin{bmatrix} a_0 \\ a_1 \\ \ldots \\ a_{k-1} \end{bmatrix}$ and $X = \begin{bmatrix} 1 & x_{1,1} & x_{1,2} & \ldots & x_{1,k-1} \\ 1 & x_{2,1} & x_{2,2} & \ldots & x_{2,k-1} \\ \ldots & \ldots & \ldots & \ldots & \ldots \\ 1 & x_{k-1,1} & x_{k-1,2} & \ldots & x_{k-1,k-1} \end{bmatrix}$. This representation references the instance when the model has a possible nonzero origin. To model the instance in which the model has a zero origin, remove the a_0 from the coefficient vector and remove the column of one's from the X matrix. In the special case when the number of observations is equivalent to the number of coefficients and the matrix X is non-singular, the solution for the coefficient vector is $\underset{\sim}{\beta} = X^{-1} \cdot \underset{\sim}{Y}$. The case in which the number of observations is less than the number of coefficients is not solvable in **n**-dimensional space. However, the treatment of generalized inverses, in the following chapter, suggests a pseudo-solution in a subspace of **n**-dimensional space. The remainder of this section focuses on the case when the number of observations is larger than the number of coefficients to be estimated.

When the number of observations exceeds the number of coefficients to be estimated, there are a number of possible models that may adequately fit the observed data. To select a model we need a criterion to define a best model. In this section, we choose the method of least squares. The method of least squares defines the best solution for the coefficient vector to be the vector $\underset{\sim}{\beta}$ that minimizes the sum of squares $(\underset{\sim}{Y} - X \cdot \underset{\sim}{\beta})^T \cdot (\underset{\sim}{Y} - X \cdot \underset{\sim}{\beta})$. The derivation of the best solution, which minimizes the sum of squares, is beyond the scope of this book. It turns out that the best solution is $\hat{\underset{\sim}{\beta}}$ such that $X^T \cdot (\underset{\sim}{Y} - X \cdot \hat{\underset{\sim}{\beta}}) = \mathbf{0}$ or $\hat{\underset{\sim}{\beta}} = (X^T \cdot X)^{-1} \cdot X^T \cdot \underset{\sim}{Y}$. Of course, we assume that the matrix $(X^T \cdot X)$ is non-singular.

6.5.1 Least Squares Coefficients With Model Having Zero Origin

In the case of the model having a zero origin, we remove the ones from the matrix an introduce the notation $X_0 = \begin{bmatrix} x_{1,1} & x_{1,2} & \dots & x_{1,k-1} \\ x_{2,1} & x_{2,2} & \dots & x_{2,k-1} \\ \dots & \dots & \dots & \dots \\ x_{k-1,1} & x_{k-1,2} & \dots & x_{k-1,k-1} \end{bmatrix}$, $SS_0 = (X_0^T \cdot X_0)$ and

$\underset{\sim}{S}_{xy} = X_0^T \cdot \underset{\sim}{Y}$. From the relationship derived above for the least square coefficients, we get $\underset{\sim}{\hat{\beta}} = (SS_0)^{-1} \cdot \underset{\sim}{S}_{xy}$. To compute the individual coefficients of the vector coefficient, we pre-multiply this result with the cartesian unit base row vector

$\underset{\sim}{E}^T_i = \begin{bmatrix} 0 & \dots & 1 & \dots & 0 & 0 \end{bmatrix}$ (The **i^th^** element has a value of one and all other elements have values of zero.). Mathematically, we write this relationship as $\hat{a}_i = \underset{\sim}{E}^T_i \cdot \underset{\sim}{\hat{\beta}}$ or equivalently $\hat{a}_i = \underset{\sim}{E}^T_i \cdot (SS_0)^{-1} \cdot \underset{\sim}{S}_{xy}$. With this relationship, we now apply the lessons we learned using the partitioning of matrices to reduce the work in applying matrix inverses.

If we define the matrix $A = \begin{bmatrix} 1 & \underset{\sim}{E}^T_i \\ \underset{\sim}{S}_{xy} & SS_0 \end{bmatrix}$, then we get the following equation: $\hat{a}_i = \underset{\sim}{E}^T_i \cdot (SS_0)^{-1} \cdot \underset{\sim}{S}_{xy} = 1 - \frac{|A|}{|SS_0|}$. If we post-multiply the matrix A by the elementary matrix E_{i-1} (subtracting column one from column **i**), we get $A \cdot E_{2-1} = \begin{bmatrix} 1 & \underset{\sim}{\mathbf{0}}^T \\ \underset{\sim}{S}_{xy} & SS_{(i-y)} \end{bmatrix}$. The matrix $SS_{(i-y)}$ is a modification

of the matrix SS_0 with the column vector $\underset{\sim}{S}_{xy}$ subtracted from column **i** of SS_0. Letting the matrix $SS_{(y \to i)}$ be a modification of matrix SS_0 by replacing column **i** with the column vector $\underset{\sim}{S}_{xy}$, we apply corollary 4.2.5 of chapter 4 to obtain the relationship $|SS_{(i-y)}| = |SS_0| - |SS_{(y \to i)}|$. Since the determinant of a type III elementary matrix is 1, we have $|A| = |SS_{(i-y)}| = |SS_0| - |SS_{(y \to i)}|$. Hence we can summarize the esti-

mation of the **i^th^** coefficient to be $\hat{a}_i = 1 - \dfrac{|A|}{|SS_0|} = \dfrac{|SS_{(y \to i)}|}{|SS_0|}$. Hence by

starting with the matrix $SS_0 = (X_0{}^T \cdot X_0)$ and replacing the **i^th^** column with the column vector $\underset{\sim}{S}_{xy}$, to obtain the matrix $SS_{(y \to i)}$, the estimate of the **i^th^** coefficient is computed as the ratio of the determinants of the matrices.

6.5.2 Least Squares Coefficients With Model Having NonZero Origin

For the nonzero origin option, the least squares solution for the coefficients of the linear model are expressed as $\begin{pmatrix} \hat{a}_0 \\ \hat{\underset{\sim}{\beta}} \end{pmatrix} = \left(\begin{pmatrix} \underset{\sim}{\mathbf{1}}^T \\ X^T{}_0 \end{pmatrix} \cdot (\underset{\sim}{\mathbf{1}}, X_0) \right)^{-1} \cdot \begin{pmatrix} \underset{\sim}{\mathbf{1}}^T \\ X^T{}_0 \end{pmatrix} \cdot \underset{\sim}{Y}$. This equation purposely separates the intercept term from the remaining coefficients to show how this assumption differs from that of the zero origin assumption. As we follow through with the matrix algebra computations, this equation takes the following form: **0**Using the notation introduced above, this equation is written as $\begin{pmatrix} \hat{a}_0 \\ \hat{\underset{\sim}{\beta}} \end{pmatrix} = \begin{bmatrix} n & n \cdot \bar{\underset{\sim}{X}}^T \\ n \cdot \bar{\underset{\sim}{X}} & \boldsymbol{SS}_0 \end{bmatrix}^{-1} \cdot \begin{pmatrix} n \cdot \bar{y} \\ \underset{\sim}{S}_{xy} \end{pmatrix}$. To evaluate the matrix inverse in this equation, we reference the results of chapter 4 for the evaluation of matrix inverses using partitions.

Applying results in chapter 4 we get $\begin{bmatrix} n & n \cdot \bar{\underset{\sim}{X}}^T \\ n \cdot \bar{\underset{\sim}{X}} & \boldsymbol{SS}_0 \end{bmatrix}^{-1} = \begin{bmatrix} b & \underset{\sim}{B}_1^T \\ \underset{\sim}{B}_2 & \boldsymbol{B}_{2,2} \end{bmatrix}$ where

$$\begin{aligned} b &= (1 + n \cdot \bar{\underset{\sim}{X}}^T \cdot \boldsymbol{B}_{2,2} \cdot \bar{\underset{\sim}{X}})/n \\ \underset{\sim}{B}_1^T &= -\bar{\underset{\sim}{X}}^T \cdot \boldsymbol{B}_{2,2} \\ \underset{\sim}{B}_2 &= -\boldsymbol{B}_{2,2} \cdot \bar{\underset{\sim}{X}} \\ \boldsymbol{B}_{2,2} &= \left(\boldsymbol{SS}_0 - n \cdot \bar{\underset{\sim}{X}} \cdot \bar{\underset{\sim}{X}}^T\right)^{-1} \end{aligned}$$

. The matrix $\left(\boldsymbol{SS}_0 - n \cdot \bar{\underset{\sim}{X}} \cdot \bar{\underset{\sim}{X}}^T\right)$ is called the covariance matrix of the independent variables of the model. If we denote this matrix as $\boldsymbol{SS}_x$ then with the above expansion of the inverse of the matrix, the estimates of the coefficients are as follows:

$\hat{a}_0 = \bar{y} \cdot (1 + n \cdot \bar{\underset{\sim}{X}}^T \cdot \boldsymbol{SS}_x^{-1} \cdot \bar{\underset{\sim}{X}}) - (\bar{\underset{\sim}{X}}^T \cdot \boldsymbol{SS}_x^{-1} \cdot \underset{\sim}{S}_{xy})$ and

$\hat{\underset{\sim}{\beta}} = \boldsymbol{SS}_x^{-1} \cdot \underset{\sim}{S}_{xy} - n \cdot \bar{y} \cdot \boldsymbol{SS}_x^{-1} \cdot \bar{\underset{\sim}{X}}$. Now we introduce another the covariant vector of the independent variables with the dependent variable. This vector is denoted as $\underset{\sim}{C}_{xy} = \underset{\sim}{S}_{xy} - n \cdot \bar{y} \cdot \bar{\underset{\sim}{X}}$. By rearranging the terms in the above matrix equations, we simplify the equations for the coefficients to the following: $\hat{a}_0 = \bar{y} - \bar{\underset{\sim}{X}}^T \cdot \boldsymbol{SS}_x^{-1} \cdot \underset{\sim}{C}_{xy}$ and $\hat{\underset{\sim}{\beta}} = \boldsymbol{SS}_x^{-1} \cdot \underset{\sim}{C}_{xy}$. Hence once we have computed the coefficients of the independent variables, we can compute the zero intercept coefficient with the relationship $\hat{a}_0 = \bar{y} - \bar{\underset{\sim}{X}}^T \cdot \hat{\underset{\sim}{\beta}}$. Now that we have a relationship similar to the relationship for the zero intercept option, we can derive a similar expression, using determinants for the nonzero intercept option. For this option, the relationship, for the independent coefficients are

$\hat{a}_i = \frac{|CS_{(y \to i)}|}{|SS_x|}$ (for **i** greater than zero). The matrix $CS_{(y \to i)}$ is the SS_x matrix with the **i**th column replaced with the column vector $\underset{\sim}{C}_{xy}$.

Exercises: Least Squares Estimates

1. Using the information in the following table, compute the least squares estimate of the coefficients in the given model.

Model: $Y = a_0 + a_1 * x + a_2 * x^2 + \varepsilon$			
SEQ.	X	X^2	Y
1	1.0	1.00	25.88
2	2.5	6.25	79.57
3	4.0	16.00	193.44
4	6.0	36.00	478.81
5	8.0	64.00	823.90
6	10.0	100.00	1314.21
7	11.0	121.00	1584.32
8	15.0	225.00	3036.74
9	20.0	400.00	5420.44
10	25.0	625.00	8512.70

2. Using the information in the following table, compute the least squares estimates of the coefficients in the stated model. For this model, a zero origin is assumed.

Model: $Y = a_1 * x + a_2 * x^2 + \varepsilon$			
SEQ.	**X**	**x^2**	**Y**
1	-5.0	25.0	146.4
2	-4.0	16.0	74.5
3	-3.0	9.0	42.3
4	-3.0	9.0	32.7
5	-1.0	1.0	7.1
6	0.0	0.0	5.1
7	2.0	4.0	17.7
8	3.0	9.0	68.1
9	4.0	16.0	127.8
10	6.0	36.0	232.2

3. Using the information in the following table, compute the least squares estimates of the coefficients in the stated model.

Model: $Y = a_0 + a_1 * x_1 + a_2 * x_2 + a_3 * x_1 * x_2 + \varepsilon$				
SEQ.	**x_1**	**x_2**	**$x_1 * x_2$**	**Y**
1	0	1	0	-1.7
2	0	5	0	-7.1
3	0	10	0	-29.6
4	1	3	3	14.8
5	1	7	7	32.2
6	1	15	15	44.7
7	1	20	20	64.3

4. Using the information in the following table, compute the least squares estimates of the coefficients in the stated model.

Model: $Ln(Y) = a_0 + a_1 * x_1 + a_2 * x_2 + \varepsilon$			
SEQ.	x_1	x_2	Y
1	0.1	-1	6.3
2	0.25	-0.5	4.4
3	0.5	-1	12.5
4	0.75	1.4	7.7
5	1	-0.25	15.9
6	1.25	0.5	23.7
7	1.5	1	41.1
8	1	0	18.9
9	0.5	-1	12.4
10	0	-0.5	3.5

5. Using the information in the following table, compute the least squares estimates of the coefficients in the stated model.

Model: $Ln(Y) = a_0 + a_1 * x_1 + a_2 * (x_1)^2 + a_3 * Ln(x_2) + \varepsilon$				
SEQ.	x_1	$(x_1)^2$	x_2	Y
1	0	0	5	4.8
2	-1	1	2	0.0
3	-2	4	3	0.0
4	1	1	2	7.5
5	2	4	1	27.8
6	3	9	3	2432
7	1	1	4	33

CHAPTER 7

Generalized Inverse

For the nonsingular matrix, we have shown in prior chapters that an inverse of the matrix exists. But for the singular matrix an inverse does not exist. This chapter shows that the singular matrix does have a counterpart to the inverse and it is called the generalized inverse. A working definition is established for the generalized inverse and properties of the generalized inverse are investigated within the following sections. For each matrix, we define a ring that is generated by the characteristic vectors or matrices of the matrix and illustrate that the generalized inverse (or inverse, if matrix is nonsingular) with the generalized identity element of the ring is contained in the ring.

A well known algorithm, the Gram-Schmidt Algorithm, is derived using the properties of idempotent matrices and properties of orthogonal vector spaces. This algorithm develops a method for deriving symmetric orthogonal vector spaces when starting with a set of linearly independent vectors.

7.1 *Working Definitions*

Revisiting properties of the matrix inverse will provide us with stimulus in determining a working definition of the generalized inverse of a matrix. Recall for a nonsingular "**n-by-n**" matrix $\boldsymbol{A}$, the inverse of the matrix is a matrix

$\boldsymbol{B}$ such that the following properties are true: $\left\{ \begin{array}{c} \boldsymbol{A} \cdot \boldsymbol{B} = \boldsymbol{B} \cdot \boldsymbol{A} = \boldsymbol{I} \\ \boldsymbol{I} \cdot \boldsymbol{A} = \boldsymbol{A} \cdot \boldsymbol{I} = \boldsymbol{A} \end{array} \right\}$.

For a singular matrix, we cannot have a matrix that multiplies with it that results in the identity matrix. We have seen in chapter 4 that a singular matrix has a determinant of zero and since the determinant of a product is the product of the determinants a singular matrix cannot produce the identity matrix when multiplied with another matrix. Hence for a generalized inverse we must provide a definition that will reduce to the above conditions when the matrix is nonsingular.

Definition 7.1.1 A generalized inverse of the matrix $\boldsymbol{A}$ is a matrix $\boldsymbol{B}$ such that the following two conditions are true

1. $\boldsymbol{A} \cdot \boldsymbol{B} \cdot \boldsymbol{A} = \boldsymbol{A}$ and
2. $\boldsymbol{A} \cdot \boldsymbol{B} = \boldsymbol{B} \cdot \boldsymbol{A}$.

With the above definition, the remainder of this section will probe the concept of the generalized inverse to show properties that might prove useful in the application requiring the use of generalized inverses. The first quantity of interest is the resultant matrix when a matrix is multiplied by its generalized inverse. Such a matrix turns out to be an idempotent matrix as seen in corollary 7.1.2 below.

Corollary 7.1.2 The matrices $\boldsymbol{A} \cdot \boldsymbol{B}$ and $\boldsymbol{B} \cdot \boldsymbol{A}$ are idempotent matrices.

Proof: By definition of generalized inverse we have $\boldsymbol{A} \cdot \boldsymbol{B} = \boldsymbol{B} \cdot \boldsymbol{A}$, hence if we prove that $\boldsymbol{A} \cdot \boldsymbol{B}$ is idempotent, it follows that $\boldsymbol{B} \cdot \boldsymbol{A}$ is idempotent since they are one and the same matrix. Squaring the matrix $\boldsymbol{A} \cdot \boldsymbol{B}$, we get $(\boldsymbol{A} \cdot \boldsymbol{B})^2 = (\boldsymbol{A} \cdot \boldsymbol{B}) \cdot (\boldsymbol{A} \cdot \boldsymbol{B})$. From the associate property of matrix multiplication we have $(\boldsymbol{A} \cdot \boldsymbol{B})^2 = (\boldsymbol{A} \cdot \boldsymbol{B} \cdot \boldsymbol{A}) \cdot \boldsymbol{B}$. Property 1 of the definition of the generalized inverse states that $\boldsymbol{A} \cdot \boldsymbol{B} \cdot \boldsymbol{A} = \boldsymbol{A}$. Making this substitution in the above, we get $(\boldsymbol{A} \cdot \boldsymbol{B})^2 = (\boldsymbol{A}) \cdot \boldsymbol{B}$. Hence we have proven the corollary to be true. The matrix $\boldsymbol{A} \cdot \boldsymbol{B}$ is an idempotent matrix.

The next question that comes to mind, is that of the generalized inverse of the matrix $\boldsymbol{B}$. The above definition only states that $\boldsymbol{B}$ is a generalized inverse of $\boldsymbol{A}$. The answer to this riddle is partially solved using corollaries 7.1.3 and 7.1.4.

Corollary 7.1.3 If the matrix $\boldsymbol{B}$ is a generalized inverse of the matrix $\boldsymbol{A}$, then $\boldsymbol{B} \cdot \boldsymbol{A} \cdot \boldsymbol{B}$ is also a generalized inverse of $\boldsymbol{A}$.

Proof: To prove this corollary it is sufficient to prove each of the items listed in the definition.

1. To prove the first property of the definition, we create the product $\boldsymbol{A} \cdot (\boldsymbol{B} \cdot \boldsymbol{A} \cdot \boldsymbol{B}) \cdot \boldsymbol{A}$. Applying the associate property of matrix multiplication, we have $\boldsymbol{A} \cdot (\boldsymbol{B} \cdot \boldsymbol{A} \cdot \boldsymbol{B}) \cdot \boldsymbol{A} = (\boldsymbol{A} \cdot \boldsymbol{B} \cdot \boldsymbol{A}) \cdot \boldsymbol{B} \cdot \boldsymbol{A}$. From property 1 of the definition, we get $\boldsymbol{A} \cdot (\boldsymbol{B} \cdot \boldsymbol{A} \cdot \boldsymbol{B}) \cdot \boldsymbol{A} = (\boldsymbol{A}) \cdot \boldsymbol{B} \cdot \boldsymbol{A}$ and hence property 1 of the definition that $\boldsymbol{B} \cdot \boldsymbol{A} \cdot \boldsymbol{B}$ is a generalized inverse of $\boldsymbol{A}$.
2. To prove the second property of the definition, we create the product $\boldsymbol{A} \cdot (\boldsymbol{B} \cdot \boldsymbol{A} \cdot \boldsymbol{B})$. Applying the associative property of matrix multiplication we have $(\boldsymbol{A} \cdot \boldsymbol{B}) \cdot (\boldsymbol{A} \cdot \boldsymbol{B})$. Using property 2 of the definition of the matrix $\boldsymbol{B}$ as a generalized inverse of the matrix $\boldsymbol{A}$ we have the following results $(\boldsymbol{A} \cdot \boldsymbol{B}) \cdot (\boldsymbol{A} \cdot \boldsymbol{B}) = (\boldsymbol{B} \cdot \boldsymbol{A}) \cdot (\boldsymbol{B} \cdot \boldsymbol{A})$. Another application of the associative property of matrix multiplication yields the results $(\boldsymbol{A} \cdot \boldsymbol{B}) \cdot (\boldsymbol{A} \cdot \boldsymbol{B}) = (\boldsymbol{B} \cdot \boldsymbol{A} \cdot \boldsymbol{B}) \cdot \boldsymbol{A}$. Hence we have verified property 2 of the definition for the matrix $(\boldsymbol{B} \cdot \boldsymbol{A} \cdot \boldsymbol{B})$ to be a generalized inverse of the matrix $\boldsymbol{A}$.

Having shown that both properties of definition 7.1.1 to be true for matrix $(B \cdot A \cdot B)$ as a generalized inverse of matrix A, this corollary is proven true.

Upon stating that B is a generalized inverse of A, you may be tempted to state that A is a generalized inverse of B. But such logic may be faulty. Later in this section, it will be shown that a second matrix may be a generalized inverse of a first matrix but the first matrix is not a generalized matrix of the second matrix. However, at this point, we turn back to the derived matrix $(B \cdot A \cdot B)$ when B is a generalized inverse of A.

Corollary 7.1.4 If the matrix B is a generalized inverse of the matrix A, then A is a generalized inverse of the derived matrix $(B \cdot A \cdot B)$.

Proof: To prove this corollary, it is sufficient to show that the two properties stated in definition 7.1.1 are true.

3. The proof of property 1 begins by establishing the product $(B \cdot A \cdot B) \cdot A \cdot (B \cdot A \cdot B)$. By applying the associate property of matrix multiplication beginning with the left-most term and the middle term, we get $(A \cdot B \cdot A) \cdot (B \cdot A \cdot B) = B \cdot A \cdot (B \cdot A \cdot B)$. Another application of the associate property yields the results $(B \cdot A \cdot B) \cdot A \cdot (B \cdot A \cdot B) = B \cdot (A \cdot B \cdot A) \cdot B$.

Applying property 1 of the definition for matrix $\boldsymbol{B}$ a generalized inverse of $\boldsymbol{A}$ we get the results

$(\boldsymbol{B} \cdot \boldsymbol{A} \cdot \boldsymbol{B}) \cdot \boldsymbol{A} \cdot (\boldsymbol{B} \cdot \boldsymbol{A} \cdot \boldsymbol{B}) = \boldsymbol{B} \cdot \boldsymbol{A} \cdot \boldsymbol{B}$.

4. This second property is obtained from corollary 7.1.3, since $(\boldsymbol{B} \cdot \boldsymbol{A} \cdot \boldsymbol{B})$ a generalized inverse of $\boldsymbol{A}$ we have from the definition that $\boldsymbol{A} \cdot (\boldsymbol{B} \cdot \boldsymbol{A} \cdot \boldsymbol{B}) = (\boldsymbol{B} \cdot \boldsymbol{A} \cdot \boldsymbol{B}) \cdot \boldsymbol{A}$.

With both properties true, we can state that $\boldsymbol{A}$ is a generalized inverse of the matrix $(\boldsymbol{B} \cdot \boldsymbol{A} \cdot \boldsymbol{B})$.

When the matrix $\boldsymbol{A}$ is singular, the matrix $\boldsymbol{A} \cdot \boldsymbol{B}$ is singular and thus the matrix $(\boldsymbol{I} - \boldsymbol{A} \cdot \boldsymbol{B})$ is not the zero matrix. If $\boldsymbol{B}$ is a generalized inverse of $\boldsymbol{A}$ and we define the matrix $\boldsymbol{M} = \boldsymbol{B} + a \cdot (\boldsymbol{I} - \boldsymbol{A} \cdot \boldsymbol{B})$ for all real values of a then $\boldsymbol{M}$ is a generalized inverse of $\boldsymbol{A}$. However, $\boldsymbol{B}$ is not a generalized inverse of the matrix $\boldsymbol{M}$. This property illustrates that for every matrix $\boldsymbol{B}$ that is a generalized inverse of $\boldsymbol{A}$, the reciprocating condition, $\boldsymbol{A}$ being a generalized inverse of $\boldsymbol{B}$ need not be true.

A final corollary in this section, that will be useful in discovering more properties about generalized inverse is given below. This corollary is proved using the definition of a generalized inverse and the associate property of matrices.

Corollary 7.1.5 If the matrix $\boldsymbol{B}$ is a generalized inverse of the matrix $\boldsymbol{A}$ and "m" is a positive integer greater than one, then we have the following relationship:

$$\boldsymbol{B} \cdot \boldsymbol{A}^m = \boldsymbol{A}^m \cdot \boldsymbol{B} = \boldsymbol{A}^{m-1}.$$

Proof: The proof for this corollary is left as an exercise for the reader.

Exercises

1. Let the matrix $\boldsymbol{B}$ be a generalized inverse of the singular matrix $\boldsymbol{A}$ and let **m** be a positive integer greater than one. Prove the assertions of corollary 7.1.5 that $\boldsymbol{B} \cdot \boldsymbol{A}^m = \boldsymbol{A}^m \cdot \boldsymbol{B} = \boldsymbol{A}^{m-1}$.

2. Let $\boldsymbol{A}$ be an idempotent singular matrix. such that $\boldsymbol{A}^2 = \boldsymbol{A}$,derive a generalized inverse of scalar multiple $\alpha \cdot \boldsymbol{A}$ for the matrix.

3. Let the matrix $\boldsymbol{B}$ be a generalized inverse of the singular matrix $\boldsymbol{A}$. For the positive integer m, derive the generalized inverse of $\boldsymbol{A}^m$.

4. Let α be a nonzero real number, let the matrix $\boldsymbol{A}$ be a singular matrix and let the matrix $\boldsymbol{B}$ be a generalized inverse of the matrix $\boldsymbol{A}$. Determine the inverse of the nonsingular matrix $\boldsymbol{A} + \alpha \cdot (\boldsymbol{I} - \boldsymbol{B} \cdot \boldsymbol{A} \cdot \boldsymbol{B})$.

5. For this exercise let's examine the case when the matrix $\boldsymbol{A}$, a singular matrix with generalized inverse $\boldsymbol{B}$, is respectively a generalized inverse of the matrix $\boldsymbol{B}$. Hence we have $\boldsymbol{A} \cdot \boldsymbol{B} \cdot \boldsymbol{A} = \boldsymbol{A}$, $\boldsymbol{B} \cdot \boldsymbol{A} \cdot \boldsymbol{B} = \boldsymbol{B}$ and $\boldsymbol{A} \cdot \boldsymbol{B} = \boldsymbol{B} \cdot \boldsymbol{A}$. Show that such a generalized inverse $\boldsymbol{B}$, is unique.

Exercises: Computing Generalized Matrices

6. Given the singular matrix $\boldsymbol{A} = \begin{bmatrix} 29 & -4 & 15 \\ 84 & -12 & 42 \\ -30 & 4 & -16 \end{bmatrix}$, determine which of the following matrices is a generalized inverse of $\boldsymbol{A}$.

a.) $\boldsymbol{B}_1 = \begin{bmatrix} 15 & -2 & 8 \\ 42 & -6 & 21 \\ -16 & 2 & -9 \end{bmatrix}$, b.) $\boldsymbol{B}_2 = \begin{bmatrix} -4 & 1 & -1.5 \\ -21 & 4 & -10.5 \\ 3 & -1 & 0.5 \end{bmatrix}$ and

c.) $\boldsymbol{B}_3 = \begin{bmatrix} 3 & 0 & 2 \\ 0 & 1 & 0 \\ -4 & 0 & -3 \end{bmatrix}$.

7. Given the singular matrix $\boldsymbol{A} = \begin{bmatrix} -5 & 4 & 4 \\ -7 & -5 & -5 \\ 1 & 0 & 0 \end{bmatrix}$, determine which of the following matrices is a generalized inverse of $\boldsymbol{A}$.

a.) $\boldsymbol{B}_1 = \begin{bmatrix} -5 & 4 & 4 \\ -7 & -5 & -5 \\ 1 & 0 & 0 \end{bmatrix}$, b.) $\boldsymbol{B}_2 = \begin{bmatrix} 3 & 2 & 2 \\ -11 & -8 & -7 \\ 8 & 6 & 5 \end{bmatrix}$ and

c.) $\boldsymbol{B}_3 = \begin{bmatrix} 3 & 2 & 2 \\ 4 & 4 & 2 \\ -7 & -6 & -4 \end{bmatrix}$.

8. Given the singular matrix $\boldsymbol{A} = \begin{bmatrix} -1 & 3 & 3 \\ 0 & 0 & -4 \\ 0 & 0 & 2 \end{bmatrix}$, determine which of the following matrices is a generalized inverse of $\boldsymbol{A}$.

a.) $\boldsymbol{B}_1 = \begin{bmatrix} -1 & 9 & 16 \\ 0 & 2 & 2 \\ 0 & 0 & 1 \end{bmatrix}$, b.) $\boldsymbol{B}_2 = \begin{bmatrix} 1 & 3 & 6 \\ 0 & 2 & 2 \\ 0 & 0 & 1 \end{bmatrix}$ and c.)

$\boldsymbol{B}_3 = \begin{bmatrix} -1 & 3 & 4.5 \\ 0 & 0 & -1 \\ 0 & 0 & 0.5 \end{bmatrix}$.

9. Given the singular matrix $\boldsymbol{A} = \begin{bmatrix} 4 & -1 & 1 \\ 0 & 0 & 0 \\ -12 & 3 & -3 \end{bmatrix}$, determine which of the following matrices is a generalized inverse of $\boldsymbol{A}$.

a.) $\boldsymbol{B}_1 = \begin{bmatrix} -7 & -1 & -3 \\ -18 & 1 & -6 \\ 18 & 3 & 8 \end{bmatrix}$, b.) $\boldsymbol{B}_2 = \begin{bmatrix} 4 & -0.25 & 1.25 \\ 4.5 & 0 & 1.5 \\ -10.5 & 0.75 & -3.25 \end{bmatrix}$ and

c.) $\boldsymbol{B}_3 = \begin{bmatrix} 1.75 & 0 & 0.5 \\ 2.25 & 0.25 & 0.75 \\ -3.75 & 0 & -1 \end{bmatrix}$.

10. Given the singular matrix $\boldsymbol{A} = \begin{bmatrix} -1 & -4 & 2 \\ -6 & -16 & 8 \\ -15 & 80 & 20 \end{bmatrix}$, determine which of the following matrices is a generalized inverse of $\boldsymbol{A}$.

a.) $\boldsymbol{B}_1 = \begin{bmatrix} 2 & 2 & -1 \\ 3 & 2 & -1 \\ 7.5 & 35 & -2.5 \end{bmatrix}$, b.) $\boldsymbol{B}_2 = \begin{bmatrix} -0.5 & -2 & 1 \\ -3 & -8 & 4 \\ -7.5 & 40 & 10 \end{bmatrix}$ and

c.) $\boldsymbol{B}_3 = \begin{bmatrix} 1 & 0 & 0 \\ 0 & -4 & 2 \\ 0 & 50 & 5 \end{bmatrix}$.

7.2 Uniqueness of kth Rank Generalized Inverse

In chapter 5 we showed that an "**n-by-n**" matrix can be expressed as a linear sum of "**n**" independent pair-wise orthogonal idempotent matrices. Each independent matrix is the binary matrix multiplication of a column vector to a row vector. Mathematically we write the following linear representation for the matrix $\boldsymbol{A}$: $\boldsymbol{A} = \sum_{i=1}^{n} x_i \cdot \boldsymbol{J}_i$; the characteristic roots of the matrix. With this terminology we now introduce a formal definition for the rank of a matrix.

Definition 7.2.1 The rank of an "**n-by-n**" matrix $\boldsymbol{A}$ is equivalent to the number of nonzero characteristic roots of the matrix which implies that it can be expressed as a linear sum of "**k**" linearly independent and pair-wise idempotent matrices $\{\boldsymbol{J}_1, \boldsymbol{J}_2, \ldots, \boldsymbol{J}_k\}$. Mathematically, this results in the linear representation $\boldsymbol{A} = \sum_{i=1}^{k} x_i \cdot \boldsymbol{J}_i$ and we write

$rank(\boldsymbol{A}) = k$.

With the matrix $\boldsymbol{A}$ expressed as a linear combination of idempotent matrices, we introduce the space S_A that represents all matrices that can be expressed as linear combinations of the "**k**" idempotent matrices that represent the characteristic matrices of the matrix $\boldsymbol{A}$. Hence for all matrices $\boldsymbol{X} = \sum_{i=1}^{k} \alpha_i \cdot \boldsymbol{J}_i$ we have $\boldsymbol{X} \in S_A$. Without going into much details concerning basic algebraic theory (beyond scope of this textbook), this space forms a ring under the binary operations of matrix addition and matrix multiplication. Additionally, if we introduce the element $J = \sum_{i=1}^{k} J_i$, we get a unity element for the ring S_A .

This is true since for every element $X \in S_A$, we have

$J \cdot X = X \cdot J = X$. However this ring is not a division ring since all elements of this ring may not have a multiplicative inverse when J is the unity element. Having established this set as a ring with a unity element, we will begin to use the properties of rings without proving each subsequent property that results from this set being a ring.

The first property of matrices, contained in this ring S_A , concerns the rank of the resultant product of any two matrices contained in the S_A .

Corollary 7.2.2 Given two matrices $X, Y \in S_A$, then we have the inequality

$$rank(X \cdot Y) \leq min(rank(X), rank(Y)) .$$

Proof: The proof of this corollary is easily shown by introducing the following notation:

11. $r_x = rank(X)$,

12. $r_y = rank(Y)$ and

13. $N(X, Y)$ = Number of zero-roots with corresponding characteristic matrices.

With this notation, we have

$rank(X \cdot Y) = k - (k - r_x) - (k - r_y) + N(X, Y)$. Literal this translates to subtracting from the maximum rank of "**k**" the number of zero roots associated with X and the number of zero roots associated with Y and then add the number of roots associated with same characteristic matrices in both matrices. This equation becomes

$rank(X \cdot Y) = r_x + r_y - (k - N(X, Y))$. The quantity $(k - N(X, Y))$ is larger than both r_x *and* r_y. Hence we get $rank(X \cdot Y) \leq rank(X)$ and $rank(X \cdot Y) \leq rank(Y)$, or equivalently $rank(X \cdot Y) \leq min(rank(X), rank(Y))$.

For the special case when one of the matrices is of rank "**k**", we have a special case of this corollary. In this special case, we have equality since $N(X, Y) = 0$; $rank(X \cdot Y) = min(rank(X), rank(Y))$.

To formalize the concept of a generalized identity element, a formal definition of the generalized identity element is warranted. Such a definition should reflect the dependence of the identity element on the original matrix A. It should also be a special case of the identity element I for all nonsingular matrices.

Definition 7.2.3 The generalized identity element of the ring S_A, spanned by the "**k**" characteristic matrices of the matrix A is the element J contained in S_A having the following properties:

14. $rank(J) = k$,

15. $J \cdot J = J$ and

16. for any element $X \in S_A$, we have $J \cdot X = X \cdot J = X$.

Previously, we proposed the generalized identity matrix $J = \sum_{i=1}^{k} J_i$, where the characteristic matrices $\{J_1, J_2, \ldots, J_k\}$ span S_A. Since all nonzero characteristic roots of this matrix have a value of one, the rank of this matrix has a value of "**k**". If we compute the square of the matrix J, we get

$J^2 = \sum_{i=1}^{k} (1)^2 \cdot J_i = \sum_{i=1}^{k} J_i = J$. Hence condition 2 of definition 7.2.3 holds for the matrix J. For condition 3, an element $X \in S_A$ implies

$X = \sum_{i=1}^{k} \alpha_i \cdot J_i$. If we pre-multiply and post-multiply X with J we get

$J \cdot X = X \cdot J = \sum_{i=1}^{k} \alpha_i \cdot J_i = X$. Hence the matrix J satisfies all conditions of the generalized identity element. Using properties of rings, the generalized inverse (synonymous with unity element of ring) is unique. I will leave the proof of this statement as an exercise for the reader. Now let's turn our attention to the properties of generalized inverses of a matrix A. The following corollary examines the rank of a generalized inverse of the matrix A.

Corollary 7.2.4 Given the "**n-by-n**" matrix B which is a generalized inverse of the matrix A and

$rank(\boldsymbol{A}) = k$, then $rank(\boldsymbol{B}) \geq k$ and furthermore $rank(\boldsymbol{A} \cdot \boldsymbol{B}) = k$.

Proof: Using the first property of the definition of a generalized inverse we have the relationship $\boldsymbol{A} \cdot \boldsymbol{B} \cdot \boldsymbol{A} = \boldsymbol{A}$. From corollary 7.2.2 we get the relationship

$rank(\boldsymbol{A} \cdot \boldsymbol{B}) \geq rank((\boldsymbol{A} \cdot \boldsymbol{B}) \cdot \boldsymbol{A}) = rank(\boldsymbol{A}) = k$ or

$rank(\boldsymbol{A} \cdot \boldsymbol{B}) \geq k$. Applying corollary 7.2.2 for the product $\boldsymbol{A} \cdot \boldsymbol{B}$ we get $rank(\boldsymbol{A}) \geq rank(\boldsymbol{A} \cdot \boldsymbol{B}) \geq k$. With this results we can conclude that $rank(\boldsymbol{A} \cdot \boldsymbol{B}) = k$. Applying corollary 7.2.2 to this equality we then get $rank(\boldsymbol{B}) \geq k$ and the statement of the corollary is true.

Another property of a generalized inverse of the matrix $\boldsymbol{A}$ is derived from the above corollary and the definition of the ring S_A. If the matrix $\boldsymbol{B}$ is a generalized inverse of the matrix $\boldsymbol{A}$ and it is an element of this ring, then we must have $rank(\boldsymbol{B}) = k$.

Corollary 7.2.5 If the matrices $\boldsymbol{X}$ and $\boldsymbol{Y}$ are contained in $\mathcal{S}_A$ with $rank(\boldsymbol{X}) = k$, then if $\boldsymbol{X} \cdot \boldsymbol{Y} = \boldsymbol{0}$ we must have that $\boldsymbol{Y} = \boldsymbol{0}$.

Proof: The proof of this corollary is easily obtained by computing the rank of the resulting product. Since $rank(\boldsymbol{X}) = k$, this is a special case of corollary 7.2.2. and therefore $rank(\boldsymbol{X} \cdot \boldsymbol{Y}) = rank(\boldsymbol{Y})$. Since $\boldsymbol{X} \cdot \boldsymbol{Y} = \boldsymbol{0}$, we must have $rank(\boldsymbol{X} \cdot \boldsymbol{Y}) = 0$. This implies $rank(\boldsymbol{Y}) = 0$ which implies that $\boldsymbol{Y} = \boldsymbol{0}$.

Corollary 7.2.6 For the matrix $\boldsymbol{A}$ and its corresponding ring $\mathcal{S}_A$, if there is a generalized inverse of the matrix $\boldsymbol{A}$ contained in $\mathcal{S}_A$ it is unique.

Proof: This corollary is derived directly from the previous corollary; corollary 7.2.5. Assume that both $\boldsymbol{B}$ and $\boldsymbol{C}$ are generalized inverses of the matrix $\boldsymbol{A}$ and that both are elements of $\mathcal{S}_A$. Then we have the equality $\boldsymbol{A} \cdot \boldsymbol{B} \cdot \boldsymbol{A} = \boldsymbol{A} \cdot \boldsymbol{C} \cdot \boldsymbol{A} = \boldsymbol{A}$. By collecting all terms on the left side of the equal sign we get $\boldsymbol{A} \cdot \boldsymbol{B} \cdot \boldsymbol{A} - \boldsymbol{A} \cdot \boldsymbol{C} \cdot \boldsymbol{A} = \boldsymbol{0}$. Using the distributive property of matrices this equation becomes the following: $(\boldsymbol{A} \cdot \boldsymbol{B} - \boldsymbol{A} \cdot \boldsymbol{C}) \cdot \boldsymbol{A} = \boldsymbol{0}$. Since all matrices are elements of $\mathcal{S}_A$ and $rank(\boldsymbol{A}) = k$, corollary 7.2.5 implies that $(\boldsymbol{A} \cdot \boldsymbol{B} - \boldsymbol{A} \cdot \boldsymbol{C}) = \boldsymbol{0}$. Applying distributive property again, we have $\boldsymbol{A} \cdot (\boldsymbol{B} - \boldsymbol{C}) = \boldsymbol{0}$ and hence we can conclude that $\boldsymbol{B} = \boldsymbol{C}$. Hence, if a generalized inverse exists in the ring S_A, it is unique.

A question that comes to mind is the following: "Given the uniqueness and existence of the generalized inverse matrix $\boldsymbol{B}$ contained in the ring S_A, how can we derive the matrix $\boldsymbol{B}$ contained in S_A". The derivation of the unique generalized inverse in S_A of $\boldsymbol{A}$ is illustrated using corollary 7.2.7 below.

Corollary 7.2.7 If both $\boldsymbol{A}$ and $\boldsymbol{B}$ are elements in S_A and $\boldsymbol{B}$ is the unique generalized inverse matrix of $\boldsymbol{A}$, then $\boldsymbol{A}$ is the unique generalized inverse matrix of $\boldsymbol{B}$ in the ring S_A.

Proof: The proof of this corollary begins with the definition of $\boldsymbol{B}$ being the generalized inverse matrix of $\boldsymbol{A}$. Applying the corollaries 7.2.4 through 7.2.6, the proof is given in the steps below.

TABLE 12. **Proof of Corollary 7.2.7**

Step	Statement	Rationale for Statement
1	$A \cdot B \cdot A = A$	Property 1 of B being the generalized inverse matrix of A
2	$B \cdot C \cdot B = B$	Assuming that C is the generalized inverse matrix of B in S_A
3	$A \cdot B \cdot C \cdot B \cdot A = A$	Substituting expression for B in step 2 into step 1
4	$A \cdot B \cdot A \cdot C \cdot B = A$	Applying commutative property of generalized inverse
5	$A \cdot C \cdot B = A$	Step 4 becomes this expression since we have $A \cdot B \cdot A = A$.
6	$A \cdot C \cdot B - A \cdot B \cdot A = 0$	Making the substitution $A \cdot B \cdot A = A$ and collecting terms
7	$A \cdot (C \cdot B - B \cdot A) = 0$	Applying distributive property of matrices
8	$C \cdot B - B \cdot A = 0$	Since A is of rank "**k**", this is derived from corollary 7.2.5
9	$(C - A) \cdot B = 0$	Using $A \cdot B = B \cdot A$ and using distributive property of matrices
10	$C - A = 0$ or $C = A$	Applying corollary 7.2.5
11	A is the unique generalized inverse of B in S_A	Applying corollary 7.2.6

Now, we are ready to derive the general form of the generalized inverse $\boldsymbol{B}$ in S_A of the matrix $\boldsymbol{A}$. There are two possibilities for the generalized inverse; it may be an element of S_A or it may not be contained in S_A. If it is contained in this ring, then we have shown in corollary 7.2.6 that it is a unique generalized inverse in S_A. When the generalized inverse in not contained, in this ring, we shall show the general form of such a matrix and describe the conditions under which such a matrix is a generalized inverse. The form of the generalized inverse is described in the theorem below.

Theorem 7.2.8 For the matrix $\boldsymbol{A}$ and its corresponding ring S_A, the generalized inverse of the matrix $\boldsymbol{A}$ is a matrix of the form

$$\boldsymbol{B} = \sum_{i=1}^{k} \left(\frac{1}{x_i}\right) \cdot \boldsymbol{J}_i + \boldsymbol{B}_0 \text{ where } \boldsymbol{A} = \sum_{i=1}^{k} x_i \cdot \boldsymbol{J}_i \text{ and all}$$

"**k**" idempotent matrices are orthogonal to matrix $\boldsymbol{B}_0$.

Proof: Beginning with the definition of a generalized inverse, the matrix $\boldsymbol{B}$ is a generalized inverse of the matrix $\boldsymbol{A}$ when $\boldsymbol{A} \cdot \boldsymbol{B} \cdot \boldsymbol{A} = \boldsymbol{A}$ and $\cdot \boldsymbol{B} \cdot \boldsymbol{A} = \boldsymbol{B} \cdot \boldsymbol{A}$. Using the commutative property of these two matrices we have the expression $\boldsymbol{A}^2 \cdot \boldsymbol{B} = \boldsymbol{A}$. By replacing $\boldsymbol{A}$ with the expression $\boldsymbol{A} = \sum_{i=1}^{k} x_i \cdot \boldsymbol{J}_i$ we get the expression

$$\sum_{i=1}^{k} (x_i)^2 \cdot \boldsymbol{J}_i \cdot \boldsymbol{B} = \boldsymbol{A}$$. Since the idempotent matrices are pair-

wise orthogonal, we get $(x_i)^2 \cdot \boldsymbol{J}_i \cdot \boldsymbol{B} = \boldsymbol{J}_i \cdot \boldsymbol{A}$ or

$(x_i)^2 \cdot \boldsymbol{J}_i \cdot \boldsymbol{B} = x_i \cdot \boldsymbol{J}_i$ when we multiply both sides of the equation by $\boldsymbol{J}_i$. Dividing both sides of the equation by $(x_i)^2$ yields the following results: $\boldsymbol{J}_i \cdot \boldsymbol{B} = \left(\frac{1}{x_i}\right) \cdot \boldsymbol{J}_i$. Since this is true for all "**k**" of the pair-wise orthogonal idempotent matrices, we can express $\boldsymbol{B}$ as the following summation: $\boldsymbol{B} = \sum_{i=1}^{k} \left(\frac{1}{x_i}\right) \cdot \boldsymbol{J}_i + \boldsymbol{B}_0$ where $\boldsymbol{B}_0$ is orthogonal to all "**k**" idempotent matrices.

If we express the matrix $\boldsymbol{B}_0$ in terms of the "**n-k**" orthogonal matrices that complements the "**k**" idempotent matrices spanning S_A, we get $\boldsymbol{B} = \sum_{i=1}^{k} \left(\frac{1}{x_i}\right) \cdot \boldsymbol{J}_i + \sum_{i=k+1}^{n} \alpha_i \cdot \boldsymbol{J}_i$. This supports corollary 7.2.4 that states the rank of the generalized inverse is greater than or equal to the rank of $\boldsymbol{A}$. If the matrix $\boldsymbol{B}_0$ is the zero-matrix, then $\boldsymbol{B}$ is contained in S_A and is unique in the ring S_A. Hence the matrix $\boldsymbol{B} = \sum_{i=1}^{k} \left(\frac{1}{x_i}\right) \cdot \boldsymbol{J}_i$ is the unique generalized inverse of the matrix $\boldsymbol{A}$ in the ring S_A. Note that when we compute the products $\boldsymbol{A} \cdot \boldsymbol{B}$ *or* $\boldsymbol{B} \cdot \boldsymbol{A}$ we get the matrix $\boldsymbol{J} = \sum_{i=1}^{k} \boldsymbol{J}_i$ that is

$A \cdot B = B \cdot A = J$. These results imply that for any matrix of rank "**k**" contained in the ring S_A, has a unique generalized inverse in S_A and the product of such a matrix with its generalized inverse is the generalized identity element J.

Exercises

The matrices A in the exercises below are singular matrices, compute the rank of each matrix and determine the idempotent matrices that span the space S_A. Using the nonzero characteristic roots of the original matrix, compute the unique generalized inverse of the matrix.

1. Let $A = \begin{bmatrix} -4 & 36 & 10 \\ 12 & -108 & -30 \\ -50 & 405 & 113 \end{bmatrix}$, compute the rank (**k**) of this matrix and derive the idempotent matrices that span the space S_A. Then compute the generalized inverse of the matrix.

2. Let matrix $A = \begin{bmatrix} 2 & -2 & -2 \\ -19 & -11 & -11 \\ 22 & 18 & 18 \end{bmatrix}$ with rank **2** and have the idempotent matrices

   $J_1 = \begin{bmatrix} -2 & -2 & -2 \\ 1 & 1 & 1 \\ 2 & 2 & 2 \end{bmatrix}$ and $J_2 = \begin{bmatrix} 3 & 2 & 2 \\ -6 & -4 & -4 \\ 3 & 2 & 2 \end{bmatrix}$, derive the generalized inverse of this matrix and derive the generalized identity matrix of the space S_A.

3. Let $A = \begin{bmatrix} 8 & -8 & 8 & 8 \\ -12 & -16 & 2 & 12 \\ -24 & 26 & -25 & -26 \\ 0 & -30 & 15 & 26 \end{bmatrix}$, compute the rank (**k**) of this matrix and derive the generalize identity matrix of the space S_A. Then compute the generalized inverse of the matrix.

4. Let matrix $A = \begin{bmatrix} 18 & -2 & 20 & -10 \\ 4 & 0 & 4 & -2 \\ 62 & -6 & 68 & -34 \\ 158 & -16 & 174 & -87 \end{bmatrix}$ with rank **2** have the idempotent matrices $J_1 = \begin{bmatrix} -7 & 1 & -8 & 4 \\ 0 & 0 & 0 & 0 \\ -21 & 3 & -24 & 12 \\ -56 & 8 & -64 & 32 \end{bmatrix}$ and $J_2 = \begin{bmatrix} 4 & 0 & 4 & -2 \\ 4 & 0 & 4 & -2 \\ 20 & 0 & 20 & -10 \\ 46 & 0 & 46 & -23 \end{bmatrix}$, derive the generalized inverse of this matrix and derive the generalized identity matrix of the space S_A.

5. Let matrix $A = \begin{bmatrix} 6 & -91 & 106 & 141 & -81 \\ -3 & 24 & -30 & -37 & 21 \\ -5 & 85 & -98 & -132 & 76 \\ 16 & -261 & 302 & 405 & -233 \\ 25 & -376 & 438 & 583 & -335 \end{bmatrix}$ with rank **3** have the idempotent matrices $J_1 = \begin{bmatrix} 0 & 31 & -34 & -47 & 27 \\ 0 & 0 & 0 & 0 & 0 \\ 0 & -31 & 34 & 47 & -27 \\ 0 & 93 & -102 & -141 & 81 \\ 0 & 124 & -136 & -188 & 108 \end{bmatrix}$, $J_2 = \begin{bmatrix} 2 & -18 & 22 & 28 & -16 \\ -2 & 18 & -22 & -28 & 16 \\ -1 & 9 & -11 & -14 & 8 \\ 4 & -36 & 44 & 56 & -32 \\ 8 & -72 & 88 & 112 & -64 \end{bmatrix}$

and $J_3 = \begin{bmatrix} 2 & -24 & 28 & 38 & -22 \\ 1 & -12 & 14 & 19 & -11 \\ -3 & 39 & -42 & -57 & 33 \\ 8 & -96 & 112 & 152 & -88 \\ 9 & -108 & 126 & 171 & -99 \end{bmatrix}$, derive the generalized inverse of this matrix and derive the generalized identity matrix of the space S_A.

7.3 Characteristic Equation and k^{th} Rank Generalized Inverse

In the previous section, a generalized inverse of the matrix $\boldsymbol{A}$ was identified and a specific form was established for computing the matrix. When the original matrix is of rank "**k**" and is expressed as the following sum $\boldsymbol{A} = \sum_{i=1}^{k} x_i \cdot \boldsymbol{J}_i$, then the generalized inverse contained in S_A is unique and has the form $\boldsymbol{B} = \sum_{i=1}^{k} \left(\frac{1}{x_i}\right) \cdot \boldsymbol{J}_i$. Following the logic used in chapter 6 of this book, there is an approach for computing the inverse of a matrix using the characteristic equation of the matrix. For a singular matrix, the characteristic equation is still valid and this section will demonstrate how to compute the generalized inverse using the characteristic equation of the matrix. Before we demonstrate this approach, there is another property of generalized inverses that must be presented. This property extends the definition of the generalized inverse and is stated in the corollary below.

Corollary 7.3.1 If $\boldsymbol{B}$ is a generalized inverse matrix of $\boldsymbol{A}$, then we have the following properties:

(1). $\boldsymbol{A} \cdot \boldsymbol{B} = \boldsymbol{B} \cdot \boldsymbol{A} = \boldsymbol{J}$

2. $\boldsymbol{A}^k \cdot \boldsymbol{B} = \boldsymbol{B} \cdot \boldsymbol{A}^k = \boldsymbol{A}^{k-1}$.

Proof: The proof of this corollary is left to the reader as an exercise.

The general form of the characteristic equation for the matrix $\boldsymbol{A}$ is $\rho_A(x) = 0$ or $\rho_A(x) = a_n \cdot x^n + a_{n-1} \cdot x^{n-1} + \ldots + a_0 = 0$. In this instance, the matrix $\boldsymbol{A}$ of rank "**k**" indicates that the multiplicity of the zero-root is equal to "**n-k**". Applying corollary 5.2.3, we can write the characteristic equation as $\rho_A(x) = (0-x)^{n-k} \cdot \rho_1(x) = 0$. The expanded form of this equation takes the form

$\rho_A(x) = a_n \cdot x^n + a_{n-1} \cdot x^{n-1} + \ldots + a_{n-k} \cdot x^{n-k} = 0$. It was also illustrated in chapter 5 that when evaluating the characteristic function using the original matrix, the characteristic equation provides a matrix equation comparable to the scalar equation for the roots. For this instance, using the matrix

$\boldsymbol{A}$, we have $\rho_A(\boldsymbol{A}) = \boldsymbol{0}$ or

$\rho_A(\boldsymbol{A}) = a_n \cdot \boldsymbol{A}^n + a_{n-1} \cdot \boldsymbol{A}^{n-1} + \ldots + a_{n-k} \cdot \boldsymbol{A}^{n-k} = \boldsymbol{0}$. If we multiply this equation by the generalized inverse of the matrix $\boldsymbol{A}$ "**n-k**" times and apply corollary 7.3.1 we obtain

$\boldsymbol{B}^{n-k} \cdot \rho_A(\boldsymbol{A}) = a_n \cdot \boldsymbol{A}^k + a_{n-1} \cdot \boldsymbol{A}^{k-1} + \ldots + a_{n-k} \cdot \boldsymbol{J} = \boldsymbol{0}$. Hence if we have a method of computing the generalized identify element of the ring $\mathcal{S}_A$, when the characteristic equation is known. From this equation, we get

$\boldsymbol{J} = (a_n \cdot \boldsymbol{A}^k + a_{n-1} \cdot \boldsymbol{A}^{k-1} + \ldots + a_{n-k+1} \cdot \boldsymbol{A})/(-a_{n-k})$. With this equation, we get two for one equation. By multiplying this equation by the generalized inverse of the matrix ($\boldsymbol{B}$) we get the following solution for the matrix $\boldsymbol{B}$:

$$\boldsymbol{B} = (a_n \cdot \boldsymbol{A}^{k-1} + a_{n-1} \cdot \boldsymbol{A}^{k-2} + \ldots + a_{n-k+1} \cdot \boldsymbol{J})/(-a_{n-k}).$$

Exercises

Applying the concepts of this section, the following problems will use the characteristic equation of the matrix $\boldsymbol{A}$ to derive the generalized inverse and the generalized identity of the space $\mathcal{S}_A$.

1. For the matrix $A = \begin{bmatrix} 3 & -10 & 8 \\ -3 & 21 & -17 \\ 6 & 22 & -18 \end{bmatrix}$ having characteristic equation $\rho_A(x) = -x^3 + x = 0$, derive the generalized inverse of A and the generalized identity element of the space S_A.

2. For the matrix $A = \begin{bmatrix} 2 & 14 & -4 \\ -3 & -3 & 0 \\ -9 & -18 & 3 \end{bmatrix}$ having characteristic equation $\rho_A(x) = -x^3 + 2 \cdot x^2 + 3 \cdot x = 0$, derive the generalized inverse of A and the generalized identity element of the space S_A.

3. For the matrix $A = \begin{bmatrix} 94 & -29 & 88 & -38 \\ -150 & 41 & -134 & 58 \\ 252 & -72 & 229 & -99 \\ 936 & -272 & 856 & -370 \end{bmatrix}$ having characteristic equation $\rho_A(x) = x^4 + 6 \cdot x^3 + 3 \cdot x^2 - 10 \cdot x = 0$, derive the generalized inverse of A and the generalized identity element of the space S_A.

4. For the matrix $A = \begin{bmatrix} -4 & -40 & -3 & 10 \\ -6 & -72 & -6 & 18 \\ -6 & -96 & -9 & 24 \\ -26 & -320 & -27 & 80 \end{bmatrix}$ having characteristic equation $\rho_A(x) = x^4 + 5 \cdot x^3 + 6 \cdot x^2 = 0$, derive the generalized inverse of A and the generalized identity element of the space S_A.

5. For the matrix $A = \begin{bmatrix} 41 & -56 & 115 & -124 & 36 \\ 72 & -91 & 158 & -173 & -51 \\ -42 & 46 & -50 & 58 & -18 \\ -21 & 32 & -79 & 84 & -24 \\ 125 & -112 & -1 & -20 & 12 \end{bmatrix}$ having characteristic equation

$$\rho_A(x) = -x^5 - 4 \cdot x^4 + x^3 + 4 \cdot x^2 = 0$$

, derive the generalized inverse of A and the generalized identity element of the space S_A.

6. Prove the statements of corollary 7.3.1.

7.4 Projections Into a Subspace and Gram-Schmidt Algorithm

In this section, we will revisit the concept of a vector space (or vector subspace) and show its relationship to the ring S_A introduced in the previous sections of this chapter.

7.4.1 Complementing Vector Spaces

Recall that the ring S_A is the set of "**n-by-n**" matrices spanned by the "**k**" matrices $\{J_1, J_2, \ldots, J_k\}$. This subspace is a subset of the n-dimensional space spanned by the superset of "**n**" matrices $\{J_1, J_2, \ldots, J_n\}$. If you remember the representation of these matrices from chapter 5, each characteristic matrix can be represented as a product of a column matrix and a row matrix (that is a column vector and a row vector); $J_i = \underset{\sim}{P}_i \cdot \underset{\sim}{Q}_i^T$. Hence if we pre-multiply an **n**-dimensional vector $\underset{\sim}{X}$ by the matrix J_i, we get $J_i \cdot \underset{\sim}{X} = (\underset{\sim}{Q}_i^T \cdot \underset{\sim}{X}) \cdot \underset{\sim}{P}_i$, a scalar multiple of the base vector $\underset{\sim}{P}_i$. Hence we define this product to be the projection of the vector $\underset{\sim}{X}$ in the direction of the vector $\underset{\sim}{P}_i$. If we post-multiply the transpose of the vector with the characteristic matrix J_i, we get $\underset{\sim}{X}^T \cdot J_i = (\underset{\sim}{X}^T \cdot \underset{\sim}{P}_i) \cdot \underset{\sim}{Q}_i^T$, a scalar multiple of the row base vector $\underset{\sim}{Q}_i^T$. (Hence this is the projection of $\underset{\sim}{X}$ along the direction of the vector $\underset{\sim}{Q}_i^T$.)

If we continue the pre-multiplication and post-multiplication for all "**k**" characteristic matrices above, we get the representation of the column and row vectors $\underset{\sim}{X}$ and $\underset{\sim}{X}^T$ in the vector subspaces spanned by $\{\underset{\sim}{P}_1, \underset{\sim}{P}_2, \ldots, \underset{\sim}{P}_n\}$ and $\{\underset{\sim}{Q}^T_1, \underset{\sim}{Q}^T_2, \ldots, \underset{\sim}{Q}^T_n\}$ respectively. For lack of a better term, I will call a pair of such vector subspaces "complementing orthogonal subspaces". Using the notation $\alpha_i = \underset{\sim}{Q}_i^T \cdot \underset{\sim}{X}$ and $\beta_i = \underset{\sim}{X}^T \cdot \underset{\sim}{P}_i$, we can represent each vector $\underset{\sim}{X}$ and $\underset{\sim}{X}^T$ in the complementing orthogonal subspaces as

$\underset{\sim}{X} = \sum_{i=1}^{n} \alpha_i \cdot \underset{\sim}{P}_i$ and $\underset{\sim}{X}^T = \sum_{i=1}^{n} \beta_i \cdot \underset{\sim}{Q}_i^T$ respectively. When projecting $\underset{\sim}{X}$ and $\underset{\sim}{X}^T$ onto the subspaces spanned by the column vectors and row vectors derived from the characteristic matrices that form the basis of the ring S_A , we can write $J \cdot \underset{\sim}{X} = \sum_{i=1}^{k} \alpha_i \cdot \underset{\sim}{P}_i$ and $\underset{\sim}{X}^T \cdot J = \sum_{i=1}^{k} \beta_i \cdot \underset{\sim}{Q}_i^T$.

For the column and row vector spaces generated from the ring S_A , as shown above, we introduce the notation $\underset{\sim}{V}(S_A)$ to represent the column vector space and $\underset{\sim}{V}^T(S_A)$ to represent the row vector space. Using this notation and the generation method above, we specify the properties in table 4 for these vector spaces.

TABLE 13. Properties of Vector spaces $\underset{\sim}{V}(S_A)$ and $\underset{\sim}{V}^T(S_A)$

Property #	Property Description
1	For any two base orthogonal idempotent matrices J_i and J_j, both of rank 1 and contained in S_A, then for an arbitrary vector $\underset{\sim}{X}$ the vectors $\underset{\sim}{X}_i = J_i \cdot \underset{\sim}{X}$ and $\underset{\sim}{X}_j = J_j \cdot \underset{\sim}{X}$ are orthogonal for $i \neq j$. $(\underset{\sim}{X}^T{}_j \cdot \underset{\sim}{X}_i = \underset{\sim}{X}^T{}_i \cdot \underset{\sim}{X}_j = 0)$
2	For an arbitrary vector $\underset{\sim}{X}$, the vector $J \cdot \underset{\sim}{X}$ is contained in $\underset{\sim}{V}(S_A)$ and the vector $\underset{\sim}{X}^T \cdot J$ is contained in $\underset{\sim}{V}^T(S_A)$.
3	If $\underset{\sim}{X} = J \cdot \underset{\sim}{X}$ or equivalently $\underset{\sim}{X}^T = \underset{\sim}{X}^T \cdot J$, then $\underset{\sim}{X}$ is contained in $\underset{\sim}{V}(S_A)$ and $\underset{\sim}{X}^T$ is contained in $\underset{\sim}{V}^T(S_A)$.
4	If each row of the matrix X is an element of $\underset{\sim}{V}^T(S_A)$ and if the generalized identity matrix J of S_A is symmetric, then the symmetric matrix given by $X^T \cdot X$ is an element of the ring S_A. Since each column is contained in $\underset{\sim}{V}(S_A)$, we have $X^T \cdot X = X^T \cdot (X \cdot J) = (X^T \cdot X) \cdot J$ and hence it is contained in the ring S_A.
5	If the symmetric matrix $X^T \cdot X$ is an element of the ring S_A, then each row of the matrix X is an element of $\underset{\sim}{V}^T(S_A)$. This property is verified using theorem 7.4.2.

Theorem 7.4.2 Let $\boldsymbol{X}$ be an **n-by-k** matrix and define the matrix $\boldsymbol{A}$ such that $\boldsymbol{A} = \boldsymbol{X}^T \cdot \boldsymbol{X}$. If the matrix $\boldsymbol{A}$ is a nonzero matrix, contained in the space $\mathcal{S}_A$ where the generalized identity matrix is $\boldsymbol{J}$ of rank m ($\boldsymbol{m} \le \boldsymbol{k}$) then each row of the matrix $\boldsymbol{X}$ is contained in the vector space $\underset{\sim}{V}^T(\mathcal{S}_A)$.

Proof: To prove this theorem, we denote each row of the matrix $\boldsymbol{X}$ as transposed vectors, namely $\boldsymbol{X} = \begin{bmatrix} \underset{\sim}{\boldsymbol{X}}_1^T \\ \underset{\sim}{\boldsymbol{X}}_2^T \\ \cdots \\ \underset{\sim}{\boldsymbol{X}}_n^T \end{bmatrix}$. Since $\boldsymbol{X}$ is an **n-by-k** matrix, each vector is a **k**-dimensional vector. From the statement of the this theorem, the matrix $\boldsymbol{A} = \boldsymbol{X}^T \cdot \boldsymbol{X}$ is contained in the space $\mathcal{S}_A$, which is of rank m ($\boldsymbol{m} \le \boldsymbol{k}$). Letting $\boldsymbol{J}$ represent the generalized inverse of this space, the projection of each vector into the space $\underset{\sim}{V}^T(\mathcal{S}_A)$ is given by $\underset{\sim}{\boldsymbol{X}}_i^T \cdot \boldsymbol{J}$. The projection of each vector in the complementing (and exclusive) space is given by $\underset{\sim}{\boldsymbol{X}}_i^T \cdot (\boldsymbol{I} - \boldsymbol{J})$. Furthermore the original vector is obtained by adding the projections as follows:

$\underset{\sim}{\boldsymbol{X}}_i^T = \underset{\sim}{\boldsymbol{X}}_i^T \cdot \boldsymbol{J} + \underset{\sim}{\boldsymbol{X}}_i^T \cdot (\boldsymbol{I} - \boldsymbol{J})$. In terms of the original matrix, we have $\boldsymbol{X} = \boldsymbol{X} \cdot \boldsymbol{J} + \boldsymbol{X} \cdot (\boldsymbol{I} - \boldsymbol{J})$. We will denote each

decomposition of the matrix as $X_0 = X \cdot J$ and
$X_1 = X \cdot (I - J)$.

Using this decomposition of X, the computation of the matrix A becomes
$X^T \cdot X = X_0^{\;T} \cdot X_0 + X_1^T \cdot X_0 + X_0^{\;T} \cdot X_1 + X_1^T \cdot X_1$.

Pre-multiplying A by J, we get $J \cdot A = X_0^{\;T} \cdot X_0 + X_0^{\;T} \cdot X_1$. Since $J \cdot A = A$ this implies that $X_1^T \cdot X_0 + X_1^T \cdot X_1 = \mathbf{0}$. With this new representation we have $A = X_0^{\;T} \cdot X_0 + X_0^{\;T} \cdot X_1$. Post multiplying A by J, we now have $A = X_0^{\;T} \cdot X_0$ and $X_0^{\;T} \cdot X_1 = \mathbf{0}$.

If we post multiply A by J first, we get $X_0^T \cdot X_1 + X_1^T \cdot X_1 = \mathbf{0}$ and then the pre-multiplication by J yields $X_1^{\;T} \cdot X_0 = \mathbf{0}$. Combining the previous equality with $X_0^{\;T} \cdot X_1 = \mathbf{0}$ also gives $X_1^{\;T} \cdot X_1 = \mathbf{0}$. Since X_1 is a real matrix, $X_1^{\;T} \cdot X_1 = \mathbf{0}$ implies that $X_1 = \mathbf{0}$. Hence we can conclude that $X = X \cdot J$ or an equivalent statement is that each row of the matrix X is an element of $\underset{\sim}{V}^T(S_A)$.

A special instance of the concept of a complementing orthogonal space/subspace pair occurs when the transpose of one vector space/subspace is its complementing orthogonal space/subspace. For the Euclidean space, the column vectors $\{\underset{\sim}{E}_1, \underset{\sim}{E}_2, \ldots, \underset{\sim}{E}_n\}$ and the row vectors $\{\underset{\sim}{E}^T{}_1, \underset{\sim}{E}^T{}_2, \ldots, \underset{\sim}{E}^T{}_n\}$ are the basis for a complementing orthogonal space. Such a pair is called a symmetric complementing orthogonal space. Recall the property of vectors discussed in chapter three; the concept of distance. For an **n**-dimensional vector in Euclidean space, the square root of the inner product of the vector with itself is defined as the distance of the vector from the origin. For a vector $\underset{\sim}{X}$ with Euclidean coordinates $\underset{\sim}{X}^T = (x_1, x_2, \ldots, x_n)$, the square of the distance from the origin is $\underset{\sim}{X}^T \cdot \underset{\sim}{X} = x_1^2 + x_2^2 + \ldots + x_n^2$. Expanding the definition for the distance from the origin for a vector, we provide the definition below.

Definition 7.4.3 Let "**k**" be a positive integer less than "**n**". Then if the vector $\underset{\sim}{X}$ is represented as $\underset{\sim}{X} = \sum_{i=1}^{k} \alpha_i \cdot \underset{\sim}{P}_i$ and its transpose is represented as $\underset{\sim}{X}^T = \sum_{i=1}^{k} \beta_i \cdot \underset{\sim}{Q}_i^T$ in the complementing orthogonal space, then the square of the distance from the origin for this vector is denoted by

$$d^2 = \underset{\sim}{X}^T \cdot \underset{\sim}{X} = \sum_{i=1}^{k} \alpha_i \cdot \beta_i .$$

A final note on the concept of distance concerns the property of invariance. When the vector is represented in Euclidean coordinates and also in complementing orthogonal vector spaces, then we have

$d^2 = \sum_{i=1}^{k} x_i^2 = \sum_{i=1}^{k} \alpha_i \cdot \beta_i$; distance is invariant regardless of the vector space. A special case of the distance formula occurs when the transpose of the column vector is the corresponding complementing orthogonal row vectors. In terms of the row and column vectors, this is given by the relationship $\underset{\sim}{Q}_i = \underset{\sim}{P}_i$. With this relationship, the distance formula becomes

$$d^2 = \underset{\sim}{X}^T \cdot \underset{\sim}{X} = \sum_{i=1}^{k} \alpha_i^2 .$$

When the complementing orthogonal vector spaces are such that $\underset{\sim}{Q}_i = \underset{\sim}{P}_i$, the complementing vector spaces are called symmetric complementing orthogonal vector spaces. For simplicity, we will refer to the basis $\{\underset{\sim}{P}_1, \underset{\sim}{P}_2, \ldots, \underset{\sim}{P}_n\}$, when the complementing orthogonal vector basis is equivalent to the transpose of the column vector basis, as an orthogonal basis. In the general case in which $\underset{\sim}{Q}_i \neq \underset{\sim}{P}_i$, the complementing vector spaces are called asymmetric complementing orthogonal vector spaces and we refer to the pair of bases $\{\underset{\sim}{P}_1, \underset{\sim}{P}_2, \ldots, \underset{\sim}{P}_n\}$ and $\{\underset{\sim}{Q}^T{}_1, \underset{\sim}{Q}^T{}_2, \ldots, \underset{\sim}{Q}^T{}_n\}$ as the complementing orthogonal bases. It is worthwhile noting that the ring S_A associated with an orthogonal basis containing only symmetric matrices generated as a linear combination of base matrices of form $J_i = \underset{\sim}{P}_i \cdot \underset{\sim}{P}_i^T$.

In general, to derive an asymmetric complementing orthogonal vector basis pair, when a single vector basis $\{\underset{\sim}{P}_1, \underset{\sim}{P}_2, \ldots, \underset{\sim}{P}_n\}$ is given, we construct a matrix and compute the inverse. Using each column vector as the column of the matrix, we get the matrix $P = (\underset{\sim}{P}_1, \underset{\sim}{P}_2, \ldots, \underset{\sim}{P}_n)$. Computing the inverse of this matrix $Q = P^{-1}$ and equating the rows of the resulting matrix to the row vectors that form the complementing orthogonal vector basis of the original column vector basis. Hence we have $Q = P^{-1} = \begin{pmatrix} \underset{\sim}{Q}^T{}_1 \\ \underset{\sim}{Q}^T{}_2 \\ \ldots \\ \underset{\sim}{Q}^T{}_n \end{pmatrix}$. The derivation of an orthogonal space is not as straightforward as the asymmetric pair, but it is derivable using the well known Gram-Schmidt algorithm. This algorithm is derived in the following section.

Exercises (Bases and Coordinate Spaces)

1. Let $A = \begin{bmatrix} -7 & 5 & 3 \\ 11 & -1 & -3 \\ -41 & 11 & 13 \end{bmatrix}$; derive the bases for vector spaces $\underset{\sim}{V}(S_A)$ and $\underset{\sim}{V}^T(S_A)$.

2. Using the matrix A , defined in exercise 1 above, derive the coordinates of the projections of the Euclidean vector $\begin{bmatrix} 1 \\ 2 \\ -1 \end{bmatrix}$ in the vector spaces $\underset{\sim}{V}(S_A)$ and $\underset{\sim}{V}^T(S_A)$.

3. Let $A = \begin{bmatrix} -16 & 24 & -10 \\ -6 & 8 & -4 \\ 3 & -6 & 1 \end{bmatrix}$; derive the bases for vector spaces $\underset{\sim}{V}(S_A)$ and $\underset{\sim}{V}^T(S_A)$.

4. Using the matrix A , defined in exercise 1 above, derive the coordinates of the projections of the Euclidean vector $\begin{bmatrix} 3 \\ 0 \\ 1 \end{bmatrix}$ in the vector spaces $\underset{\sim}{V}(S_A)$ and $\underset{\sim}{V}^T(S_A)$.

5. Let $A = \begin{bmatrix} -35 & -29 & -5 \\ 54 & 42 & 6 \\ -92 & -68 & -8 \end{bmatrix}$; derive the bases for vector spaces $\underset{\sim}{V}(S_A)$ and $\underset{\sim}{V}^T(S_A)$.

6. Using the matrix A , defined in exercise 1 above, derive the coordinates of the projections of the Euclidean vector $\begin{bmatrix} 2 \\ 4 \\ 1 \end{bmatrix}$ in the vector spaces $\underset{\sim}{V}(S_A)$ and $\underset{\sim}{V}^T(S_A)$.

7. Let $A = \begin{bmatrix} -1 & 13 & 8 \\ 12 & -30 & -22 \\ -17 & 47 & 34 \end{bmatrix}$; derive the bases for vector spaces $\underset{\sim}{V}(S_A)$ and $\underset{\sim}{V}^T(S_A)$.

8. Using the matrix A , defined in exercise 1 above, derive the coordinates of the projections of the Euclidean vector $\begin{bmatrix} 10 \\ 3 \\ 1 \end{bmatrix}$ in the vector spaces $\underset{\sim}{V}(S_A)$ and $\underset{\sim}{V}^T(S_A)$.

9. Let $A = \begin{bmatrix} 46 & 29 & -47 & -18 \\ 148 & 71 & -128 & -48 \\ 98 & 70 & -109 & -42 \\ 100 & 4 & -40 & -13 \end{bmatrix}$; derive the bases for spaces $\underset{\sim}{V}(S_A)$ and $\underset{\sim}{V}^T(S_A)$.

10. Using the matrix A , defined in exercise 1 above, derive the coordinates of the projections of the Euclidean vector $\begin{bmatrix} 1 \\ -1 \\ 1 \end{bmatrix}$ in the vector spaces $\underset{\sim}{V}(S_A)$ and $\underset{\sim}{V}^T(S_A)$.

Exercises

For exercises 11 through 15 below, use the given matrix denoted as X in each exercise to compute $A = X^T \cdot X$ and derive the generalized identity matrix J of the space S_A. Then verify the relationship $X = X \cdot J$.

11. $X = \begin{bmatrix} 1 & 1 & 1 & 1 \\ 1 & 1 & 0 & 2 \\ 1 & 2 & 1 & -1 \end{bmatrix}$

12. $X = \begin{bmatrix} 1 & 2 & 1 & -1 & 3 \\ 1 & 0 & 0 & 0 & 4 \\ 1 & 2 & 1 & -1 & -3 \end{bmatrix}$

13. $X = \begin{bmatrix} 1 & 0 & 2 & 4 & 6 \\ 1 & 1 & 3 & 5 & 1 \end{bmatrix}$

14. $X = \begin{bmatrix} 1 & 2 & 4 \\ 1 & 2 & 4 \\ 1 & 0 & 1 \end{bmatrix}$

15. $X = \begin{bmatrix} 1 & 0 & 1 & 2 & 1 \\ 1 & 1 & 1 & 3 & 4 \\ 1 & 0 & 0 & 4 & 6 \\ 1 & 1 & 0 & 5 & 8 \end{bmatrix}$

7.4.2 Gram-Schmidt Algorithm

Having shown the properties associated with orthogonal bases, it would be nice to convert "**n**" arbitrary nonorthogonal independent vectors to "**n**" independent orthogonal vectors to form an orthogonal basis. The Gram-Schmidt algorithm shows a sequential approach for constructing an orthogonal basis. This algorithm is given in forms of a theorem.

Before proving the Gram-Schmidt Algorithm, we introduce the following notation for a series of rings generated from "**n**" idempotent orthogonal matrices $\{J_1, J_2, \ldots, J_n\}$:

1. S_n is the ring generated by the "**n**" orthogonal matrices $\{J_1, J_2, \ldots, J_n\}$.
2. S_{n-1} is the ring generated by the "**n-1**" orthogonal matrices $\{J_2, \ldots, J_n\}$.
3. S_{n-k} is the ring generated by the "**n-k**" orthogonal matrices $\{J_{k+1}, \ldots, J_n\}$.

The corresponding complementing rings for the above rings are designated using the following notation:

1. τ_1 is the ring generated by the single idempotent matrix J_1 of rank one. It is composed of all matrices that is a scalar multiple of J_1 and it is orthogonal to all matrices contained in S_{n-1}.
2. τ_2 is the ring generated by the two idempotent and orthogonal matrices $\{J_1, J_2\}$. Every matrix contained in this ring is orthogonal to every matrix contained in S_{n-2}.
3. τ_k is the ring generated by the "**k**" idempotent and orthogonal matrices $\{J_1, J_2, \ldots, J_k\}$ .Every matrix contained in this ring is orthogonal to every matrix contained in S_{n-k}.

Theorem - Gram-Schmidt Algorithm Let the sequence of vectors $\{\underset{\sim}{P}_1, \underset{\sim}{P}_2, \ldots, \underset{\sim}{P}_n\}$ be vectors forming a basis for $\boldsymbol{R}^n$, then there is a sequence of idempotent matrices such that for each "i" we have $\underset{\sim}{u}_i = \mathsf{M}_i \cdot \underset{\sim}{P}_i$ and the vectors $\{\frac{\underset{\sim}{u}_1}{|\underset{\sim}{u}_1|}, \frac{\underset{\sim}{u}_2}{|\underset{\sim}{u}_2|}, \ldots, \frac{\underset{\sim}{u}_n}{|\underset{\sim}{u}_n|}\}$ form an orthogonal basis for $\boldsymbol{R}^n$. Furthermore, the representation of each of the derived orthogonal vectors is given as $\underset{\sim}{u}_1 = \underset{\sim}{P}_1$ and

$$\underset{\sim}{u}_i = \underset{\sim}{P}_i - \sum_{k=1}^{i-1} \frac{\underset{\sim}{P}_i^T \cdot \underset{\sim}{u}_k}{(\underset{\sim}{u}_k^T \cdot \underset{\sim}{u}_k)} \cdot \underset{\sim}{u}_k \text{ for } i > 1 .$$

Proof: The proof of this theorem estimates the "**n**" orthogonal idempotent matrices $\{\boldsymbol{J}_1, \boldsymbol{J}_2, \ldots, \boldsymbol{J}_n\}$ in the table below. Using these matrices, we compute the sequence of identity elements for the rings $\mathsf{S}_n, \mathsf{S}_{n-1}, \ldots, \mathsf{S}_1$. With the identity matrices, we compute the sequence of matrices M_i and the sequence of orthogonal vectors $\underset{\sim}{u}_i = \mathsf{M}_i \cdot \underset{\sim}{P}_i$.

TABLE 14. **Proof for Gram-Schmidt Algorithm**

Step	Statement	Rationale for Statement
1	We begin with "**n**" orthogonal idempotent and symmetric matrices $\{J_1, J_2, \ldots, J_n\}$ that forms basis for ring S_n.	Assume that all matrices of form $\sum_{i=1}^{n} \alpha_i \cdot J_i$ are contained in ring S_n.
2	The identity matrix satisfies relationship $I = \sum_{i=1}^{n} J_i$.	The identity matrix is contained in S_n and since these are **n-by-n** matrices the sum of the idempotent matrices is the identity matrix.
3	As a starting point, set the vector $\underset{\sim}{u}_1 = \underset{\sim}{P}_1$ and define the idempotent matrix J_1 to be the matrix $\frac{(\underset{\sim}{u}_1 \cdot \underset{\sim}{u}_1^T)}{(\underset{\sim}{u}_1^T \cdot \underset{\sim}{u}_1)}$.	The starting point is completely arbitrary. We could choose any one of the original vectors as a starting point. The selection of J_1 was selected so that $J_1^2 = J_1$.
4	Defining matrix $M_2 = I - J_1$, we have $\underset{\sim}{u}_2 = M_2 \cdot \underset{\sim}{P}_2$ is orthogonal to the vector $\underset{\sim}{u}_1$.	The matrix $M_2 = I - J_1$ is the identity element of the ring S_{n-1}. The vectors $\underset{\sim}{u}_2 = M_2 \cdot \underset{\sim}{P}_2 \in \underset{\sim}{V}(S_{n-1})$ and $\underset{\sim}{u}_1 \in \underset{\sim}{V}(\tau_1)$ are orthogonal.

TABLE 14. **Proof for Gram-Schmidt Algorithm**

Step	Statement	Rationale for Statement
5	Using the results of step 4, we define the idempotent matrix J_2 to be the matrix $\dfrac{(\underset{\sim}{u}_2 \cdot \underset{\sim}{u}_2^T)}{(\underset{\sim}{u}_2^T \cdot \underset{\sim}{u}_2)}$.	With this selection, we get $J_2^2 = J_2$ and J_1 is orthogonal to J_2. Hence we have defined the first two symmetric idempotent and orthogonal matrices.
6	Defining $M_3 = M_2 - J_2$, we have $\underset{\sim}{u}_3 = M_3 \cdot \underset{\sim}{P}_3$ is orthogonal to the vectors $\underset{\sim}{u}_1$ *and* $\underset{\sim}{u}_2$.	The matrix $M_3 = I - J_1 - J_2$ is the identity element of the ring S_{n-2}. The vectors $\underset{\sim}{u}_3 = M_3 \cdot \underset{\sim}{P}_3 \in \underset{\sim}{V}(S_{n-2})$ and $\underset{\sim}{u}_1, \underset{\sim}{u}_2 \in \underset{\sim}{V}(\tau_2)$ are orthogonal.
7	The general form of the sequence of the idempotent matrices is $M_i = I - \sum_{k=1}^{i-1} \dfrac{\underset{\sim}{u}_k \cdot \underset{\sim}{u}_k^T}{(\underset{\sim}{u}_k^T \cdot \underset{\sim}{u}_k)}$ and $\underset{\sim}{u}_i = M_i \cdot \underset{\sim}{P}_i$ is orthogonal to the vectors $\underset{\sim}{u}_1, \underset{\sim}{u}_2, \ldots, \underset{\sim}{u}_{i-1}$.	Using mathematical induction and sequentially estimating each idempotent matrix J_i, we construct the identity matrix M_i for the ring $S_{n-(i-1)}$. The vectors $\underset{\sim}{u}_i = M_i \cdot \underset{\sim}{P}_i \in \underset{\sim}{V}(S_{n-(i-1)})$ and $\underset{\sim}{u}_1, \underset{\sim}{u}_2, \ldots, \underset{\sim}{u}_{i-1} \in \underset{\sim}{V}(\tau_{i-1})$ are orthogonal.

TABLE 14. **Proof for Gram-Schmidt Algorithm**

Step	Statement	Rationale for Statement
8	$\underset{\sim}{u}_i = \underset{\sim}{P}_i - \sum_{k=1}^{i-1} \frac{\underset{\sim}{u}_k \cdot \underset{\sim}{u}_k^T}{(\underset{\sim}{u}_k^T \cdot \underset{\sim}{u}_k)} \cdot \underset{\sim}{P}_i$	Substituting the value for $\mathbf{M}_i$ and performing the multiplication, we get this expression for the vector $\underset{\sim}{u}_i$.
9	The final form of the vector representation is $\underset{\sim}{u}_1 = \underset{\sim}{P}_1$ and $\underset{\sim}{u}_i = \underset{\sim}{P}_i - \sum_{k=1}^{i-1} \frac{\underset{\sim}{P}_i^T \cdot \underset{\sim}{u}_k}{(\underset{\sim}{u}_k^T \cdot \underset{\sim}{u}_k)} \cdot \underset{\sim}{u}_k$ for $i > 1$.	Use the commutative property of matrix multiplication to get this form of the algorithm.

Exercises

For exercises 1 through 5 below, apply the Gram-Schmidt Algorithm to derive a set of orthogonal vectors from the independent vectors given.

1. Given the set of independent vectors $\left\{ \begin{bmatrix} 42 \\ 74 \\ 234 \end{bmatrix}, \begin{bmatrix} 8 \\ 12 \\ 42 \end{bmatrix}, \begin{bmatrix} -10 \\ -17 \\ -55 \end{bmatrix} \right\}$, derive a set of orthogonal vectors that form a basis for the three dimensional vector space.

2. Given the set of independent vectors $\left\{ \begin{bmatrix} -109 \\ -330 \\ -642 \end{bmatrix}, \begin{bmatrix} -6 \\ -25 \\ -38 \end{bmatrix}, \begin{bmatrix} 22 \\ 70 \\ 131 \end{bmatrix} \right\}$, derive a set of orthogonal vectors that form a basis for the three dimensional vector space.

3. Given the set of independent vectors $\left\{ \begin{bmatrix} -2 \\ 1 \\ 1 \\ -2 \end{bmatrix}, \begin{bmatrix} 6 \\ 7 \\ -2 \\ 8 \end{bmatrix}, \begin{bmatrix} -18 \\ -28 \\ 3 \\ -22 \end{bmatrix}, \begin{bmatrix} -5 \\ -5 \\ 3 \\ -9 \end{bmatrix} \right\}$, derive a set of orthogonal vectors that form a basis for the four dimensional vector space.

4. Given the set of independent vectors $\left\{ \begin{bmatrix} 18 \\ -20 \\ 10 \\ 70 \end{bmatrix}, \begin{bmatrix} -20 \\ 12 \\ -19 \\ -58 \end{bmatrix}, \begin{bmatrix} 0 \\ 2 \\ 1 \\ -4 \end{bmatrix}, \begin{bmatrix} -10 \\ 8 \\ -8 \\ -33 \end{bmatrix} \right\}$, derive a set of orthogonal vectors that form a basis for the four dimensional vector space.

5. Given the set of independent vectors $\left\{ \begin{bmatrix} 10 \\ 10 \\ 32 \\ -74 \\ 59 \end{bmatrix}, \begin{bmatrix} 12 \\ 20 \\ 62 \\ -146 \\ 113 \end{bmatrix}, \begin{bmatrix} -36 \\ -58 \\ -169 \\ 392 \\ -314 \end{bmatrix}, \begin{bmatrix} -15 \\ -22 \\ -68 \\ 159 \\ -125 \end{bmatrix}, \begin{bmatrix} -2 \\ 0 \\ -6 \\ 16 \\ -9 \end{bmatrix} \right\}$, derive a set of orthogonal vectors that form a basis for the five dimensional vector space.

7.4.3 Q R Decomposition

A benefit derived from the Gram-Schmidt algorithm is the representation of a nonsingular matrix as the product of orthogonal matrix (product of matrix with its transpose matrix equals identity matrix) and a triangular matrix. This is easily seen by the transformation of the arbitrary independent columns of the nonsingular matrix using the Gram-Schmidt algorithm. Suppose we have a matrix represented as $\boldsymbol{P} = (\underset{\sim}{\boldsymbol{P}}_1, \underset{\sim}{\boldsymbol{P}}_2, \ldots, \underset{\sim}{\boldsymbol{P}}_n)$. Using the derivation in the orthogonal vectors in the Gram-Schmidt algorithm, we get $\underset{\sim}{v}_k = \underset{\sim}{\boldsymbol{u}}_k / |\underset{\sim}{\boldsymbol{u}}_k|$ where

$\underset{\sim}{u}_1 = \underset{\sim}{P}_1$ and $\underset{\sim}{u}_i = \underset{\sim}{P}_i - \sum_{k=1}^{i-1} \frac{\underset{\sim}{P}_i^T \cdot \underset{\sim}{u}_k}{(\underset{\sim}{u}_k^T \cdot \underset{\sim}{u}_k)} \cdot \underset{\sim}{u}_k$. Using this representation and strategically replacing the **u**-vectors with the base vectors we have

$\underset{\sim}{v}_i \cdot |\underset{\sim}{u}_i| = \underset{\sim}{P}_i - \sum_{k=1}^{i-1} \frac{\underset{\sim}{P}_i^T \cdot \underset{\sim}{u}_k}{|\underset{\sim}{u}_k|} \cdot \underset{\sim}{v}_k$. Solving for the vector $\underset{\sim}{P}_i$, we can express the independent vectors of the matrix as $\underset{\sim}{P}_i = \underset{\sim}{v}_i \cdot |\underset{\sim}{u}_i| + \sum_{k=1}^{i-1} \frac{\underset{\sim}{P}_i^T \cdot \underset{\sim}{u}_k}{|\underset{\sim}{u}_k|} \cdot \underset{\sim}{v}_k$. You will note that the ith vector of the matrix P, is expressed as a linear combination of the first "i" orthogonal vectors derived using the Gram-Schmidt algorithm. If we represent this relationship in matrix terminology, we have the following:

$$P = (\underset{\sim}{v}_1, \underset{\sim}{v}_2, \ldots, \underset{\sim}{v}_n) \cdot \begin{bmatrix} a_{1,1} & a_{1,2} & \ldots & a_{1,n} \\ 0 & a_{2,2} & \ldots & a_{2,n} \\ \ldots & \ldots & \ldots & \ldots \\ 0 & 0 & \ldots & a_{n,n} \end{bmatrix}$$

. Hence we have a product representation of the original matrix as the product of an orthogonal matrix and a triangular matrix. With this representation of the matrix, the determinant of the matrix is simply the product of the values along the diagonal of the triangular matrix times the determinant of the orthogonal matrix. The determinant of the orthogonal matrix is either one or negative one.

In the case with the matrix P being a singular matrix, the sequential approach of the Gram-Schmidt algorithm allows us to create a product until we exhaust the independent column vectors.

Exercises

For exercises 1 through 5 below, apply the Gram-Schmidt algorithm to compute the orthogonal matrix and derive the Q-R decomposition of the original matrix.

1. Given the matrix $\begin{bmatrix} 19 & 6 & 8 \\ -45 & -14 & -23 \\ 0 & 0 & 3 \end{bmatrix}$, use the independent column vectors of this matrix to compute an orthogonal matrix and derive the Q-R decomposition of the original matrix.

2. Given the matrix $\begin{bmatrix} -4 & 0 & 0 \\ 0 & -4 & 0 \\ 5 & -5 & 1 \end{bmatrix}$, use the independent column vectors of this matrix to compute an orthogonal matrix and derive the Q-R decomposition of the original matrix.

3. Given the matrix $\begin{bmatrix} -11 & -54 & 50 & 40 \\ -8 & -39 & 38 & 30 \\ 8 & 54 & -47 & -38 \\ -24 & -138 & 126 & 101 \end{bmatrix}$, use the independent column vectors of this matrix to compute an orthogonal matrix and derive the Q-R decomposition of the original matrix.

4. Given the matrix $\begin{bmatrix} 1 & 17 & -9 & 8 \\ 3 & 33 & -17 & 16 \\ 19 & 11 & -15 & 8 \\ 26 & -64 & 18 & -26 \end{bmatrix}$, use the independent column vectors of the matrix to compute an orthogonal matrix and derive the Q-R decomposition of the original matrix.

5. Given the matrix $\begin{bmatrix} -8 & 9 & -24 & 2 & 7 \\ 44 & -30 & 45 & -20 & -26 \\ 42 & -18 & 21 & -18 & -20 \\ -132 & 60 & -72 & 57 & 64 \\ 108 & -27 & -3 & -48 & -41 \end{bmatrix}$, use the independent column vectors of the matrix to compute an orthogonal matrix and derive the Q-R decomposition of the original matrix.

7.4.4 Gram-Schmidt & Characteristic Root of Multiplicity > 1

Earlier we introduced the concept of a pair of asymmetric complementing orthogonal vector spaces. When dealing with "**n**" independent vectors such as $\{\underset{\sim}{P}_1, \underset{\sim}{P}_2, \ldots, \underset{\sim}{P}_n\}$. It turned out that the corresponding complementing and orthogonal row vectors are computed by computing the inverse of the matrix given by $\boldsymbol{P} = (\underset{\sim}{P}_1, \underset{\sim}{P}_2, \ldots, \underset{\sim}{P}_n)$. Hence if $\boldsymbol{Q} = \boldsymbol{P}^{-1}$ and the rows of the inverse are $\begin{bmatrix} \underset{\sim}{Q}^T{}_1 \\ \underset{\sim}{Q}^T{}_2 \\ \ldots \\ \underset{\sim}{Q}^T{}_n \end{bmatrix}$, then the vectors $\{\underset{\sim}{Q}^T{}_1, \underset{\sim}{Q}^T{}_2, \ldots, \underset{\sim}{Q}^T{}_n\}$ are the complementing orthogonal vectors of the original independent vectors. Hence for "n" independent vectors, the computation of symmetric complementing orthogonal vector spaces is trivial.

Now we examine the situation in which we have "**k**" independent vectors and $k < n$. Since we are not dealing with a basis for an **n**-dimensional vector space, we must define the space that the vectors span. Beginning with the identity element J of a ring S_A of rank "**k**", we can apply the same approach shown for the proof of the Gram-Schmidt algorithm; when developing the asymmetric complementing orthogonal vectors contained in the vector spaces $\underset{\sim}{V}(S_A)$ and $\underset{\sim}{V}^T(S_A)$. We formulate the following theorem to extend the Gram-Schmidt algorithm to that of a ring of rank "**k**".

Theorem 7.4.5 Given "**k**" independent vectors $\{\underset{\sim}{P}_1, \underset{\sim}{P}_2, \ldots, \underset{\sim}{P}_k\}$ and the identity element J of a ring S_A, which has rank "**k**", then there is a sequence of idempotent matrices such that for each "**i**" we have $\underset{\sim}{u}_i = \mathsf{M}_i \cdot \underset{\sim}{P}_i$ and $\underset{\sim}{v}^T_i = \underset{\sim}{P}^T_i \cdot \mathsf{M}_i$. The vectors $\{\frac{\underset{\sim}{u}_1}{|\underset{\sim}{u}_1|}, \frac{\underset{\sim}{u}_2}{|\underset{\sim}{u}_2|}, \ldots, \frac{\underset{\sim}{u}_k}{|\underset{\sim}{u}_k|}\}$ and $\{\frac{\underset{\sim}{v}^T_1}{|\underset{\sim}{v}_1|}, \frac{\underset{\sim}{v}^T_2}{|\underset{\sim}{v}_2|}, \ldots, \frac{\underset{\sim}{v}^T_k}{|\underset{\sim}{v}_k|}\}$ form asymmetric complementing orthogonal vector bases for $\underset{\sim}{V}(S_A)$ and $\underset{\sim}{V}^T(S_A)$. The representation of each of the derived orthogonal column vectors is given as

$$\underset{\sim}{u}_1 = J \cdot \underset{\sim}{P}_1 \text{ and } \underset{\sim}{u}_i = J \cdot \underset{\sim}{P}_i - \sum_{j=1}^{i-1} \frac{\underset{\sim}{v}^T_j \cdot \underset{\sim}{P}_i}{(\underset{\sim}{v}^T_j \cdot \underset{\sim}{u}_j)} \cdot \underset{\sim}{u}_j \quad \text{for}$$

$i > 1$. The representation of each of the derived orthogonal row vectors is given as $\underset{\sim}{v}^T_1 = \underset{\sim}{P}^T_1 \cdot J$ and

$$\underset{\sim}{v}^T_i = \underset{\sim}{P}^T_i \cdot J - \sum_{j=1}^{i-1} \frac{\underset{\sim}{P}^T_i \cdot \underset{\sim}{u}_j}{(\underset{\sim}{v}^T_j \cdot \underset{\sim}{u}_j)} \cdot \underset{\sim}{v}^T_j \quad \text{for } i > 1.$$

Proof: The proof for this theorem is similar to the proof for the Gram-Schmidt algorithm. The steps for this proof are shown in table 7.6 below.

TABLE 15. Proof for Theorem 7.4.4

Step	Statement	Rationale For Statement
1	We begin with "**k**" orthogonal idempotent matrices $\{\boldsymbol{J}_1, \boldsymbol{J}_2, \ldots, \boldsymbol{J}_k\}$ that forms basis for ring S_A.	Assume that all matrices of form $\sum_{i=1}^{k} \alpha_i \cdot \boldsymbol{J}_i$ are contained in ring S_A. We will derive these matrices using the given "**k**" independent vectors $\{\underset{\sim}{\boldsymbol{P}}_1, \underset{\sim}{\boldsymbol{P}}_2, \ldots, \underset{\sim}{\boldsymbol{P}}_k\}$.
2	As a starting point, set the vector $\underset{\sim}{\boldsymbol{u}}_1 = \boldsymbol{J} \cdot \underset{\sim}{\boldsymbol{P}}_1$ and set the row vector to be $\underset{\sim}{\boldsymbol{v}}_1^T = \underset{\sim}{\boldsymbol{P}}_1^T \cdot \boldsymbol{J}$. The idempotent matrix $\boldsymbol{J}_1$ we define to be the matrix $\dfrac{\left(\underset{\sim}{\boldsymbol{u}}_1 \cdot \underset{\sim}{\boldsymbol{v}}_1^T\right)}{\left(\underset{\sim}{\boldsymbol{v}}_1^T \cdot \underset{\sim}{\boldsymbol{u}}_1\right)}$.	The starting point is completely arbitrary. We could choose any one of the original vectors as a starting point. The selection of $\boldsymbol{J}_1$ was selected so that $\boldsymbol{J}_1^2 = \boldsymbol{J}_1$.
3	Defining matrix $\mathsf{M}_2 = \boldsymbol{J} - \boldsymbol{J}_1$, we have $\underset{\sim}{\boldsymbol{u}}_2 = \mathsf{M}_2 \cdot \underset{\sim}{\boldsymbol{P}}_2$ and $\underset{\sim}{\boldsymbol{v}}^T_2 = \underset{\sim}{\boldsymbol{P}}^T_2 \cdot \mathsf{M}_2$ are respectively orthogonal to the vectors $\underset{\sim}{\boldsymbol{v}}^T_1$ and $\underset{\sim}{\boldsymbol{u}}_1$.	The matrix $\mathsf{M}_2 = \boldsymbol{J} - \boldsymbol{J}_1$ is orthogonal to the matrix $\boldsymbol{J}_1$ and hence to the vectors $\underset{\sim}{\boldsymbol{v}}^T_1$ and $\underset{\sim}{\boldsymbol{u}}_1$.

TABLE 15. Proof for Theorem 7.4.4

Step	Statement	Rationale For Statement
4	We define the idempotent matrix J_2 to be the matrix $\dfrac{(u_2 \cdot v^T_2)}{(v^T_2 \cdot u_2)}$.	With this selection, we get $J_2^2 = J_2$ and J_1 is orthogonal to J_2 . Hence we have defined the first two symmetric idempotent and orthogonal matrices.
5	Defining $M_3 = M_2 - J_2$, we have $u_3 = M_3 \cdot P_3$ and $v^T_3 = P^T_3 \cdot M_3$ are respectively orthogonal to the vectors (v^T_1, v^T_2) and (u_1, u_2).	The matrix $M_3 = M_2 - J_2$ is orthogonal to the matrices J_1 and J_2 . Hence it is orthogonal to the vectors (v^T_1, v^T_2) and (u_1, u_2).
6	The general form of the sequence of the idempotent matrices is $M_i = J - \sum_{k=1}^{i-1} \dfrac{u_k \cdot v_k^T}{(v_k^T \cdot u_k)}$ and $u_i = M_i \cdot P_i$ is orthogonal to the vectors $u_1, u_2, \ldots, u_{i-1}$.	Using mathematical induction and sequentially estimating each idempotent matrix J_i, we construct the identity matrix M_i for the sub-ring generated by matrices $\{J_i, \ldots, J_k\}$. The vectors u_i and v^T_i are respectively orthogonal to $\{v^T_1, v^T_2, \ldots, v^T_{i-1}\}$ and $\{u_1, u_2, \ldots, u_{i-1}\}$.

TABLE 15. Proof for Theorem 7.4.4

Step	Statement	Rationale For Statement
7	$\underset{\sim}{u}_i = J \cdot \underset{\sim}{P}_i - \sum_{k=1}^{i-1} J_k \cdot \underset{\sim}{P}_i$ and $\underset{\sim}{v}_i^T = \underset{\sim}{P}_i^T \cdot J - \sum_{k=1}^{i-1} \underset{\sim}{P}_i^T \cdot J_k$	Substituting the value for M_i and performing the multiplication, we get this expression for the vectors $\underset{\sim}{u}_i$ and $\underset{\sim}{v}_i^T$.
8	The final form of the derived orthogonal column vectors are given as $\underset{\sim}{u}_1 = J \cdot \underset{\sim}{P}_1$ and $\underset{\sim}{u}_i = J \cdot \underset{\sim}{P}_i - \sum_{j=1}^{i-1} \frac{\underset{\sim}{v}_j^T \cdot \underset{\sim}{P}_i}{(\underset{\sim}{v}_j^T \cdot \underset{\sim}{u}_j)} \cdot \underset{\sim}{u}_j$ for $i > 1$. The representation of each of the derived orthogonal row vectors is given as $\underset{\sim}{v}^T{}_1 = \underset{\sim}{P}_1^T \cdot J$ and $\underset{\sim}{v}^T{}_i = \underset{\sim}{P}_i^T \cdot J - \sum_{j=1}^{i-1} \frac{\underset{\sim}{P}_i^T \cdot \underset{\sim}{u}_j}{(\underset{\sim}{v}_j^T \cdot \underset{\sim}{u}_j)} \cdot \underset{\sim}{v}_j^T$ for $i > 1$.	Use the commutative property of matrix multiplication to get this form of the algorithm.

With theorem 7.4.4, we derive the matrix $\mathsf{M} = \sum_{j=1}^{k} \frac{\underset{\sim}{u}_j \cdot \underset{\sim}{v}_j^T}{(\underset{\sim}{v}_j^T \cdot \underset{\sim}{u}_j)}$. Properties of this matrix are $\mathsf{M} = \mathsf{M} \cdot \boldsymbol{J} = \boldsymbol{J} \cdot \mathsf{M}$ and $\mathsf{M}^2 = \mathsf{M}$. By rearranging terms, we have the equation $\mathsf{M}^2 - \mathsf{M} \cdot \boldsymbol{J} = \mathbf{0}$ or $\mathsf{M} \cdot (\mathsf{M} - \boldsymbol{J}) = \mathbf{0}$. If the matrix M is of rank "**k**" and is an element of the ring for which $\boldsymbol{J}$ is the identity element, then theorem 7.2.6 implies that $\mathsf{M} = \boldsymbol{J}$. Hence we have $\boldsymbol{J} = \sum_{j=1}^{k} \frac{\underset{\sim}{u}_j \cdot \underset{\sim}{v}_j^T}{(\underset{\sim}{v}_j^T \cdot \underset{\sim}{u}_j)}$. This application, of the Gram-Schmidt algorithm, gives us an alternative approach for spectral decomposition of the matrix $\boldsymbol{A}$, when each characteristic matrix is associated with the same characteristic root value. In other words, if a given characteristic root has multiplicity "**k**" where $k > 1$.

If the characteristic equation for the matrix $\boldsymbol{A}$ is $\rho_A(x) = 0$ and the root x_0 is of multiplicity "**k**", then we can define the polynomial $\rho_1(x) = \rho_A(x) / (x - x_0)^k$. From corollary 5.6.4, the sum of the **k** characteristic matrices associated with the root x_0 is $\boldsymbol{J} = \rho_1(\boldsymbol{A}) / \rho_1(x_0)$. Once we have computed this matrix, we apply theorem 7.4.4 to decompose it into characteristic matrices associated with the root x_0. To apply this theorem, we need at least one column vector $\underset{\sim}{\boldsymbol{P}}_i$ (or possibly k independent column vectors) such that $\underset{\sim}{u}_i = \mathsf{M}_i \cdot \underset{\sim}{\boldsymbol{P}}_i \neq \varnothing$.

Example

Given the **3-by-3** matrix $\boldsymbol{A} = \begin{bmatrix} -5 & -9 & -4 \\ 0 & -2 & 0 \\ 3 & 9 & 2 \end{bmatrix}$, with characteristic equation

$\rho_A(x) = x^3 + 5 \cdot x^2 + 8 \cdot x + 4 = 0$ and roots -1 and -2 (multiplicity 2); apply theorem 7.4.4 to decompose the idempotent matrix associated with root -2.

6. To apply corollary 5.6.4 we derive the polynomial $\rho_1(x) = \rho_A(x) / (x+2)^2 = (x+1)$. The associated idempotent matrix associated with the root -2 of multiplicity 2 is defined to be $(\boldsymbol{A} + \boldsymbol{I}) / (-1)$.

7. From step 1, the idempotent matrix is computed to be $\boldsymbol{J} = \begin{bmatrix} 4 & 9 & 4 \\ 0 & 1 & 0 \\ -3 & -9 & -3 \end{bmatrix}$. Select $\underset{\sim}{\boldsymbol{P}}_1 = \begin{bmatrix} 1 \\ 0 \\ 0 \end{bmatrix}$; that is $\underset{\sim}{\boldsymbol{E}}_1$, since diagonal element at (1,1) is nonzero. This will insure that both $\underset{\sim}{\boldsymbol{u}}_1 = \boldsymbol{J} \cdot \underset{\sim}{\boldsymbol{P}}_1$ and $\underset{\sim}{\boldsymbol{v}}^T{}_1 = \underset{\sim}{\boldsymbol{P}}_1^T \cdot \boldsymbol{J}$ are nonzero.

8. Using the results in step 2 we get $\underset{\sim}{u}_1 = \begin{bmatrix} 4 \\ 0 \\ -3 \end{bmatrix}$ and

$\underset{\sim}{v}^T{}_1 = \begin{bmatrix} 4 & 9 & 4 \end{bmatrix}$ .The first idempotent matrix in the decomposi-

tion is $J_1 = \dfrac{\left(\underset{\sim}{u}_1 \cdot \underset{\sim}{v}_1{}^T\right)}{\left(\underset{\sim}{v}_1{}^T \cdot \underset{\sim}{u}_1\right)}$ or $J_1 = \begin{bmatrix} 4 & 9 & 4 \\ 0 & 0 & 0 \\ -3 & -\frac{27}{4} & -3 \end{bmatrix}$.

9. Since the root -2 has a multiplicity of 2, we can compute the sec-

ond characteristic matrix as $J_2 = J - J_1$ or $J_2 = \begin{bmatrix} 0 & 0 & 0 \\ 0 & 1 & 0 \\ 0 & -\frac{9}{4} & 0 \end{bmatrix}$.

Hence we the have decomposition of the matrix

$$J = \begin{bmatrix} 4 & 9 & 4 \\ 0 & 0 & 0 \\ -3 & -\frac{27}{4} & -3 \end{bmatrix} + \begin{bmatrix} 0 & 0 & 0 \\ 0 & 1 & 0 \\ 0 & -\frac{9}{4} & 0 \end{bmatrix}.$$

Exercises

For each exercise, 1 through 5 below, a matrix and its characteristic equation is shown. Decompose the associated idempotent matrix with the characteristic root with multiplicity greater than one.

1. For the matrix $A = \begin{bmatrix} 1 & -2 & -1 \\ 4 & 7 & 2 \\ -6 & -6 & 0 \end{bmatrix}$ with characteristic equation

$\rho_A(x) = -x^3 + 8 \cdot x^2 - 21 \cdot x + 18 = 0$ and characteristic roots $\{2, 3, 3\}$, decompose the associated idempotent matrix of the characteristic root 3.

2. For the matrix $A = \begin{bmatrix} 97 & -108 & 96 & -69 \\ -69 & 76 & -84 & 57 \\ 393 & -432 & 448 & -309 \\ 794 & -876 & 892 & -618 \end{bmatrix}$ with characteristic equation

$\rho_A(x) = x^4 - 3 \cdot x^3 - 24 \cdot x^2 + 80 \cdot x = 0$ and characteristic roots $\{-5, 0, 4, 4\}$, decompose the associated idempotent matrix of the characteristic root 4.

3. For the matrix $A = \begin{bmatrix} -4 & -6 & 18 & -6 \\ -24 & -22 & 72 & -24 \\ -51 & -35 & 127 & -43 \\ -120 & -72 & 276 & -94 \end{bmatrix}$ with characteristic equation

$\rho_A(x) = x^4 - 7 \cdot x^3 + 6 \cdot x^2 + 28 \cdot x - 40 = 0$ and characteristic roots $\{-2, 5, 2, 2\}$, decompose the associated idempotent matrix of the characteristic root 2.

4. For the matrix $A = \begin{bmatrix} -9 & 33 & 3 & 37 & 14 \\ 3 & -50 & -14 & -49 & -16 \\ 3 & 43 & 17 & 41 & 12 \\ -6 & -13 & -11 & -12 & -2 \\ 6 & 149 & 59 & 145 & 43 \end{bmatrix}$ with characteristic equation

$\rho_A(x) = -x^5 - 11 \cdot x^4 - 41 \cdot x^3 - 49 \cdot x^2 + 30 \cdot x + 72 = 0$ and characteristic roots $\{-4, -3, -3, -2, 1\}$, decompose the associated idempotent matrix of the characteristic root 3.

5. For the matrix $A = \begin{bmatrix} -12 & -22 & -130 & 68 & -34 \\ -24 & -8 & 36 & -24 & 12 \\ 72 & 72 & 292 & -144 & 72 \\ 136 & 130 & 478 & -232 & 118 \\ 16 & -2 & -134 & 76 & -34 \end{bmatrix}$ with characteristic roots

$\{-4, -2, 4, 4, 4\}$, decompose the associated idempotent matrix of the characteristic root 4.

7.5 *Application Of Generalized Inverse*

In section 6.5 of chapter 6, we derived estimates of the coefficients of a linear equation using the theory of least squares. One requirement for the existence of the solution for the coefficient estimates is that the matrix $A = X^T \cdot X$ be nonsingular. This section reexamines the solution of the least squares equation; namely $(X^T \cdot X) \cdot \hat{\underset{\sim}{\beta}} = X^T \cdot \underset{\sim}{Y}$. But this time, we relax the restriction of the symmetric matrix $A = X^T \cdot X$ being nonsingular.

When the matrix $A = X^T \cdot X$ is singular, it has a unique generalized inverse contained in S_A. We shall denote this generalized inverse as $A^G = (X^T \cdot X)^G$. In keeping with the notation of rings as discussed throughout this book, the generalized identity element of the ring S_A is denoted as J. (Note that if A is nonsingular, the generalized inverse becomes the traditional identity element I.) Outside the vector space $\underset{\sim}{V}(S_A)$, when the matrix A is singular, the least squares equation has many possible solutions. By restricting our attention to the vector subspace $\underset{\sim}{V}(S_A)$, a unique solution is obtained. The unique solution is obtained by multiplying both sides of the equation by $(X^T \cdot X)^G$. Hence the results is

$$J \cdot \hat{\underset{\sim}{\beta}} = (X^T \cdot X)^G \cdot X^T \cdot \underset{\sim}{Y} .$$

Recall property 2 of vector spaces $\underset{\sim}{V}(S_A)$ in section 7.4. From this property, the estimate $J \cdot \hat{\underset{\sim}{\beta}}$ is the projection of the least squares estimate into the vector space $\underset{\sim}{V}(S_A)$. Hence conclude that when the matrix is singular, that all linear transformations of the coefficient vector contained in the vector space $\underset{\sim}{V}(S_A)$ are estimable. Some typical estimable transformations are listed below.

1. For any matrix B , contained in the space S_A , then an estimate of the transformation $B \cdot \underset{\sim}{\beta}$ is possible. This is easily proved since $B = B \cdot J$ (by definition of $B \in S_A$) and hence $B \cdot \underset{\sim}{\beta} = B \cdot J \cdot \underset{\sim}{\beta}$.

2. Property 5 of the properties of the vector spaces $\underset{\sim}{V}(S_A)$ and $\underset{\sim}{V}^T(S_A)$ shows that the rows of the matrix X are elements of the vector space $\underset{\sim}{V}^T(S_A)$ when $X^T \cdot X$ belongs to S_A . Hence we have $X = X \cdot J$ and therefore the quantity

$$Y = X \cdot \underset{\sim}{\beta} \quad \text{is estimable.}$$

3. For any given vector $\underset{\sim}{\alpha}^T \in \underset{\sim}{V}^T(S_A)$ we have $\underset{\sim}{\alpha}^T \cdot J = \underset{\sim}{\alpha}^T$ and therefore $\underset{\sim}{\alpha}^T \cdot J \cdot \underset{\sim}{\beta} = \underset{\sim}{\alpha}^T \cdot \underset{\sim}{\beta}$. Hence conclude that $\underset{\sim}{\alpha}^T \cdot \underset{\sim}{\beta}$ is an estimable linear transformation of $\underset{\sim}{\beta}$.

Example

This example illustrates the estimation of a linear combination of coefficients when the matrix $A = X^T \cdot X$ is singular.

1. For this example, we have the model

$$y = \beta_0 + \beta_1 \cdot x_1 + \beta_2 \cdot x_2 + \beta_3 \cdot x_3 + \beta_4 \cdot x_4 .$$

2. The response values associated with the independent values are listed in this table.

k	X_1	X_2	X_3	X_4
1	-1	1	2	3
2	-1	1	2	3
3	-1	2	2	5
4	1	2	0	7
5	1	2	2	-5

Since the model assumes a nonzero intercept, the matrix X is expressed as

$$X = \begin{bmatrix} 1 & -1 & 1 & 2 & 3 \\ 1 & -1 & 1 & 2 & 3 \\ 1 & -1 & 2 & 2 & 5 \\ 1 & 1 & 2 & 0 & 7 \\ 1 & 1 & 2 & 2 & -5 \end{bmatrix} .$$

3. The matrix $\boldsymbol{A} = \boldsymbol{X}^T \cdot \boldsymbol{X}$ is $\begin{bmatrix} 5 & -1 & 8 & 8 & 13 \\ -1 & 5 & 0 & -4 & -9 \\ 8 & 0 & 14 & 12 & 20 \\ 8 & -4 & 12 & 16 & 12 \\ 13 & -9 & 20 & 12 & 117 \end{bmatrix}$. It has a characteristic equation of

$$\rho_A(x) = -x^5 + 157 \cdot x^4 - 4146 \cdot x^3 + 21584 \cdot x^2 - 4160 \cdot x = 0 .$$

4. Using the properties of characteristic equations and defining the polynomial $\rho_1(x) = \rho_A(x) / (4160 \cdot x) + 1$, we have

$\rho_1(x) = \dfrac{-x^4 + 157 \cdot x^3 - 4146 \cdot x^2 + 21584 \cdot x}{4160}$ and yields an explicit formula for the generalized matrix; $\boldsymbol{J} = \rho_1(\boldsymbol{A})$. Applying this formula yields

$$\boldsymbol{J} = \begin{bmatrix} 0.51 & 0.31 & -0.12 & 0.37 & 0.06 \\ 0.31 & 0.81 & 0.08 & -0.23 & -0.04 \\ -0.12 & 0.08 & 0.97 & 0.09 & 0.02 \\ 0.37 & -0.23 & 0.09 & 0.72 & -0.05 \\ 0.06 & -0.04 & 0.02 & -0.05 & 0.99 \end{bmatrix} .$$

5. Since $\begin{bmatrix} 0 & 0 & 0 & -1 & 6 \end{bmatrix} \cdot \boldsymbol{J} = \begin{bmatrix} 0 & 0 & 0 & -1 & 6 \end{bmatrix}$, the linear combination $6 \cdot \beta_4 - \beta_3$ is estimable.If we are interested in estimating the single coefficient β_0, then we must verify that the post multiplication of the vector $\begin{bmatrix}1 & 0 & 0 & 0 & 0\end{bmatrix}$ by $\boldsymbol{J}$ must yield the original vector $\begin{bmatrix}1 & 0 & 0 & 0 & 0\end{bmatrix}$. Upon multiplying by $\boldsymbol{J}$, we get $\begin{bmatrix}1 & 0 & 0 & 0 & 0\end{bmatrix} \cdot \boldsymbol{J} = \begin{bmatrix}0.51 & 0.31 & -0.12 & 0.37 & 0.06\end{bmatrix}$. Hence we are not able to estimate the coefficient β_0 given the above data.

Exercises

For the linear models listed in the exercises below, compute the matrices $A = X^T \cdot X$ and J then determine if the linear combination of the coefficients are estimable.

1. The model of interest is $y = \beta_0 + \beta_1 \cdot x_1 + \beta_2 \cdot x_2 + \beta_3 \cdot x_3 + \beta_4 \cdot x_4$ and the data is contained in the following table:

k	X_1	X_2	X_3	X_4
1	1	1	1	1
2	2	4	3	6
3	3	9	2	6

Determine which of the following linear combinations of the coefficients are estimable using this data.

a. $\beta_0 + \beta_1 + \beta_2 + \beta_3 + \beta_4$

b. β_0

c. $\beta_1 + 5 \cdot \beta_2 - \beta_3$

2. The model of interest is $y = \beta_1 \cdot x_1 + \beta_2 \cdot x_2 + \beta_3 \cdot x_3$ and the data is contained in the following table:

k	X_1	X_2	X_3
1	2	-1	0
2	0	2	2
3	2	1	2
4	2	0	1

Determine which of the following linear combinations of the coefficients are estimable using this data.

a. $\beta_1 + \beta_2 + \beta_3$

b. $\beta_2 + \beta_3$

c. β_1

3. The model of interest is $y = \beta_1 \cdot x_1 + \beta_2 \cdot (x_1)^2 + \beta_3 \cdot x_3 + \beta_4 \cdot x_4$ and the data is contained in the following table:

k	X_1	$(X_1)^2$	X_3	X_4
1	-1	1	5	0
2	0	0	10	0
3	1	1	20	1

Determine which of the following linear combinations of the coefficients are estimable using this data.

a. $-\beta_1 + \beta_2$

b. β_3

c. $\beta_1 + \beta_2 + \beta_3 + \beta_4$

4. The model of interest is $y = \beta_0 + \beta_1 \cdot x_1 + \beta_2 \cdot x_2 + \beta_3 \cdot x_3 + \beta_4 \cdot x_4$ and the data is contained in the following table:

k	X_1	X_2	X_3	X_4
1	0	0	0	2
2	0	0	1	2
3	1	1	1	2
4	1	1	0	5
5	1	1	0	5
6	1	1	0	5

Determine which of the following linear combinations of the coefficients are estimable using this data.

a. $\beta_1 + \beta_2$

b. β_3

c. $\beta_1 + \beta_2 + \beta_3 + \beta_4$

5. For the model $y = \beta_0 + \beta_1 \cdot x_1 + \beta_2 \cdot x_2 + \beta_3 \cdot x_3 + \beta_4 \cdot x_4 + \beta_5 \cdot x_5$ and the data contained in the following table:

k	x_1	x_2	x_3	x_4	x_5
1	1	1	0	0	2
2	5	0	10	0	0
3	10	-1	5	1	2

determine which of the following linear combinations of the coefficients are estimable using this data.

a. $0.2 \cdot \beta_0 + \beta_2 + 2 \cdot \beta_4$

b. $\beta_0 + \beta_1 + \beta_2 + \beta_3 + \beta_4 + \beta_5$

c. $\beta_0 + \beta_2 + 15 \cdot \beta_3 - \beta_4 - 2 \cdot \beta_5$

This Page is Intentionally Left Blank

Bibliography

1. **Matrix Calculus** by E. Bodewig, 1959; North-Holland Publishing Company, Amsterdam, second revised

2. **Linear Algebra** by A. H. Lightstone, 1969; Appleton-Century-Crofts of Meredith Corporation, New York

3. **Elementary Linear Algebra** by Bernard Kolman, 1970; The Macmillan Company of Collier-Macmillan Limited of London

4. **Linear Algebra With Applications** by Leonard E. Fuller, 1968; Dickenson Publishing Company Inc., Belmont California, Third Printing

5. **Introduction To Modern Algebraic Concepts** by Max D. Larsen, 1969; Addison-Wesley Publishing Company, Reading Massachusetts

A. Installation and Abbreviated User's Guide

For me the excitement of 'e-learning' in mathematics is using the computer to animate mathematical concepts. The computer presents the user with challenging exercises that requires user interaction while applying learned concepts. Such an approach gives real world applicability to the accompanying printed text. In short, '**e-learning**' in mathematics reinforces mathematical concepts by giving the user hands-on experience. This system, most often referred to as a learning management system (LMS), uses a completely modular approach. Each chapter of the book is presented as a learning vignette which allows the user to interactively explore the mathematical concepts presented in each chapter. The lms is accessible over the internet from any location.

A.1 Functional Overview

A functional overview of this particular learning management system is shown in Figure A.1 below. This LMS uses a browser (such as Internet Explorer, Netscape, Mozilla,....) as its primary graphical user interface. Using HTML and Java applets, an interactive graphical user interface is implemented that is portable to any operational platform (Windows, Unix, Solaris and others). graphical user interface is implemented that is portable to any operational platform (Windows, Unix, Solaris and others).

As the title of Figure A.1 indicates, the LMS is a frame control system. Hence, when the user invokes the LMS through a browser, frames are generated by the tool. The frame is the principle input mechanism and it is controlled by the frame controller. Messages are sent between the controller and the frame(s) to implement the various functions of the LMS. Depending upon the inputs or responses by the user, the message sent to the frame controller may be sent to the server or may cause the frame controller to take an appropriate action on the client. Each frame communicates with the user by creating panels that are specified using a string of commands specified in the HTML interface.

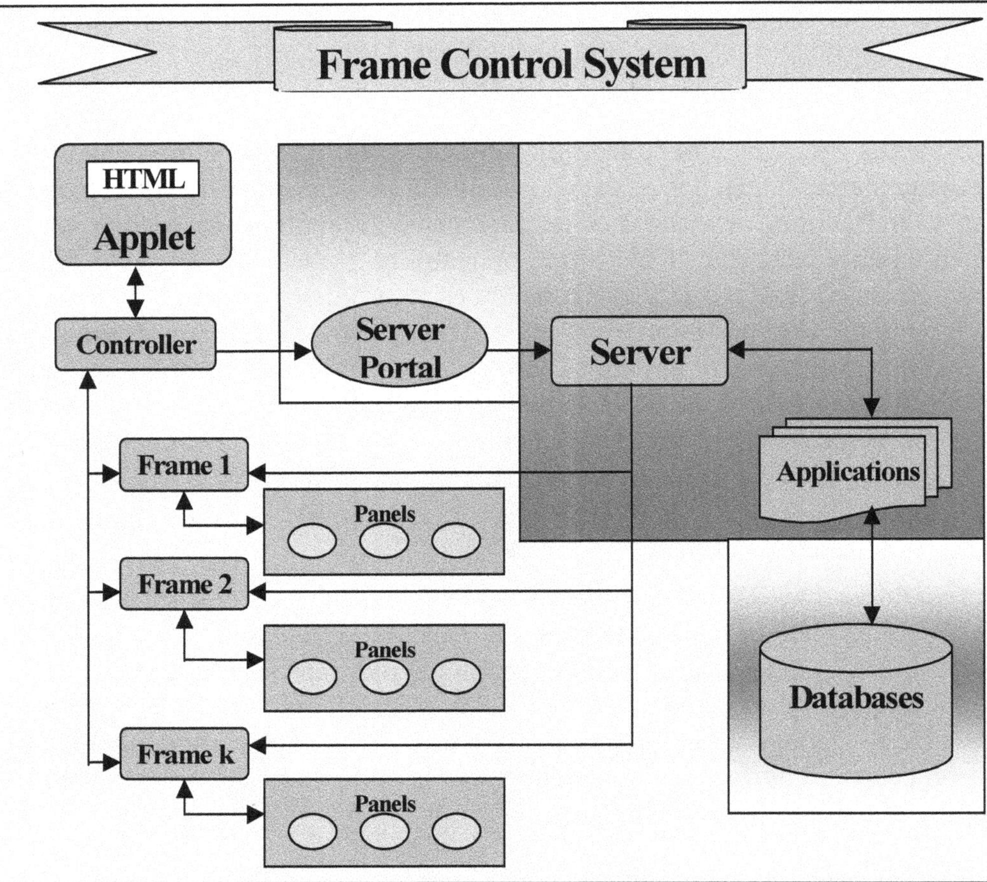

Figure A.1: Functional Architecture for Linear Algebra LMS

The graphical interface or the HTML page may be designed by the user, but a default page is provided for the Linear Algebra e-learning system and has the appearance shown in Figure A.2.

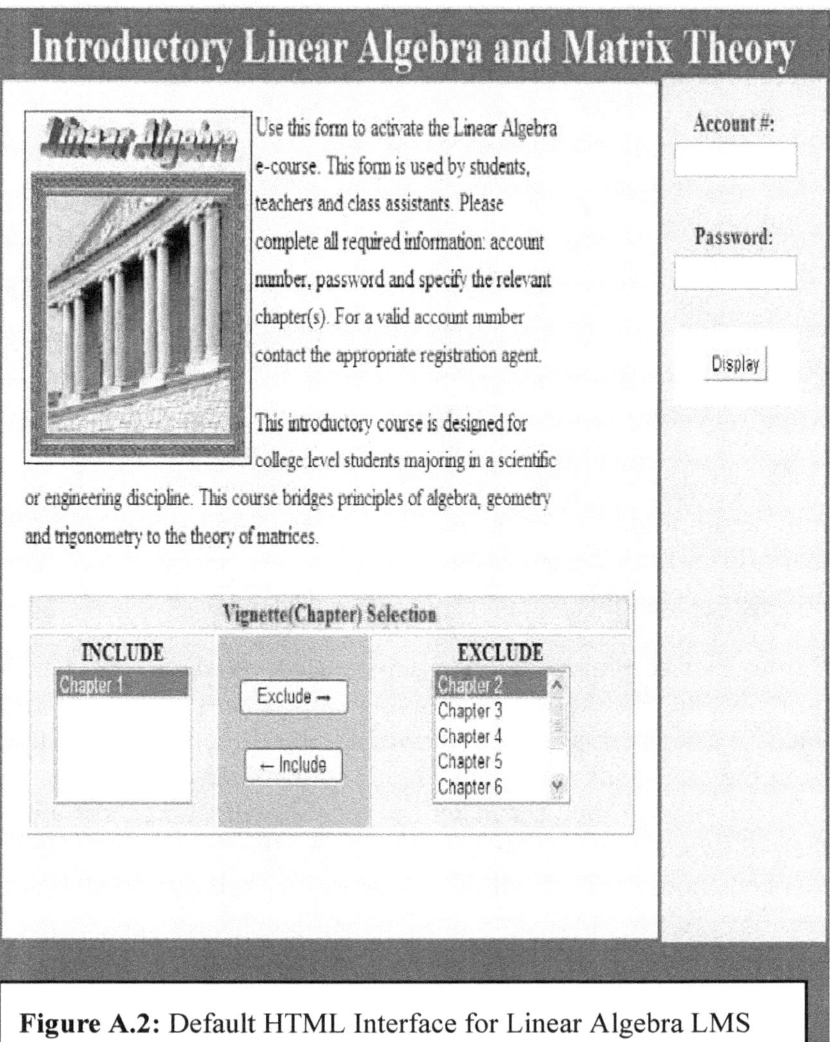

Figure A.2: Default HTML Interface for Linear Algebra LMS

Using this interface, the user selects the chapter of interests and clicks the select button to initiate the chapter vignette of interest. For example, in the "Vignette Selection" section choose "Chapter 1" in the "EXCLUDE" box positioned on the right. After making the selection, click on the right facing arrow (<--) to move the choice to the "INCLUDE" box positioned on the left in the Vignette Selection area. A discussion of each chapter vignette will follow in subsequent sections of this appendix.

After selecting the chapters or vignettes to be activated in the LMS, the user should enter his/her account number. The account number is assigned by the administrator of the remote LMS. The remote administrator also assigns the password for the user account. The LMS can only be activated by the appropriate account number and password combination.

A.2 Installation

Two possible installation choices are available for the linear algebra LMS. They are client LMS and server LMS. The client LMS is used by the student or a remote instructor (remote relative to the hosting server). Each client interacts with the server through a TCP/IP connection using a browser. The server LMS is used by the hosting instructor or institution to provide centralized resources to the client LMS's. Figure A.3, shown below, depicts the architecture of a fully functioning LMS system. The server LMS is installed in an environment that allows the clients access to its files and directories within a prescribed directory location.

Before a successful installation can be accomplished, an existing Java installation must be present on the computer. If it is not present, the installer will produce an error message with accompanying directions for installing Java (jdk) from the "http://java.sun.com" site. The installer will stop the installation process and request the user to return upon downloading Java and installing it to their computer. With Java installed on the computer, the installer, when invoked should process all files successfully and complete the installation process. Part of the installation process will include the alteration of the Java.Policy file as described in the "Security" section below.

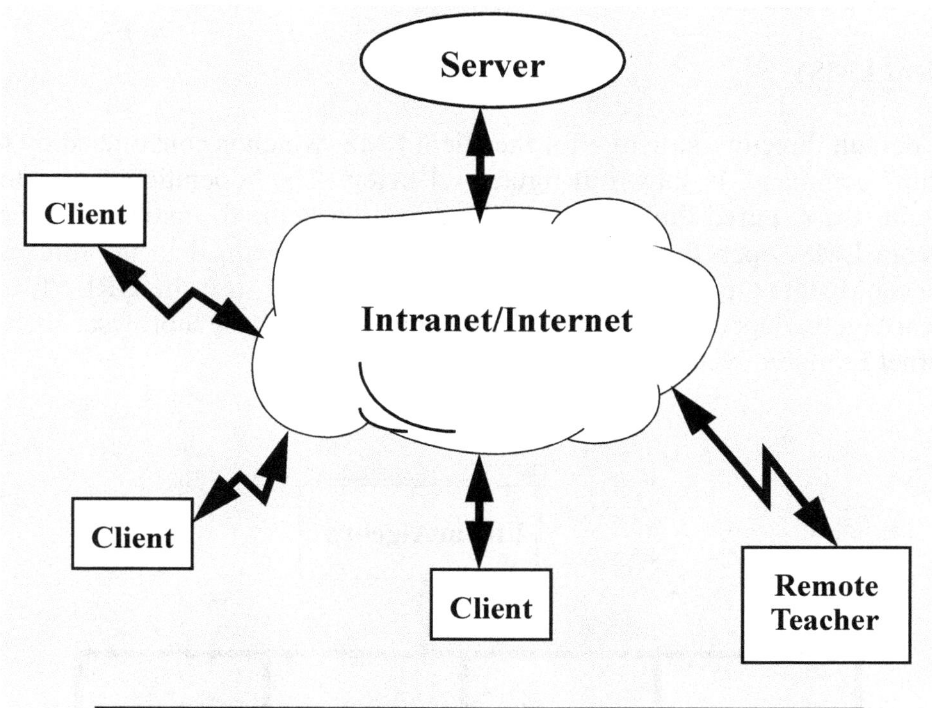

Figure A.3: Functional Architecture for Linear Algebra LMS

The installation for both the client and server LMS will be identical. The CD labeled "Linear Algebra Installation Software" should be inserted in the appropriate CD driver and the relevant operating system directory selected by the installer.

For the Windows operating system, the user should run the command "setup" under the directory labeled "windowsOS" on the CD drive. For the Unix or Linux operating system, the user should run the command "setup" under the directory labeled "UnixOS" and run the command "setup".

Not that for each client installation a directory structure is created for holding working files and supporting files. The file tcpServer.class file is copied to the

"Bin" directory shown in Figure A.4 This represents the minimal setup and can be accomplished manually.

(Client LMS)

The default directory structure for the client LMS, which is constructed by the "setup" command, is shown in figure A.4 below. The "operations" directory contains the required files for a fully functional system. To invoke the Linear Algebra LMS, open the document "LinAlg.html" contained in the directory "\LinearAlgebra\Operations\HTML". In a local setting, use the URL "file:/// LinearAlgebra/Operations/HTML/LinAlg.html" when using a browser such as Internet Explorer, Netscape or Mozilla.

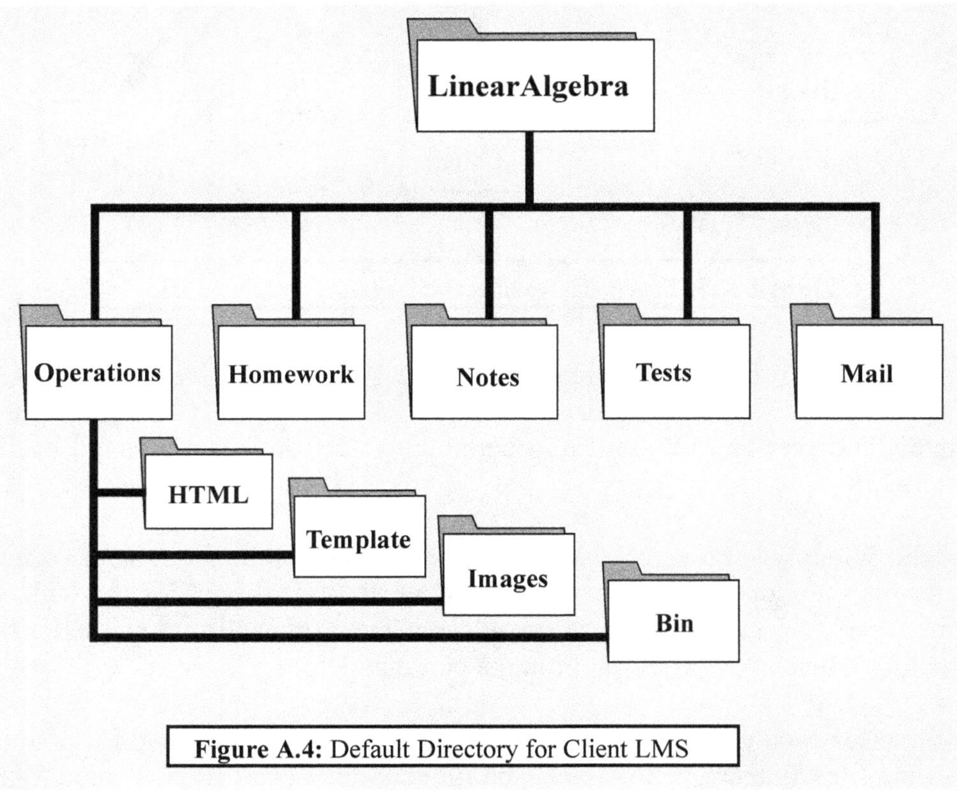

Figure A.4: Default Directory for Client LMS

The LMS software uses Java applets to implement the graphical user (GUI) logic and all client functionality. For the file handling capability, Java applications are available. Security related issues prohibit the browser from directly accessing the local computer disk files. To circumvent these restrictions, the Java applets communicate with a locally installed Java application using TCP sockets. The Java application is then able to access local files or communicate remotely with a host using a TCP socket without raising any security exceptions.

For simplicity, a single entry is used for the Java applications. Upon invoking the client via the browser, the user should invoke the tcpServer application. This application is contained in the Bin directory as shown in Figure A.4 above. A typical invocation of this program is the following: **java \LinearAlgebra\Operations\Bin\tcpServer**.

For a Windows operating environment, the user will need to open an MS-DOS window and then enter the tcpServer command described in the previous paragraph. Typically the MS-DOS window can be reached by clicking on the 'Start' menu, the Programs menu, the Accessories menu and finally the MS-DOS Command Prompt button. In my test environment, I change to the "C:\LinearAlgebra" directory, then I perform the invocation of the tcpServer command as shown in Figure A.5 below. To end this command, make the MS-DOS window the current window and press the "Ctrl" and the "C" simultaneously. This will stop the program and keep the window open. Of course, you could always exit the MS-DOS window and this will interrupt program as well as cause the window to disappear. Once the program has been stopped, the client machine will not be able to implement "Local" commands.

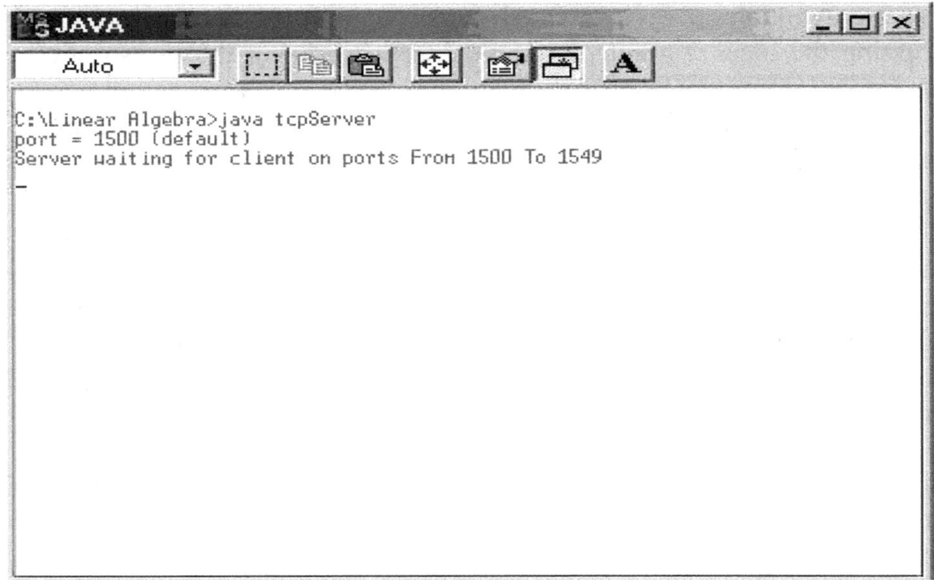

Figure A.5: MS-DOS Window for Invocation of tcpServer

Once this command has been invoked, it will allow the browser to communicate with the server through TCP sockets (sockets 1500 through 1549[for larger servers and larger class sizes this number may be increased]). It receives a command string from the client browser and sends the command to the LMS server. The LMS server then processes the command and returns the results to the client. In most cases, the client software provides the business applications to perform the requested activities requested by each message stream. The primary purpose of the server software is to provide access to a common database and as a message interpreter.

The tcpServer command is used on the client machine to communicate with the underling operating system. As commands in the "Local" categories are executed, the LMS will communicate with this program to implement the appropriate commands.

(Server LMS)

The default directory structure for the server LMS is similar to that of the client LMS directory structure. The only difference is the addition of an "Administration" directory that will be used to administer student records and user accounts/passwords. This difference is seen in Figure A.5 below.

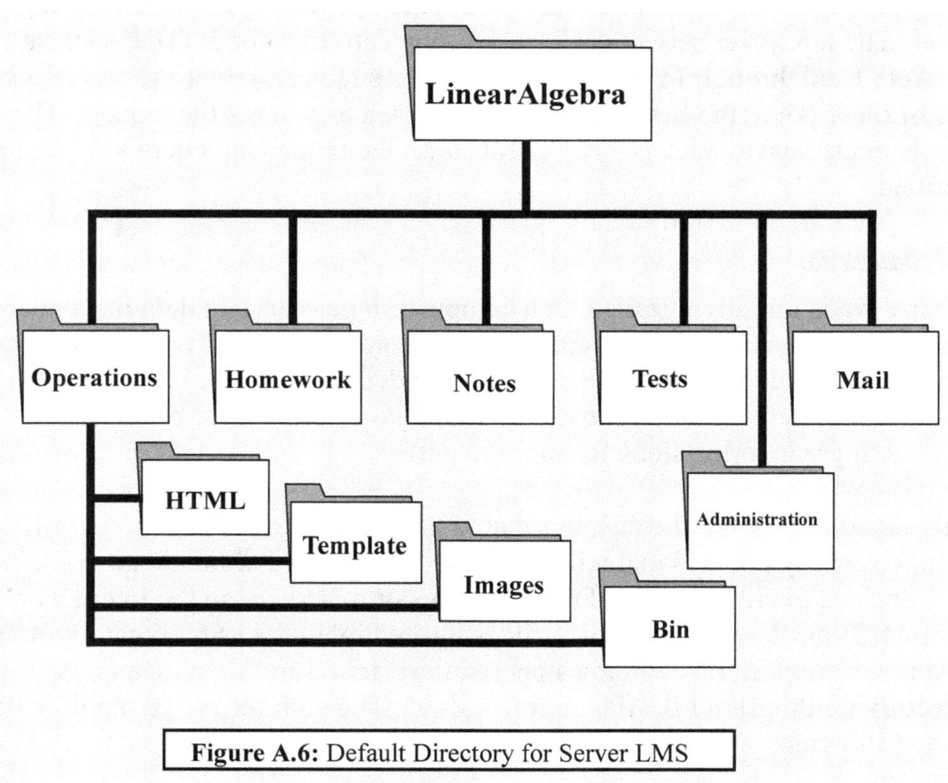

Figure A.6: Default Directory for Server LMS

The server LMS will be installed in a similar manner as the client LMS is installed. Additional functionality will be included to help administrators or teachers build tests, assignments, reference material and to send messages to the registered students for the linear algebra class. When using a remote server to access the Linear Algebra LMS, the user will enter a command such as http://server-address/LinAlg.html. The server address refers to the URL of the server and the directory location of the "LinAlg.html file within the server. When using a remote server to host the Linear Algebra LMS, the client will reference files on the server using the 'Remote' menu command. It will reference files on the client machine using the 'Local' menu command.

To negotiate TCP type sockets with the clients, invoke the tcpServer command using the following command: **java \LinearAlgebra\Operations\Bin\tcpServer**. The tcpServer servlet performs polling activities for 50 TCP sockets (sockets 1500 through 1549) for availability. As requests enter the server on one of these ports, the first available port is used to process the request. The port is made unavailable to other clients until the requesting client's request is fulfilled.

A.3 Security

When Java is initially installed on a computer, it has security defaults that are cautiously protective of the operating and file system. Before an LMS client can communicate with an LMS server, these defaults must be overridden. To override the these defaults, entries in the file java.policy must be altered or entered to grant permissions for the java.net.

This section describes the changes that must be made to the java.policy file to change security policies that will allow the browser to communicate through TCP sockets to either the local machine files or to servers on the internet or intranet. Typically the java.policy file is located in the security directory of the java installation. In my Windows operating system, I installed java in the directory C:\Program Files\Java\j2re1.4.2_05. The path for my java.policy file is the following:

C:\Program Files\Java\j2re1.4.2_05\lib\security\java.policy.

For my installation, I have an LMS server located at address 192.168.1.100 of an intranet and one client that is located at address 192.168.1.120. The command string that is included in the java.policy file of the client machine is as follows:

```
grant {
   permission java.net.SocketPermission "localhost:1024-","listen";
   permission java.net.SocketPermission "192.168.1.100:1024-","connect,accept";
   permission java.net.SocketPermission "192.168.1.120:1024-",
"listen,connect,accept";
};
```

With the above command, the server (192.168.1.100) is given permission to connect to a TCP socket and to accept connections on sockets 1024 and higher. The client (192.168.1.120) is given permission to listen, connect and accept connections on sockets 1024 and higher on the client machine. Specifying the actual addresses of the server and the client affords the client machine some additional security, since it will not open up the client machine to all users on the internet or intranet.

For the server LMS machine, the java.policy file may need to include a permission statement for each client that will have access to the server. Hence, some preliminary research may be involved before giving clients access to the server. In the special case of the server on an intranet and all clients on the same intranet, the serve can specify each client machine by IP address.

A.4 User's Guide: Chapter 1

Upon clicking the 'Display' button on the main page of the LMS (See figure A.3), the chapter 1 menu panel is displayed (User Selection). Since the command lines sent to the browser are user-specifiable, the actual appearance of each menu panel is determined by the user. The default command lines for the Linear Algebra LMS are discussed in this appendix.

The commands for chapter 1 are given in the menu screens shown below. As shown in the menu above, there are six (6) major categories of menus for each chapter. They are the following:

- Chapter 1
- File
- Local File

- Class
- Local Class
- Tools

For chapter number "k", the major category "**Chapter 1**" will be replaced with the category "**Chapter k**" , where k = 1, 2, ..., 7.

(Chapter 1 Command)

When using the LMS, the first menu or menu that should be first requested by the user is labeled "Chapter 1". For the **Linear Algebra and Matrix Theory** text book, this menu corresponds to the chapter (Chapter 1) that supports the basic mathematical concepts and definitions such as elementary functions, number systems and such concepts as continuity. This menu also provides the user access to the other chapters in the learning management system.

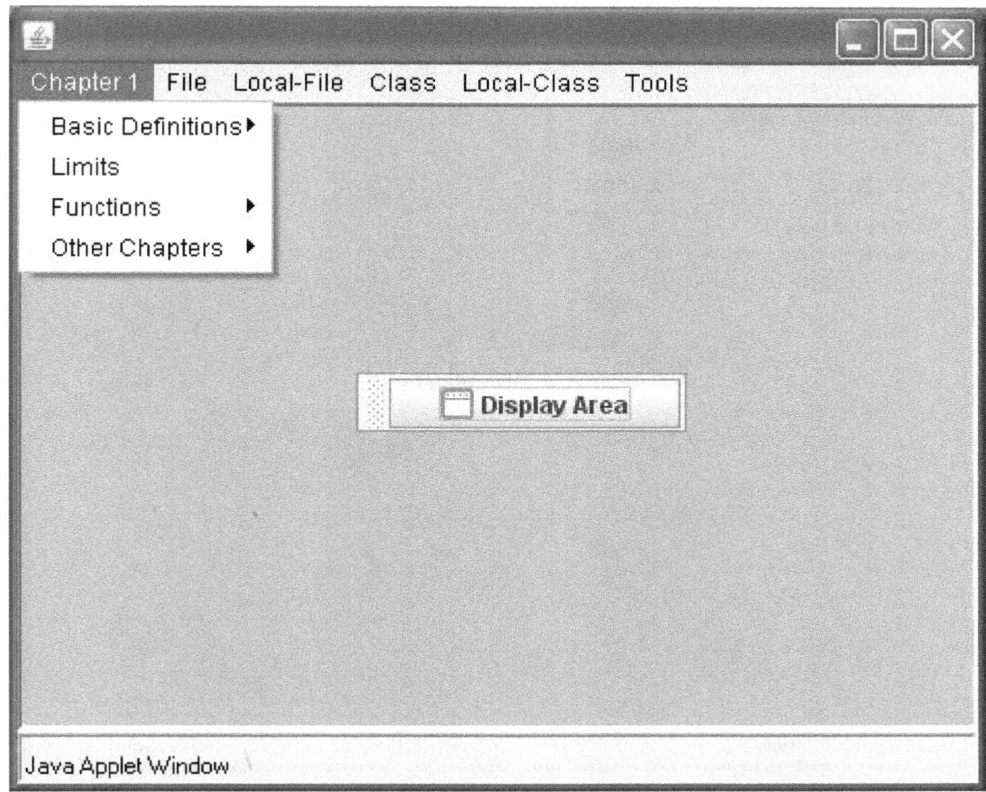

Figure A.7: Chapter 1 Menu Command

The menu item labeled "Chapter 1", provides synopses for the various sub-sections contained in chapter 1 of the book. As we examine this menu item, the first command item is labeled "Basic Definitions" and provides the functionality listed below.

Basic Definitions - This command is used for displaying the basic definitions for (a) Integer, (b) Rational and (c) Irrational numbers. See Figure below.

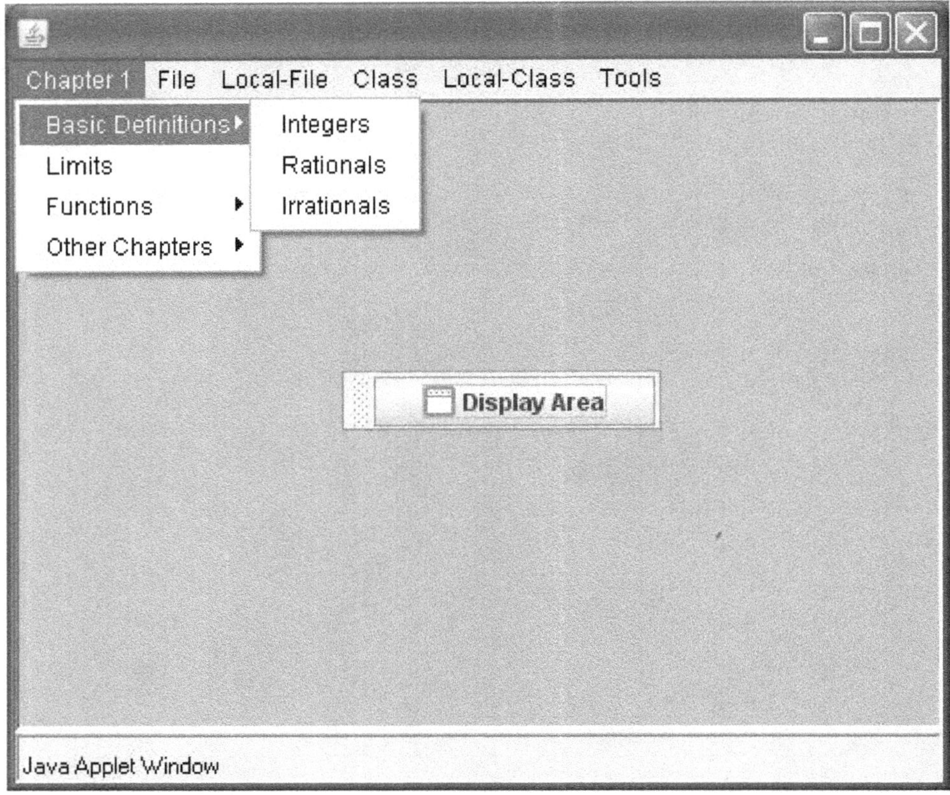

Figure A.8: Chapter 1 Basic Definitions Commands

The next command is the Limits command is has the following description.

Limits - This command displays the basic definition of limits and shows mathematical nomenclature for limits of continuous functions and limits of sequences.

The third command is the Functions command and has the following description.

Functions - This command reveals additional commands for describing (a) Linear functions, (b)Exponential functions, (c)Logarithmic functions and (d)Trigonometric functions. See figure below for visual presentation of commands.

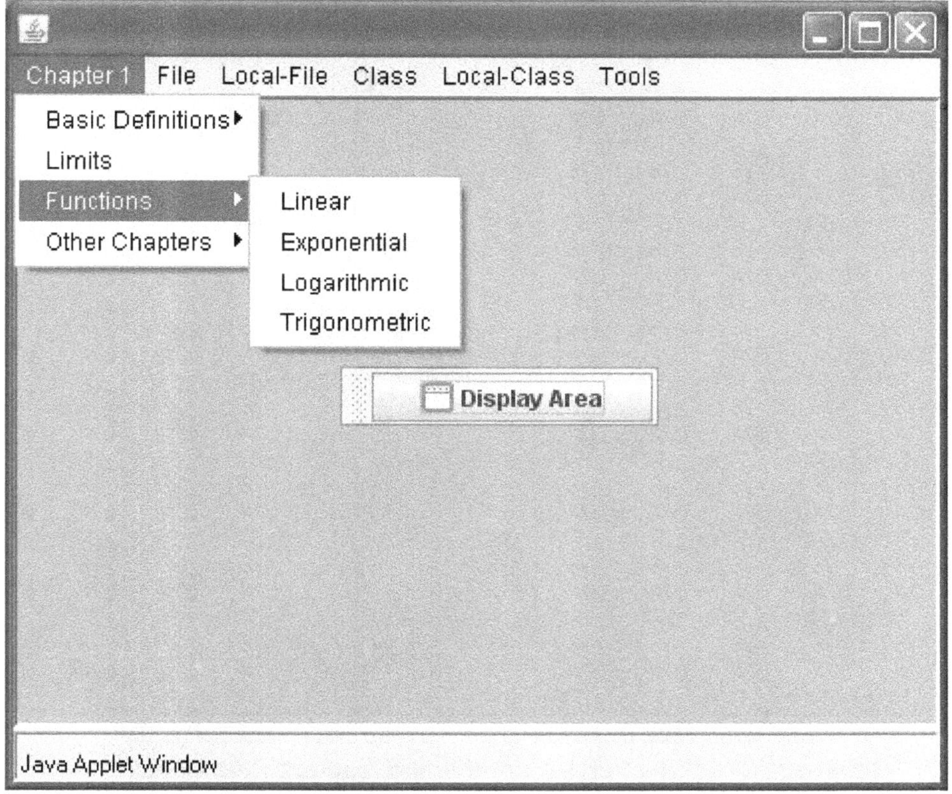

Figure A.9: Chapter 1 Functions Command

To invoke other main menus associated with other chapters, use the "Other Chapters" command described below.

Other Chapters - This command displays chapters 2 through 7 and the appendix. A visual display of this menu is shown below.

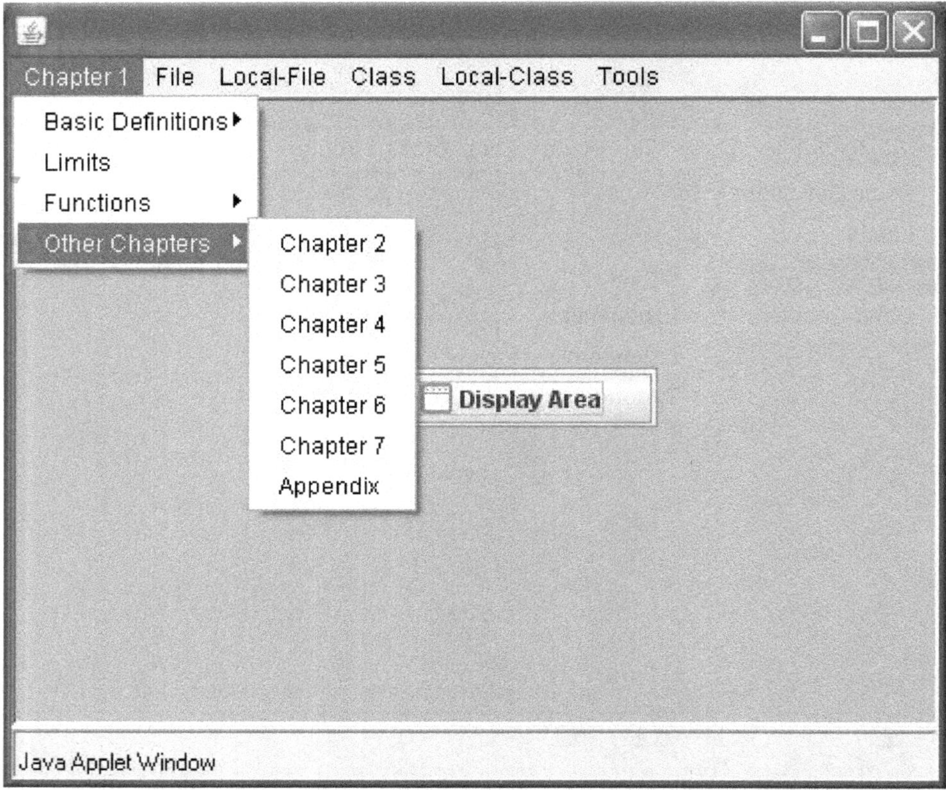

Figure A.10: Chapter 1 Menu Other Chapters Command

Note, that the appendix menu will give addition commands that will support computations in chapters 1 through 7.

(File Menu)

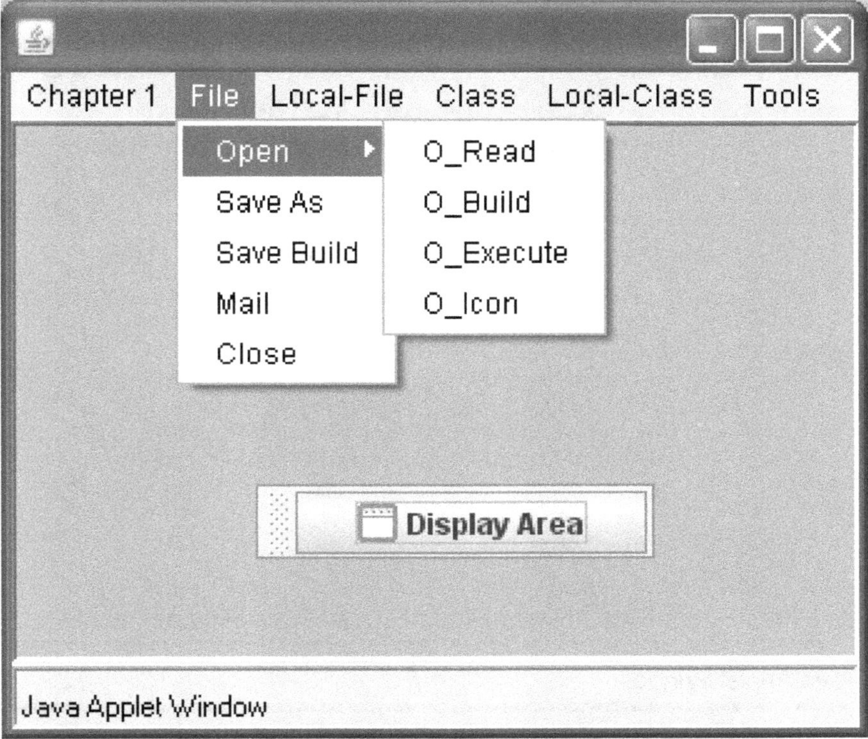

Figure A.11: File Menu Category Commands

The commands associated with this category retrieves files from the server or stores files to the server. For example to retrieve panels and build a panel window use the Build subcommand of the Open command. This provides a convenient mechanism for retrieving new panels created by the instructor or administrator. Use the Save Build command to store any open panel to a directory on the remote server.

(Local - File Menu)

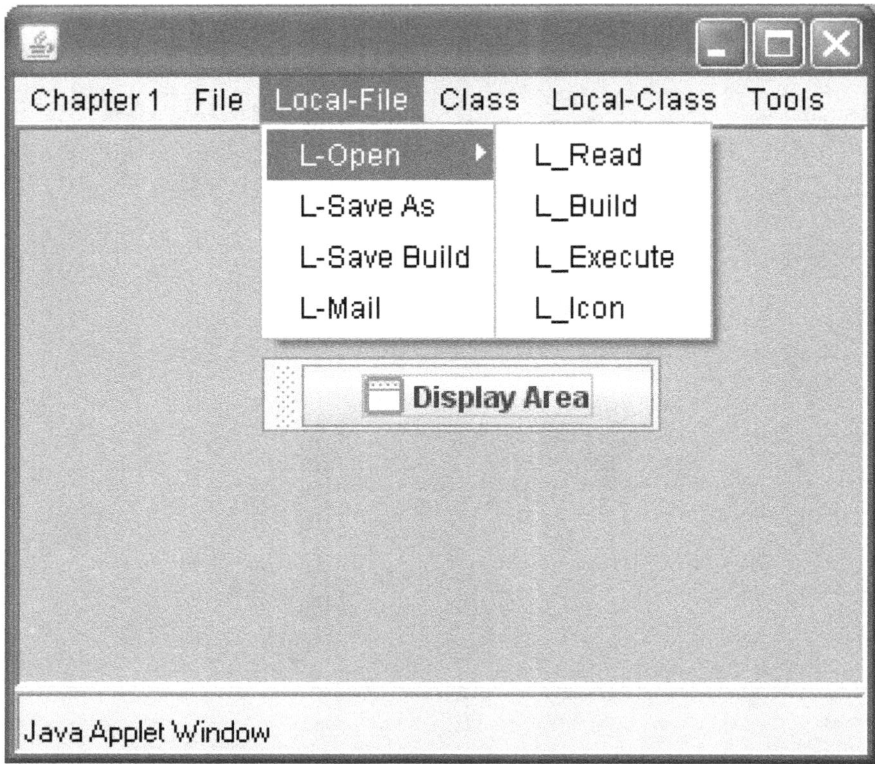

Figure A.12: Local-File Menu Category Commands

Similar to the "File" menu, the "Local-File" menu retrieves files or stores files to the local client machine's directory. For reading commands into the menu buffer, use the Read command. For executing commands contained in a file on the local computer, use the Execute command. The Icon command retrieves icon files and displays the content in the menu frame. To store the contents of the menu buffer, use the "Save As" command. To save an open (it must not be minimized) panel to the local compute, use the "Save Build". In conjunction with the "File" menu (previously mentioned), these two menus will allow the transfer of files between the local and remote systems.

(Class Menu)

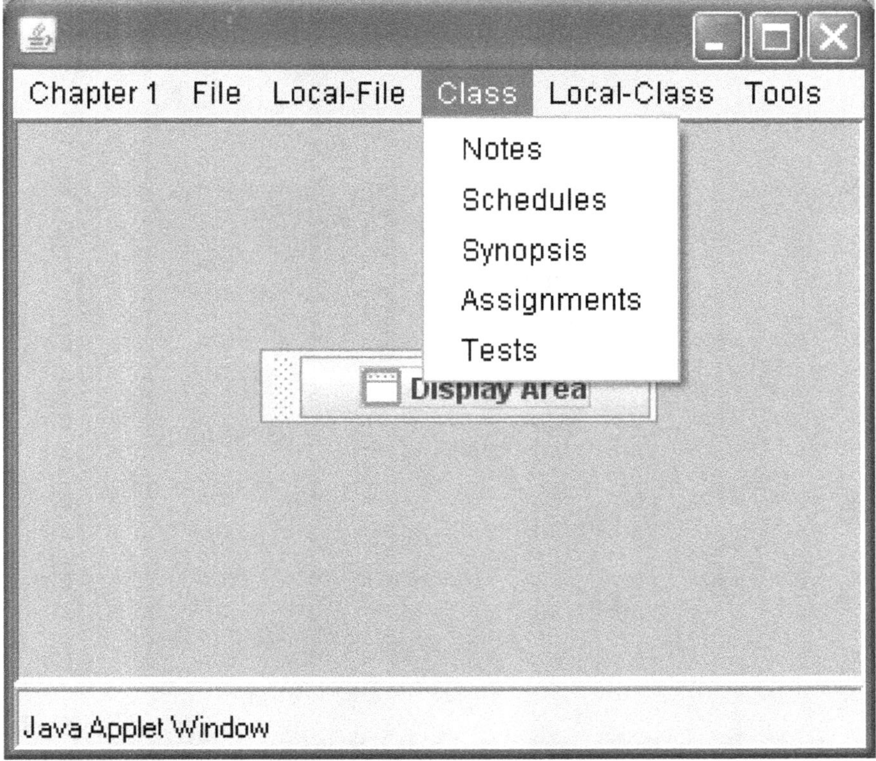

Figure A.13: Class Menu Category Commands

The commands in the "Class" menu represents standard menus that are placed on the server for retrieval by the client. The instructor (or administrator) places notes, schedules, course synopsis, assignments and tests on the server to be retrieved by the user. As the course evolves, the instructor can change the content in these areas to enhance the user's experience. Of course, a communication plan is required to inform the user of the sequence for retrieving notes, synopsis, assignments and tests. This communication may be accomplished with the "Schedules" command or a sequence of e-mails.

(Local - Class Menu)

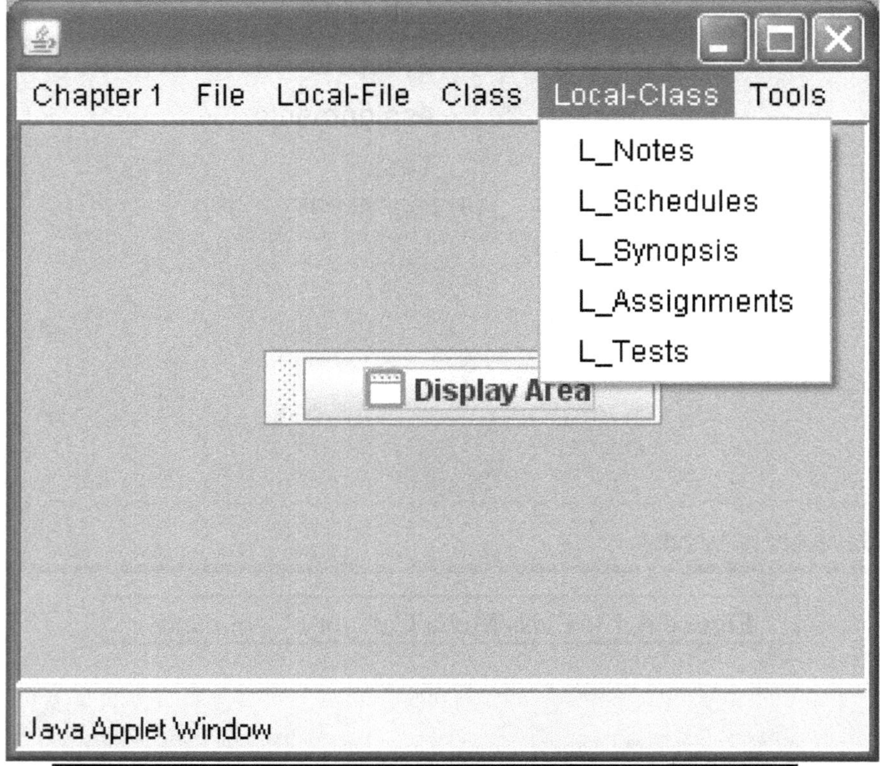

Figure A.14: Local-Class Menu Category Commands

As the user retrieves notes, schedules, synopsis, assignments and tests from the server; It is helpful if they have the ability to store them locally. These commands allow the client to retrieve the information locally, which allows them to work increase their response time for retrieving files in heavy internet traffic.

(Tools Menu)

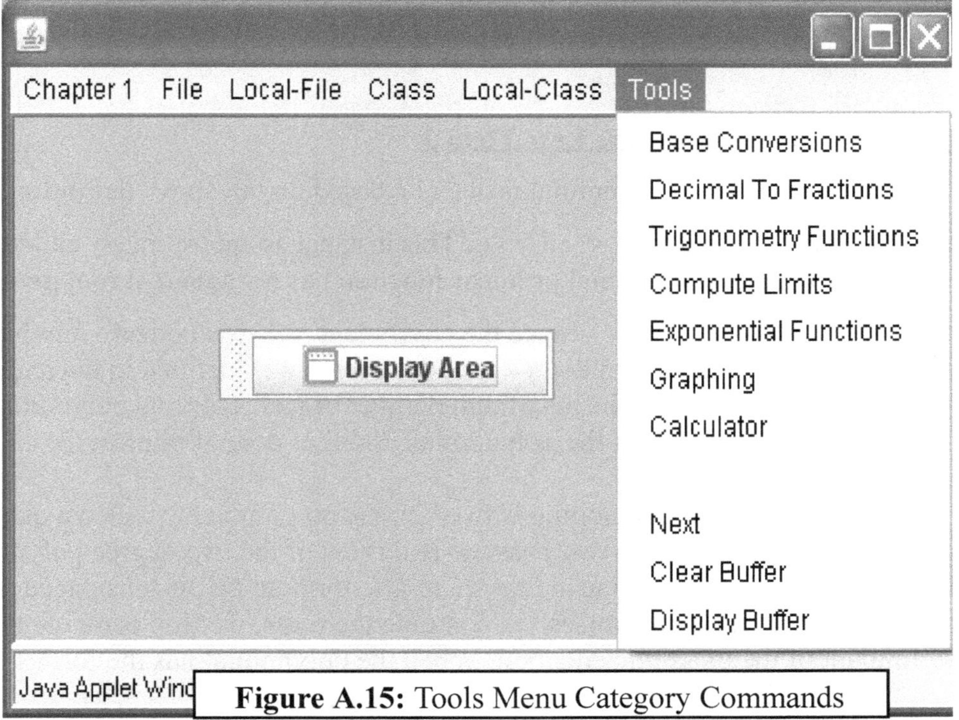

Figure A.15: Tools Menu Category Commands

Commands that are relevant to chapter 1 of the text book are included in this menu category. To convert one number from one base to another, use the "Base Conversions" command. This command will cause a panel to be displayed in the display area of menu frame.Once the panel is displayed, the user should adjust the dimensions of the panel window, using their mouse, for an acceptable visual presentation.

B. Solving for Roots of Polynomials

We begin the discussion of polynomials of a real value "x" by giving a working definition. A polynomial is a linear combination of non-negative integer powers of the real number "x". The degree of the polynomial is the highest integer power of the term "x" with a non-zero coefficient. Mathematically, we represent an **n**th degree polynomial with the following summation:

$c_n \cdot x^n + c_{n-1} \cdot x^{n-1} + \ldots + c_1 \cdot x + c_0$.

B.1 Polynomial with Degree Less Than 3

When the degree of the polynomial is one (1), based on the above definition, the polynomial has the form $c_1 \cdot x + c_0$. This instance is appropriately called a linear function. This polynomial or linear function has a single real root given by the formula $x_1 = -c_0/c_1$. Since the coefficient c_1 is a nonzero number, by definition of polynomial of degree one, this root is well defined in the realm of real numbers. In general, the maximum number of real roots of a polynomial is equivalent to the degree of the polynomial. Another note of interest for all

When the degree of the polynomial is two (2), the polynomial is called a quadratic. An expression for the two possible real roots of the two degree polynomial (if they exist) is derived using the quadratic formula that is introduced in most introductory algebra courses. To compute the roots, we first compute the discriminant of the quadratic equation. When the polynomial has the form $c_2 \cdot x^2 + c_1 \cdot x + c_0$, the discriminant function is defined to be

$D = (c_1)^2 - 4 \cdot c_2 \cdot c_0$. When the discriminant function is non-negative, at least one real root exists for the quadratic polynomial. If the discriminant function equals zero, exactly one root exist. If the discriminant function is positive, then two distinct roots exist for the polynomial. The expressions for the roots of a quadratic polynomial are $\dfrac{-c_1-\sqrt{(c_1)^2-4\cdot c_2\cdot c_0}}{2\cdot c_2}$ and $\dfrac{-c_1+\sqrt{(c_1)^2-4\cdot c_2\cdot c_0}}{2\cdot c_2}$.

If we were to continue this approach with a cubic polynomial (polynomials of form $c_3\cdot x^3+c_2\cdot x^2+c_1\cdot x+c_0$), there is much information in the literature on solving cubic polynomials. Such approaches as Cardano's method, Lagrange resolvents and Chebyshev radicals may be used to find the roots. Rather than continuing with this line of reasoning and developing a myriad of equations and discriminant functions (to test for real roots), a simpler numerical approach is suggested. The next section discusses the numerical approach presented in chapter four (4) and describes a computer based tool for computing the real roots of polynomials.

B.2 Polynomials of Degree Greater Than Two

In chapter five (5), a numerical approach for the computation of real roots for the third degree polynomial is discussed. Using the first derivative of the polynomial, local minima and local maxima are found for the polynomial function. The derivative of a third degree polynomial is a second degree polynomial. Hence the roots of the derivative polynomial yields the local minimum and local maximum of the original three degree polynomial. The roots of the derivative function is obtained using the quadratic formula.

Converting the polynomial to the form $x^3+c_2\cdot x^2+c_1\cdot x+c_0$, the polynomial evaluates to negative values as x becomes more negative. The polynomial evaluates to positive values as x becomes more positive. Hence choosing a large negative number, the local minimum, the local maximum and a large positive number; we organize these four numbers from smallest to largest. With these four points, intermediate ranges are evaluated to search for zero values between adjacent selected points. Using a binary searching algorithm and starting from the extreme negative point, we search for a zero root of the polynomial. After searching this range, we move to the next range to find a possible zero root in the next adjacent range. We continue this process until all three ranges have been searched. The sequential equations for this approach are displayed in table 2 of section 5.3.2 (**Solving for Roots of a "3-by-3" Matrix**).

For polynomials of degrees greater than three (3), a similar numerical approach is used as shown for three degree polynomial. To facilitate the derivation of roots for these higher degree polynomial, a computer routine is made available. The computer routine is assessable in the 'Chapter 5' menu frame under the "Tool" command menu. The appearance of the "Polynomial Roots" panel is shown in figure B.1 below.

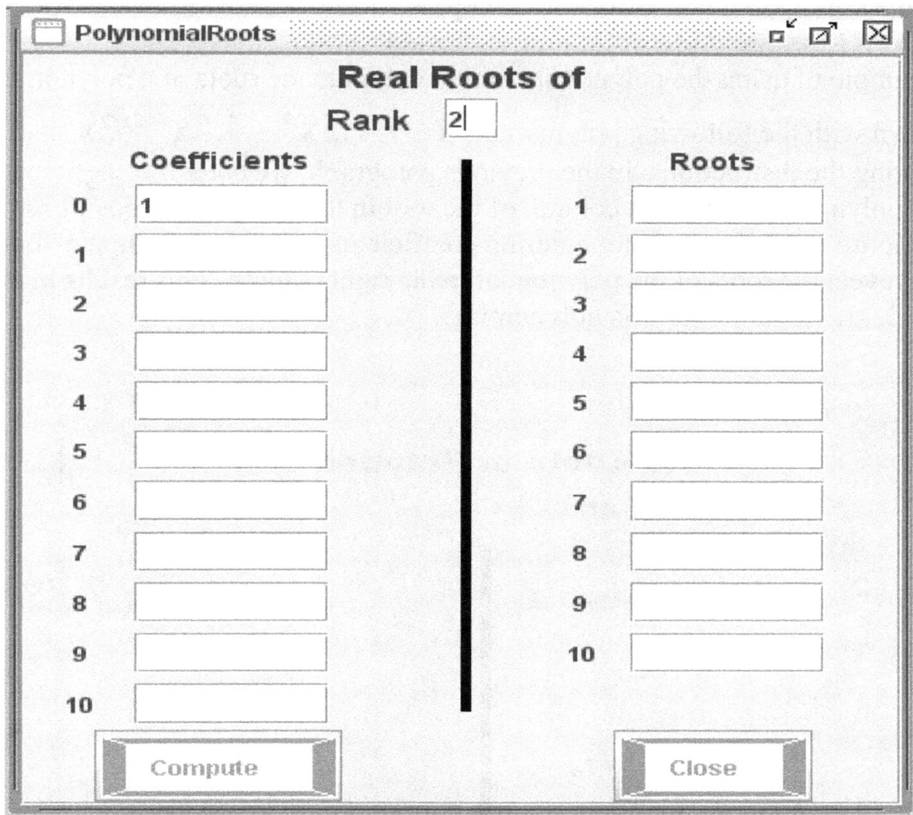

Figure B.1: Polynomial Roots Tool Panel

To use this tool, enter the rank (or degree) of the polynomial in the text box labeled “Rank”. Then enter the coefficients of each power term of the polynomial in the column labeled “Coefficients”. After entering the kth coefficient next to the corresponding number in the column, you are now ready to compute the roots of the polynomial. To compute the roots, click the “Compute” button. All real roots should appear in the column labeled “Roots”. For additional polynomials, re-enter the rank and the polynomial coefficients and repeat the procedure. As can be seen in the above figure, this tool is limited to a polynomial of rank 10. To exit the tool, click the "Close" button. The "Polynomial Roots" panel will close and disappear.

An example of using the polynomial tool to compute the roots of a polynomial is shown with the following polynomial: $x^4 + 1.5x^3 - 4.5x^2 - 2x + 2$. Following the instructions, in the previous paragraph, we enter the coeficients of the polynomial in the left column of the tool in their respective positions (See Figure B.2 below.). After entering coefficients, click the "Compute" button to reveal the roots of the polynomial in the right column. This results in the roots -2, -1, 0.5 and 2 for this polynomial.

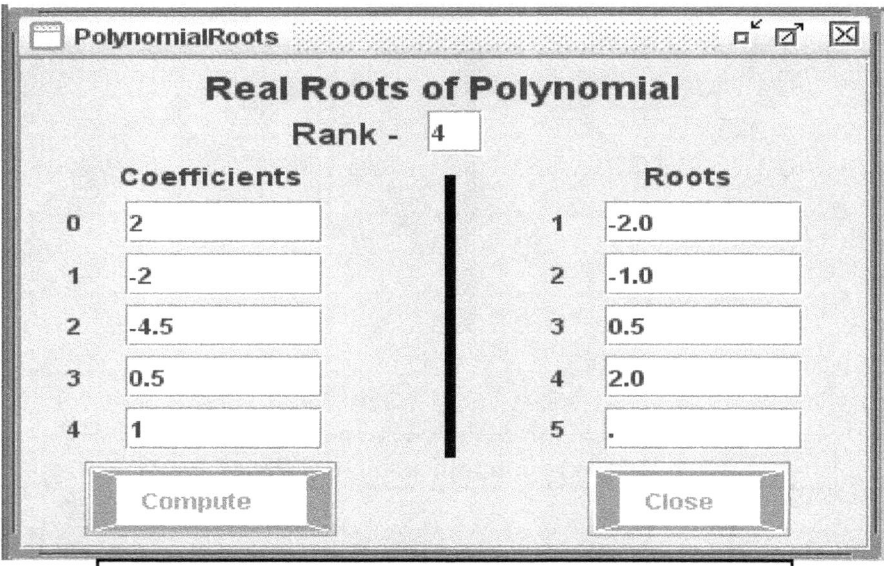

Figure B.2: Computing Polynomial Roots

C. Spreadsheet for Matrix Computations

To enhance the learning experience for matrix theory and linear algebra, I also provide a Microsoft Excel™ Spreadsheet that enables the user to practise some

	A	B	C	D	E	F	G	H	I
1	Characteristic Coefficients	Roots	4.00			Initial Matrix			
2	-8.00	-4.0000	38.00	-19.00	-19.00	-11.00			
3	-6.00	-2.0000	99.00	-52.00	-48.00	-27.00			
4	7.00	-1.0000	-226.00	118.00	110.00	62.00			
5	6.00	1.0000	365.00	-187.00	-179.00	-102.00			
6	1.00								
7									
8									
9									
10									
11									
12						Idempotent Characteristic Matrix by Root			
13									
14									
15									
16									
17									
18									
19									
20									
21									
22									
23						Matrix Accumulator			
24			1.00	0.00	0.00	0.00			
25			0.00	1.00	0.00	0.00			
26	XT times X		0.00	0.00	1.00	0.00			
27			0.00	0.00	0.00	1.00			
28	Build Matrix								
29									
30									
31	Gram-Schmidt								
32									
33									
34	Generalized Identity								
35									

Figure C.1: Matrix Computation Spreadsheet

advanced concepts associated with chapters 5 through 7. The spreadsheet uses visual basic macros to implement the mathematical computation with the push of a single button in some instances. Since the matrix computation tool was designed using Excel™ 97, it is recommended that the user have access to Excel™ 97 or a later version.

The appearance of the spreadsheet is shown in Figure C.1. The functions associated with the spreadsheet are described in the following sections.

C.1 Random Generation of Matrices and Computing Characteristic Roots

The first capability associated with the matrix computation tool is the generation of square matrices for practicing some of the more advanced matrix concepts. The matrices are generated in the spreadsheet cells enclosed by the rectangular area with diagonal end points **C2** and **L11**. This area will hold at most a maximum a 10 by 10 square matrix. To generate a matrix, enter a number between 2 and 10 inclusively in the cell **C1**. Upon entering a value in the cell **C1**, you should erase any value in cell **C12**. The importance of this cell will be explained later. Now you can generate a new random matrix by clicking the button labeled "**Build Matrix**". This button spans cells **A28** and **A29**.

Once the "**Build Matrix**" button is clicked, a square matrix will appear in the rectangular area described above. Additionally, the "**B**" column is populated with numbers in rows 2 through 11. As the heading of the column suggests, these are the real roots of the characteristic equation of the generated matrix. Column "**A**" contains the coefficients of the characteristic equation and the correspondence of coefficients to spreadsheet cells is the following:

C0 ....**A2**
C1 ....**A3**
C2 ....**A4**
C3 ....**A5**
C4 ....**A6**
C5 ....**A7**
C6 ....**A8**
C7 ....**A9**
C8 ....**A10**

Figure C.2: Entry Area for X-Transpose Times X Matrix

C9 .... **A11**
C10 .... **A12**

If the generated matrix is an "**N**" by "**N**" matrix, then only $\mathbf{C_0}$ through $\mathbf{C_N}$ coefficients are displayed. Remember that the value of "**N**" must be between 2 and 10 inclusive.

Having generated the matrix, the roots and the coefficients, a task that may be required of the user is to derive the associated idempotent matrix for each distinct root of characteristic equation. To accomplish this, enter the number associated with the root (1 through "**N**") into cell "**C12**" and click the "**Build Matrix**" button. This action places the associated idempotent matrix in square area of the spreadsheet with diagonal endpoints of "**C13**" and "**L22**".

Another button that is visible in column "**A**" is the "**Generate Root/Coefs.**" button. Use this button to find the characteristic roots of the matrix specified in the area labeled "**Initial Matrix**". Before clicking this button, insure that the size of the matrix is specified in cell "**C1**" and the appropriately sized matrix is placed in the "**Initial Matrix**" area. Once the button is clicked, the characteristic coefficients and the characteristic roots of the matrix are displayed in columns "**A**" and "**B**" respectively. This capability gives the user the ability to compute roots of user specified matrices and to derive the associated idempotent matrices. To derive the idempotent matrices, use the approach described in the previous paragraph.

In columns "**C**" through "**L**", note that there is another rectangular collection of cells labeled "**Matrix Accumulator**". Each time an idempotent matrix is generated, the idempotent matrix is also added to the previous content in this section. It is important to remember to clear this area before the first idempotent matrix is generated. If a value is placed in cell "**C23**", this value is multiplied to the idempotent matrix before it is added to the "**Matrix Accumulator**". This functionality allows the user to compute the inverse of the matrix using idempotent matrices and the associated roots. When the matrix is singular, this allows an alternate approach to computing the generalized inverse of the matrix. Other uses of this capability will become apparent as you become familiar with the more advanced concepts described in this book.

C.2 X-Transpose Times X

The "**XT times X**" button is used to compute the product of the transpose of a matrix with the matrix itself. This enables the user to user to compute covariance matrices in statistics and for the computation of matrices in the solving of least squares estimation in multivariable linear regression applications. To compute the product you begin by entering the original matrix in the rectangular section of cells labeled "**X Matrix**" in the Polynoms.xls EXCEL workbook.

.

To use this capability, enter the matrix in the "**X Matrix**" area (You are restricted to a maximum of a 10 columns and 51 rows.) After specifying the matrix, click the "**XT times X**" button. This will cause the product of the transpose of the matrix with itself to be displayed in the "**Initial Matrix**" section and the characteristic coefficients and the roots of the resulting matrix product are displayed in columns "**A**" and "**B**" respectively. You will also note that the size of newly created square matrix is placed in the cell "**C1**".

Upon generating this matrix, the user may perform analysis such as determining the inverse, computing idempotent matrices and other functional analysis associated with linear combinations of the idempotent matrices.

C.3 Generalized Identity

When the square matrix (**M**), displayed in the "Initial Matrix" section of the spreadsheet is a non-singular matrix, the only idempotent matrix **J**, such that **J*M = M*J = M**, is the identity matrix. However, when the matrix is singular, there exist a unique idempotent matrix of minimal rank such that this condition is also true. This idempotent matrix, as described in chapter 7, is called the generalized identity matrix associated with the matrix **M**.

To generate the associated generalized identity matrix, click the "**Generalized Identity**" button after entering the square matrix of interest in the "**Initial Matrix**" section of the spreadsheet. When the button is clicked, the generalized identity matrix is placed in the "**Matrix Accumulator**" section of the spreadsheet. Manually (assuming all characteristic roots are real), this operation is equivalent to adding all the Idempotent matrices associated with each nonzero root of the matrix. Hence an equivalent operation would be to enter the integers one through the number of roots shown in column "**B**" into cell "**C12**" and click the "**Build Matrix**" button. To compute the generalized inverse of the matrix in the "**Initial Matrix**" section follow this same manual procedure, but enter the reciprocal of the nonzero root into the cell "**C23**" prior to clicking the "**Build Matrix**" button. Other functions of the original matrix can be computed by placing the square of each root into the cell "**C23**" before pressing the button. As other functions are required the roots can be modified as the required. Future revisions of this spreadsheet may automate many of these functions using a series of push buttons.

C.4 Gram-Schmidt Algorithm

As shown in chapter 7 of this text book, the Gram-Schmidt algorithm gives us the ability to represent a non-singular matrix as the product of an orthogonal matrix (**Q**) and a triangular matrix (**R**). If the non-singular matrix is denoted as M; we get the relationship $\mathbf{M=QR}$ where $\mathbf{Q}^T\mathbf{Q} = \mathbf{I}$. If the user clicks the "**Gram-Schmidt**" button, the orthogonal matrix **Q** is placed in the "**Idempotent Characteristic Matrix by Root**" section of the spreadsheet and the triangular matrix **R** is placed in the "**Matrix Accumulator**" section of the spreadsheet.

Upon clicking the button, try multiplying the resultant matrix **Q** times the matrix **R**. You should obtain the original matrix **M**. However, since the round off error with the Gram-Schmidt algorithm is large, you may get an approximation to the original matrix. This is especially true when the rank of the matrix exceeds 5. To carry out this multiplication using Excel ™ (we will assume for a five-by-five matrix), select a 5-by-5 square area of the spreadsheet such as **$N13:$R17**. Upon selecting this region, enter the following command: "**=MMULT($C13:$G17, $C23:$G27)**". Since this is an array defining function, you should press **Ctrl+Shift+Enter** to register this command.

This Page is Intentionally Left Blank

This Page is Intentionally Left Blank

www.ingramcontent.com/pod-product-compliance
Lightning Source LLC
Chambersburg PA
CBHW080123220326
41598CB00032B/4937